# Linear Operators in Function Spaces

12th International Conference
on Operator Theory
Timişoara (Romania)
June 6–16, 1988

**Volume Editors**

**H. Helson
B. Sz.-Nagy
F.-H. Vasilescu**

**Managing Editor
Gr. Arsene**

1990

**Birkhäuser Verlag
Basel · Boston · Berlin**

Volume Editorial Office:

Department of Mathematics
INCREST
Bd. Păcii 220
79622 Bucharest
Romania

**CIP-Titelaufnahme der Deutschen Bibliothek**

**Linear operators in function spaces** / 12th Internat. Conference
on Operator Theory, Timişoara (Romania), June 6–16, 1988.
Vol. ed. H. Helson ... Managing ed. Gr. Arsene. – Basel ;
Boston ; Berlin ; Birkhäuser, 1990
  (Operator theory ; Vol. 43)
ISBN-13: 978-3-0348-7252-2      e-ISBN-13: 978-3-0348-7250-8
DOI: 10.1007/978-3-0348-7250-8

NE: Helson, Henry [Hrsg.]; International Conference on Operator
  Theory <12, 1988, Timişoara>; GT

# CONTENTS

# FOREWORD

The Operator Theory conferences, organized by the Department of Mathematics of INCREST and the Department of Mathematics of the University of Timişoara, are conceived as a means to promote cooperation and exchange of information between specialists in all areas of operator theory. This book comprises carefully selected papers on theory of linear operators and related fields. Original results of new research in fast developing areas are included. Several contributed papers focus on the action of linear operators in various function spaces. Recent advances in spectral theory and related topics, operators in indefinite metric spaces, dual algebras and the invariant subspace problem, operator algebras and group representations as well as applications to mathematical physics are presented.

The research contacts of the Department of Mathematics of INCREST with the National Committee for Science and Technology of Romania provided means for developing the research activity in mathematics; they represent the generous framework of these meetings too.

It is our pleasure to acknowledge the financial support of UNESCO which also contributed to the success of this meeting.

We are indebted to Professor Israel Gohberg for including these Proceedings in the OT Series and for valuable advice in the editing process. Birkhauser Verlag was very cooperative in publishing this volume.

Camelia Minculescu, Irén Némethi and Rodica Stoenescu dealt with the difficult task of typing the whole manuscript using a Rank Xerox 860 word processor; we thank them for this exellent job.


Organizing Committee,

Head of Mathematics Department of INCREST,          For Organizers,


Zoia Ceausescu                                       F.-H. Vasilescu

## LIST OF PARTICIPANTS[*]

| | |
|---|---|
| **ADAMJAN, V.M.** | - University of Odessa, USSR |
| **ALBRECHT, Ernst** | - University of Saarland, West Germany |
| **AMBROZIE, Călin** | - INCREST, Bucharest |
| **AROV, D.Z.** | - Pedagogical Institute, Odessa, USSR |
| **ARSENE, Grigore** | - INCREST, Bucharest |
| **ARSU, Gruia** | - INCREST, Bucharest |
| **ARZUMANIAN, V.** | - Institute of Mathematics, Yerevan, USSR |
| **AXLER, Sheldon** | - Michigan State University, USA |
| **AZIZOV, T.Ya.** | - State University of Voronezh, USSR |
| **BACALU, Ion** | - Politechnical Institute, Bucharest |
| **BAKONYI, Mihai** | - INCREST, Bucharest |
| **BÁLINT, Ştefan** | - University of Timişoara |
| **BEZNEA, Lucian** | - INCREST, Bucharest |
| **BIRÁUŞ, Silviu** | - University of Timişoara |
| **BOCA, Florin** | - INCREST, Bucharest |
| **BOYADZIEV, K.N.** | - Institute of Mathematics, Sofia, Bulgaria |
| **BUCUR, Gheorghe** | - INCREST, Bucharest |
| **BUJUKLIEV, Nikolay** | - University of Sofia, Bulgaria |
| **CARTIANU, Dan** | - ICECHIM, Bucharest |
| **CEAUŞESCU, Zoia** | - INCREST, Bucharest |
| **CEAUŞU, Traian** | - University of Timişoara |
| **CHEVREAU, Bernard** | - University of Bordeaux, France |
| **CLARK, Douglas N.** | - University of Georgia, USA |
| **COFAN, Nicolae** | - University of Timişoara |
| **CONSTANTINESCU, Tiberiu** | - INCREST, Bucharest |
| **CORNEA, Emil** | - INCREST, Bucharest |
| **CORNEA, Florin** | - University of Bucharest |
| **CRAIOVEANU, Mircea** | - University of Timisoara |
| **CURTO, Raul-Enriques** | - University of Iowa, USA |
| **DĂDĂRLAT, Marius** | - INCREST, Bucharest |
| **DEACONU, Valentin** | - INCREST, Bucharest |
| **ECKSTEIN, Gheorghe** | - University of Timişoara |
| **ESCHMEIER, Jörg** | - University of Münster, West Germany |
| **EXNER, George** | - Oberlin College, USA |
| **FRAGOULOPOULOU, Maria** | - University of Athens, Greece |
| **FRUNZĂ, Ştefan** | - University of Iaşi |
| **GADIDOV, Radu** | - INCREST, Bucharest |
| **GAŞPAR, Dumitru** | - University of Timişoara |
| **GĂVRUŢĂ, Pasc** | - University of Timişoara |
| **GEORGESCU, Vladimir** | - IFIN, Bucharest |
| **GEORGIEV, Vladimir** | - Institute of Mathematics, Sofia, Bulgaria |
| **GHEONDEA, Aurelian** | - INCREST, Bucharest |

---

[*] Romanian participants are listed only with the name of their institution and the city.

| | |
|---|---|
| GHEORGHE, Gabriela | - ICECHIM, Bucharest |
| GODINI, Gliceria | - INCREST, Bucharest |
| GOLOGAN, Radu | - INCREST, Bucharest |
| GRIGORIAN, S. | - Institute of Mathematics, Yerevan, USSR |
| HALANAY, Andrei | - Polytechnical Institute, Bucharest |
| HĂRĂGUȘ, Dumitru | - University of Timișoara |
| HELSON, Henry | - University of California, Berkeley, USA |
| HELEMSKII, Alexander Ya. | - University of Moscow, USSR |
| HERRERO, Domingo | - University of Arizona, Tempe, USA |
| IONAȘCU, Eugen | - INCREST, Bucharest |
| IONAȘCU, Ileana | - INCREST, Bucharest |
| IONESCU, Adrian | - INCREST, Bucharest |
| IONESCU, Valentin | - INMT, Bucharest |
| IVANOV, Alexandru | - ICECHIM, Bucharest |
| IVANOVSKI, N. | - Faculty of Mathematics, Skopjie, Yugoslavia |
| JANAS, Jan | - Institute of Mathematics, Krakow, Poland |
| JEBELEAN, Petru | - Polytechnical Institute, Bucharest |
| JOIȚA, Maria | - INCREST, Bucharest |
| JONAS, Peter | - Institute of Mathematics, Berlin, GDR |
| KYRIASIS, Athanasios | - University of Athens, Greece |
| LEVY, Roni N. | - Institute of Mathematics, Sofia, Bulgaria |
| MARTIN, Mircea | - INCREST, Bucharest |
| MARZOUK, Ahmed | - University of Bordeaux, France |
| MASANI, Pesi R. | - University of Pittsburgh, USA |
| MATACHE, Valentin | - University of Timișoara |
| MEGAN, Mihail | - University of Timișoara |
| MIHALACHE, Georgeta | - INCREST, Bucharest |
| MIHALACHE, Nicolae | - INCREST, Bucharest |
| MÜLLER, Vladimir | - Institute of Mathematics, Prague, Czechoslovakia |
| NAGY, Gabriel | - INCREST, Bucharest |
| NAKAZI, Takahiko | - Hokkaido University, Sapporo, Japan |
| NENCIU, Gheorghe | - IFIN, Bucharest |
| NICA, Alexandru | - INCREST, Bucharest |
| NICULESCU, Constantin | - University of Craiova |
| NIKOLOV, Krasimir | - Institute of Mathematics, Sofia, Bulgaria |
| NISTOR, Victor | - INCREST, Bucharest |
| NIȚICĂ, Viorel | - INCREST, Bucharest |
| OJA, Eve | - Tartu State University, USSR |
| PASCU, Mihai | - INCREST, Bucharest |
| PASNICU, Cornel | - INCREST, Bucharest |
| PĂUNESCU, Doru | - University of Timișoara |
| PETRESCU, Steliana | - INCREST, Bucharest |
| POP, Florin | - INCREST, Bucharest |
| POPA, Nicolae | - INCREST, Bucharest |
| POPESCU, Gelu | - INCREST, Bucharest |
| POPESCU, George | - ITCI, Bucharest |
| PRĂJITURĂ, Gabriel | - High School, Slatina |
| PRUNARU, Bebe | - INCREST, Bucharest |
| PTÁK, Vlastimil | - Institute of Mathematics, Prague, Czechoslovakia |
| PUTA, Mircea | - University of Timișoara |
| PUTINAR, Mihai | - INCREST, Bucharest |
| RADULESCU, Florin | - INCREST, Bucharest |
| REGHIȘ, Mircea | - University of Timișoara |

| | |
|---|---|
| **RICKER, Werner** | – University of Canberra, Australia |
| **ROVNYAK, James L.** | – University of Virginia, Charlottesville, USA |
| **RUDOL, Krzysztof** | – Institute of Mathematics, Krakow, Poland |
| **ŞABAC, Mihai** | – University of Bucharest |
| **SHTRAUS, Vladimir** | – Polytechnical Institute, Celiabinsk, USSR |
| **STAN, Ilie** | – University of Timişoara |
| **STĂNĂŞILĂ, Octavian** | – Polytechnical Institute, Bucharest |
| **ŞTEFAN, Tudor** | – Polytechnical Institute, Bucharest |
| **STRĂTILĂ, Şerban** | – INCREST, Bucharest |
| **SUCIU, Ion** | – INCREST, Bucharest |
| **SUCIU, Nicolae** | – University of Timisoara |
| **SZAFRANIEC, Franciszek H.** | – Jagiellonian University, Krakow, Poland |
| **SZÖKEFALVI-NAGY, Béla** | – University of Szeged, Hungary |
| **TERESCENCO, Alexandru** | – ITCI, Timişoara |
| **TIMOTIN, Dan** | – INCREST, Bucharest |
| **TODOROV, Todor** | – Institute of Mathematics, Sofia, Bulgaria |
| **TOPUZU, Paul** | – University of Timişoara |
| **TÖRÖK, Andrei** | – INCREST, Bucharest |
| **TOTOLICI, Ioan** | – University of Galaţi |
| **TSEKANOVSKI, E.R.** | – Donetsk State University, USSR |
| **TURNPU, Heino** | – Tartu State University, Tartu, USSR |
| **VALUŞESCU, Ilie** | – INCREST, Bucharest |
| **VASILESCU, Florian-Horia** | – INCREST, Bucharest |
| **VOLANSCHI, Constantin** | – ICECHIM, Bucharest |
| **VRBOVÁ, Pavla** | – Institute of Mathematics, Prague, Czechoslovakia |
| **VUZA, Dan** | – INCREST, Bucharest |
| **WAELBROECK, Lucien** | – Free University of Brusseles, Belgium |
| **WOLLENBERG, Manfred** | – Institute of Mathematics, Berlin, GDR |
| **ZAJAC, Michal** | – Institute of Mathematics, Bratislava, Czechoslovakia |
| **ZAROUF, Fouad** | – University of Bordeaux, France |
| **ZOUAKIA, Fouad** | – University of Bordeaux, France |

**Secretariat**

**Minculescu Camelia**

**Némethi Irén**

# PROGRAMME OF THE CONFERENCE

## Tuesday, June 7

### Joint Session

*Chairman: D. Herrero*

9.00- 9.45   **H. Helson:** Large analytic functions.

9.50-10.35   **E. Albrecht:** Multicyclic n-tuples of commuting operators.

10.50-11.35   **J. Eschmeier:** Operators with rich invariant subspace lattices.

11.40-12.25   **M. Putinar:** On the division of vector valued distributions by linear functions.

### Section A

*Chairman: Gr. Arsene*

16.00-16.30   **R. Gologan:** Another proof of the subadditive ergodic theorem.

16.35-17.05   **W.J. Ricker:** Boolean algebras of projections and spectral measures in dual spaces.

17.20-17.50   **Şt. Frunză:** The extension problem for n-coadjoint and n-isometric operators on Hilbert spaces.

17.55-18.25   **F. Zouakia:** Stabilité des semigroupes d'opérateurs.

### Section B

*Chairman: V. Pták*

16.00-16.30   **S. Bálint:** Linear operators, several complex variables and power systems.

16.35-17.05   **M. Craioveanu, M. Puta:** Basic de Rham cohomology.

17.20-17.35   **H. Turnpu:** Inclusion with speed of summability methods in the class of functional series.

17.40-17.55   **C. Volanschi:** A proof for reconstructing matrix conjecture.

## Wednesday, June 8

### Joint Session

*Chairman: J. Rovnyak*

9.00- 9.45   **D. Herrero:** All about triangular operators.

9.50-10.35   **R.E. Curto:** Operator factorizations and quasisimilarity orbits.

10.50-11.35   **B. Sz.-Nagy:** Operators similar to isometries.

11.40-12.25   **V. Pták:** Intertwining relations.

### Section A

*Chairman: I. Suciu*

16.00-16.30   **F.H. Szafraniec:** The multiplication operator over **C**. An explicit example.

16.35-17.05   **K. Rudol:** Toeplitz operators on the Bergman space of infinitely many variables.

17.20-17.50   **T. Nakazi:** Commuting dilations and uniform algebras.

17.55-18.10   **N. Ivanovski:** Quasisimilarity of bilateral operator weighted shifts.

18.15-18.30   **F. Pop :** Some results on nest-subalgebras of von Neumann algebras.

### Section B

*Chairman: R.E. Curto*

16.00-16.30   **V.A. Shtraus:** Functional models for operators acting in indefinite inner product spaces.

16.35-17.05   **A. Terescenco:** Categories of quotient Banach spaces.

17.20-17.35   **A. Marzouk:** Fonctions généralisées solutions d'equations algèbriques.

## Thursday, June 9

### Joint Session

*Chairman: D.Z. Arov*

9.00-9.45   **M. Wollenberg:** On causal nets of causal algebras.

9.50-10.35   **V. Georgescu:** Conjugate operators and limiting absorbtion principle.

10.50-11.35   **Gh. Nenciu:** Asymptotic invariant subspaces for evolution operators. (Adiabatic theorem of Quantum Mechanics).

11.40-12.10   **V. Georgiev:** RAGE theorem for power bounded operators.

### Section A

*Chairman: D. Gaşpar*

16.00-16.30 **E.R. Tsekanovski:** Accretive and sectorial extensions and characteristic operator functions.

16.35-17.05 **J. Janas:** Inductive limit of Toeplitz operators.

17.20-17.35 **R. Gadidov:** Choice sequences and subisometric dilations.

17.40-17.55 **V. Ionescu:** Processes with parameter periodicities.

18.00-18.15 **P. Găvruţă, D. Păunescu:** Moment problems for operators.

### Section B

*Chairman: M. Reghiş*

16.00-16.30 **M. Pascu:** Asymptotic completeness of the wave operators for simply characteristic operators and long-range potentials by Enss method.

16.35-17.05 **G. Arsu:** Spectral analysis for simply characteristic operators by Mourre's method.

17.20-17.35 **K.N. Boyadzhiev:** Some notes on the Pincus principal function and on the Krein spectral shift function.

17.40-17.55 **T.S. Todorov:** On irreducibility properties of unbounded operator $*$-algebras.

18.00-18.15 **M. Puta:** The integrability problem at quantic level.

### Friday, June 10

#### Joint Session

*Chairman: H. Helson*

9.00- 9.45 **J.Rovnyak:** Contractive substitution transformations.

9.50-10.35 **V.M. Adamyan:** Asymptotic relations for the positive Toeplitz matrices and operators.

10.50-11.35 **Gr. Arsene, Zoia Ceauşescu, T. Constantinescu:** On some completion problems.

11.40-12.25 **T. Constantinescu:** Consequences of some completion problems.

#### Section A

*Chairman: B.Sz.-Nagy*

16.00-16.30 **Zoia Ceauşescu, I. Suciu:** Extreme points in the set of contractive intertwining dilations.

14

16.35-17.05  **D. Gaspar, N. Suciu, I. Valusescu:** Factorizations and Szegö theorem for semispectral measures on $\mathbf{T}^2$.

17.20-17.35  **G. Popescu:** Multi-analytic operators and some factorization theorems.

17.40-17.55  **M. Zajac:** Pairs of commuting contractions.

18.00-18.15  **M. Bákonyi:** Spectral factors and analytic completion.

18.30-19.30  **J. Rovnyak: Seminar.**

### Section B

*Chairman: N. Popa*

16.00-16.30  **G. Godini:** Operators in normed almost linear spaces.

16.35-17.05  **C. Niculescu:** Alfsen-Effros structures associated to a vector norm.

17.20-17.35  **A. Ivanov:** Compatibility and commutativity in orthomodular lattices.

17.40-17.55  **E. Oja:** M-ideals of compact operators.

18.00-18.15  **D. Vuza:** Characterization of Carleman operators.

### Saturday, June 11

### Joint Session

*Chairman: L. Waelbroeck*

9.00- 9.45  **P. Vrbova:** Some remarks on lifting intertwining relation for contractions.

9.50-10.35  **P.R. Masani:** Stationary measures over L.C.A. groups $\Gamma$.

10.50-11.35  **N. Popa:** Isomorphism theorms for dyadic Hardy spaces $H_\infty$.

11.40-12.25  **D. Vuza:** Operators of convolution type and their applications to a problem of H.H. Schaefer.

### Monday, June 13

### Joint Session

*Chairman: P.R. Masani*

9.00- 9.45  **A.Ya. Helemskii:** Results and problems on the homology of the "Algebras in Analysis".

9.50-10.35  **F. Rădulescu:** On smooth extensions for odd dimensional spheres.

10.50-11.35  **M. Dădărlat, A. Némethi:** Shape theory for $C^*$-algebras.

11.40-12.25  **V. Nistor:** Homotopy properties of automorphisms of AF-algebras.

### Section A

*Chairman: A.Ya. Helemskii*

16.00-16.30  **R.N. Levy**: Chern character for n-tuples of operators.

16.35-17.05  **M. Martin**: Algebraic manifolds associated with convolution subalgebras of a $C^*$-algebra.

17.20-17.35  **A. Kyriazis**: Tensor product algebra bundles.

17.40-17.55  **A. Nica**: Some remarks on the $C^*$-algebras of the positive cones of $H_3$ and $H_4$.

18.00-18.15  **A. Török**: AF-algebras with unique trace.

18.20-18.35  **Maria Fragoulopoulou**: Tensor products of locally $C^*$-algebras and applications.

18.40-19.40  **D.Z. Arov**: Seminar.

### Tuesday, June 14

#### Joint Session

*Chairman: E. Albrecht*

9.00- 9.45  **L. Waelbroeck**: Holomorphic functions taking their values in quotient spaces.

9.50-10.35  **F.-H. Vasilescu**: A multioperator spectral theory in quotient Fréchet spaces.

10.50-11.35  **S. Axler**: Hankel operators on Bergman spaces.

#### Section A

*Chairman: Ş. Strătilă*

16.00-16.30  **V. Arzumanian, S. Grigorian**: Noncommutative uniform algebras.

16.35-17.05  **C. Pasnicu, M. Dădărlat**: Inductive limits of C(X)-modules and continuous fields of AF-algebras.

17.20-17.35  **V. Deaconu**: Automorphisms of AF-algebras.

17.40-17.55  **F. Boca, V. Niţică**: Combinatorial properties of simple groups and $C^*$-algebras with unique trace.

18.00-18.15  **G. Nagy**: On lifting invertible elements from quotient $C^*$-algebras.

18.30-19.30  **V.M. Adamyan**: Seminar.

**Wednesday, June 15**

**Joint Session**

*Chairman: V.M. Adamyan*

9.00- 9.45   **P. Jonas:** On compact perturbations of definitizable operators in Krein space.

9.50-10.35   **D.Z. Arov:** Regular J-inner matrix-functions and corresponding extension problems.

10.50-11.35   **D.N. Clark:** The Krein space $\Phi H^2$ .

11.40-12.25   **T.Ia. Azizov:** Operator theory in indefinite inner product spaces and operator pencils.

14.45-15.45   **D.N. Clark:** Seminar.

**Section A**

*Chairman: D.N. Clark*

16.00-16.30   **A. Gheondea:** A geometric question concerning the strong duality of neutral spaces.

16.35-17.05   **F. Zarouf:** Sur les opérateurs sous-normaux de la classe A.

17.20-17.35   **B. Prunaru:** Rational approximation and invariant subspaces.

17.40-17.55   **A. Halanay:** On the existence of invariant subspaces for some contractions with spectrum dominating an arc on the unit circle.

18.00-18.15   **V. Nistor:** A $C^*$-algebra which is not the $C^*$-algebra of a grupoid.

**Thursday, June 16**

**Joint Session**

*Chairman: F.-H. Vasilescu*

9.00- 9.45   **G. Exner:** Analytic invariant subspaces and the class A.

9.50-10.35   **B. Chevreau:** Contractions with isometric functional calculus are reflexive.

10.50-11.20   **V. Müller:** The numerical radius of commuting products.

11.30-12.30   **B. Chevreau:** Seminar.

Operator Theory:
Advances and Applications, Vol. 43
© 1990 Birkhäuser Verlag Basel



# ASYMPTOTIC PROPERTIES FOR POSITIVE AND
# TOEPLITZ MATRICES AND OPERATORS

**V.M. Adamjan**

## 1.

In this paper it is presented an elementary method to obtain some asymptotic properties for certain classes of positive definite matrices and operators. This method is based, besides the Cauchy - Schwarz inequality, on the following inequality for positive operators.

Let $A$ be a continuous positive operator with continuous inverse, defined on a Hilbert space $H$, let $H_1$ be an arbitrary nontrivial subspace of $H$ with the orthogonal complement $H_2$; $P_1$ and $P_2$ denote the orthogonal projections on $H_1$ and $H_2$, respectively. With respect to the representation of $H$ as the orthogonal sum $H_1 \oplus H_2$, $A$ is represented as a block-matrix

$$A = \begin{bmatrix} A_1 & C \\ C^* & A_2 \end{bmatrix}$$

$A_i = P_i A \,|\, H_i$, $i = 1, 2$ ; $C = P_1 A \,|\, H_2$. Then

$$(1) \qquad \begin{bmatrix} A_1^{-1} & 0 \\ 0 & 0 \end{bmatrix} \leq A^{-1},$$

which means that for the quadratic forms of the operators $A_1^{-1}$ and $A^{-1}$, the inequality

$$(A_1^{-1} P_1 f, f) \leq (A^{-1} f, f),$$

holds for every vector $f \in H$.

In particular, inequality (1) holds (by comparing the corresponding quadratic forms) if $A$ is a matrix of order $N'$ which generates a positive definite quadratic form on $\mathbf{C}^{N'}$, and $A_1$ is its left upper block of order $N < N'$.

For proving (1), put $Q = P_2 A^{-1} | H_2$. Since A is a continuous positive operator with continuous inverse, the operator Q has continuous inverse on $H_2$. We have

$$(2) \qquad A_1^{-1} P_1 = A^{-1} - A^{-1} P_2 Q^{-1} P_2 A^{-1},$$

which can be proved multiplying both sides by $P_2$ on the left, and by $A_1 P_1$ on the right. Since the operator $A^{-1} P_2 Q^{-1} P_2 A^{-1}$ is nonnegative, (1) follows from (2).

Let us present now an important consequence of the inequality (1). An infinite matrix $J = (J_{jk})_1^\infty$ will be called *positive* if all its blocks $J_N = (J_{jk})_1^N$, $N = 1, 2, \ldots$, are matrices associated to hermitian positive definite quadratic forms. We associate to an infinite positive matrix J the sequence of sums

$$d_N(J) = \sum_{j,k=1}^n (J_N^{-1})_{j,k} \, ,$$

where $J_N^{-1}$ is the inverse of $J_N$. Inequality (1) implies that the sequence $d_N(J)$ is non decreasing. To see this, let $e_N$ be the vector in $\mathbf{C}^N$ which has all coordinates equal to the unity. For the block $J_N$ of the matrix J, the sum $d_N(J)$ can be written using quadratic forms as:

$$d_N(J) = (J_N^{-1} e_N, e_N).$$

If $N < N'$, then from (1) we infer

$$d_N(J) = (J_N^{-1} e_N, e_N) = \left( \begin{matrix} J_N^{-1} & 0 \\ 0 & 0 \end{matrix} e_{N'}, e_{N'} \right) \le (J_{N'}^{-1} e_{N'}, e_{N'}) = d_{N'}(J).$$

The aim of this paper is to study the asymptotic behaviour of the numbers $d_N(J)$ as $N \to \infty$, for an infinite positive Toeplitz matrix J $(J_{j,k} = J_{j-k})$, and also the corresponding asymptotics for multi-dimensional and continuous analogues of positive Toeplitz matrices. This problem was stated for Toeplitz matrix by V.S. Vladimirov and I.V. Volovich in connection with some problems arising from the study of the Gaussian model of Statistical Physics, see [1]. For this kind of matrix we compute below the limit $d_J$ of the sequence $d_N(J)$, and, in the case when $d_J = +\infty$, under certain conditions, we also compute the limit $\chi_J$ of the sequence $\chi_N(J) = (1/N) d_N(J)$. Using the construction necessary for the computation of $\chi_J$, it is obtained the expression of the limit

$$S_J = \lim_{N \to \infty} \frac{1}{N} \operatorname{Tr} J_N^{-1}$$

and, as a corollary, the Szegö formula for the limit

$$D_J = \lim_{N \to \infty} \frac{1}{N} \ln \det J_N.$$

Analogous relations are obtained for arbitrary dimensional positive Toeplitz matrices and operators, generated by hermitian positive functions.

## 2.

We are concerned now with the asymptotic behaviour of the numbers $d_N(J)$ for an arbitrary positive matrix $J = (J_{j,k})_1^\infty$. Let us put:

$$m_N(J) = (1/N^2) \sum_{j,k=1}^{N} J_{j,k} = (1/N^2)(J_N e_N , e_N).$$

For any continuous positive operator $A$ with continuous inverse acting on a Hilbert space $H$ and for any vector $f \in H$, we have

$$(f,f) \leq \sqrt{(Af,f)(A^{-1}f,f)}.$$

Therefore

(3)
$$N^2 = [(e_N , e_N)]^2 \leq (J_N e_N , e_N) \cdot (J_N^{-1} e_N , e_N),$$

or

$$m_N^{-1}(J) \leq d_N(J).$$

Let

$$d_J = \lim_{N \to \infty} d_N(J), \quad m_J = \varliminf_{N \to \infty} m_N(J).$$

Thus we proved:

**THEOREM 1.** *For an arbitrary infinite positive matrix $J$ the sequence $\{d_N(J)\}_1^\infty$ is nondecreasing and its limit satisfies*

(4)
$$d_J \geq m_J^{-1}.$$

From Theorem 1 it follows that for an infinite positive matrix $J$, we have $d_J = +\infty$ if $m_J = 0$. In particular, if $J$ is the matrix of a continuous positive operator $\hat{J}$ on $\ell^2$, then

$$m_N(J) = (1/N^2)(J_N e_N , e_N) \leq (\| J \| /N^2)\| e_N \|^2 = (\| J \| /N) \xrightarrow[N \to \infty]{} 0$$

and, consequently, $d_J \to \infty$.

Let us note that, for an arbitrary positive matrix $J$, the limits $m_J^{-1}$ and $d_J$ may not coincide. For example, let us consider the infinite matrix $G$ with entries

$$G_{j,k} = 1 + (2j - 1)\delta_{j,k}, \quad \text{where } \delta_{j,k} = \begin{cases} 1 & j = k \\ 0 & j \neq k. \end{cases}$$

This matrix is positive since for any non-zero vector $\xi = (\xi_j)_1^N$ from $\mathbf{C}^N$ we have

$$(G_N\xi,\xi) = \left| \sum_{j=1}^{N} \xi_j \right|^2 + \sum_{j=1}^{N} (2j - 1)|\xi_j|^2 > 0.$$

Moreover, we infer that:

$$m_N(G) = 2, \quad m_G^{-1} = \frac{1}{2};$$

$$(G_N^{-1})_{j,k} = (1/(2j - 1))\delta_{j,k} - \left[1 + \sum_{s=1}^{N} 1/(2s - 1)\right]^{-1}(1/(2j - 1))\cdot(1/(2k - 1)),$$

$$d_N(G) = \sum_{s=1}^{N} (1/(2s - 1))/\left(1 + \sum_{\ell=1}^{N} (1/(2\ell - 1))\right) \xrightarrow[N \to \infty]{} d_G = 1.$$

However, there exists a family of positive matrices for which we have equality in (4). Let us denote by Q the infinite matrix with all entries equal to one, and by I the infinite unit matrix.

**THEOREM 2.** *If the positive matrix* J *is such that* $m_J \geq 0$ *and for some* $\lambda \geq 0$, *the matrix*

$$J_\lambda = J - m_J Q + \lambda I$$

*is also positive, then we have*

$$d_J = m_J^{-1}.$$

PROOF. Since the case $m_J = 0$ was already considered, we can assume that $m_J > 0$. Denote, for short, the matrix $J + \lambda I$ by $\tilde{J}_\lambda$. Suppose that the hypothesis of the theorem is fulfilled for $\lambda > \lambda_o \geq 0$. Since

$$\lim_{N \to \infty} m_N(\tilde{J}_\lambda) = \lim_{N \to \infty} [m_N(J) + (1/N)\cdot\lambda - m_J] \xrightarrow[N \to \infty]{} 0,$$

then, using Theorem 1 for $\lambda > \lambda_o$, we infer

(5) $$\lim_{N \to \infty} d_N(\tilde{J}_\lambda) = m_{\tilde{J}}^{-1} = +\infty.$$

As for $\xi \in \mathbf{C}^N$

$$Q(\xi) = (\xi, e_N)e_N, \qquad J_{\lambda,N}\xi = \tilde{J}_{\lambda,N}\xi + m_J(\xi, e_N)e_N,$$

then

$$J_{\lambda,N}^{-1} = \tilde{J}_{\lambda,N}^{-1}\xi - m_J(1/(1 + m_J d_N(\tilde{J}_\lambda)))(\xi, \tilde{J}_{\lambda,N}^{-1}e_N)\tilde{J}_{\lambda,N}^{-1}e_N,$$

(6)

$$(J_{\lambda,N}^{-1}e_N, e_N) = d_N(J_\lambda) = d_N(\tilde{J}_\lambda)/(1 + m_J d_N(\tilde{J}_\lambda)).$$

From (5) and (6) it follows that, for $\lambda > \lambda_o$,

(7)
$$\lim_{N\to\infty} d_N(J_\lambda) = m_J^{-1}.$$

When $\lambda_o = 0$, the theorem is proved. Suppose now that $\lambda_o > 0$. For the positive matrix $J$, the functions of complex variable $z$

$$d_N(J_z) = ((J_N + zI_N)^{-1}e_N, e_N)$$

are rational and have positive real part on the closed right half-plane. The rational functions

$$\phi_N(z) = [d_N(J_z) - m_J^{-1}][d_N(J_z) + m_J^{-1}]^{-1}$$

are contractions on the right half-plane (i.e., $|\phi_N(z)| < 1$ for **Re** $z > 0$). Using Montel Theorem, for any positive $\rho < \lambda_o$ one can find a subsequence $\phi_{N_s}(z)$ of $\phi_N(z)$ which is uniform convergent in the disc $|z - \lambda_o| < \rho$. Form (7) it follows that the limit of this subsequence (which is a holomorphic function in the disc $|z - \lambda_o| < \rho$) equals zero on the interval $[\lambda_o, \lambda_o + \rho)$ of the real axis, and therefore is identical zero. In particular,

$$\lim_{N_s\to\infty} \phi_{N_s}(\lambda) = 0,$$

for $\lambda \in (\lambda_o - \rho, \lambda_o)$. Taking into account the monotonicity of the sequence $d_N(J_\lambda)$, we obtain

$$\lim_{N_s\to\infty} d_{N_s}(J_\lambda) = \lim_{N\to\infty} d_N(J_\lambda) = m_J^{-1}.$$

Thus, the relation (7) is true for every $\lambda > 0$. By Theorem 1 it follows that the sequence $d_N(J_\lambda)$ is non-decreasing and therefore, for $\lambda > 0$ we get

(8)
$$d_N(J_\lambda) \leq m_J^{-1}.$$

By continuity, the inequality (8) also holds for $\lambda = 0$. Consequently

$$(9) \qquad \lim_{N \to \infty} d_N(J) \leq m_J^{-1}.$$

Using (4), the relation (9) becomes an equality.

**3.**

The set of infinite positive matrices which satisfy the conditions of Theorem 2 is large enough. It contains, in particular, the infinite positive Toeplitz matrices, and the Hankel matrices with uniformly bounded elements.

By Carathéodory - Toeplitz Theorem, an infinite Toeplitz matrix $J = (J_{j-k})_1^\infty$ is positive if and only if the terms of the sequence $(J_k)_0^\infty$ can be represented as

$$J_k = \int_{-\pi}^{\pi} e^{-ik\theta} d\sigma(\theta),$$

where $\sigma(\theta)$ is a non-decreasing function with an infinite number of points of growth in the interval $[-\pi,\pi]$, see [3]. For the positive Toeplitz matrix generated by such a non-decreasing function $\sigma$, we have:

$$m_N(J) = (1/N^2) \sum_{j,k=0}^{N-1} \int_{-\pi}^{\pi} e^{-i(j-k)\theta} d\sigma(\theta) = (1/N^2) \int_{-\pi}^{\pi} (\sin^2(N\theta/2)/\sin^2(\theta/2)) d\sigma(\theta).$$

It is easy to see that

$$\lim_{N \to \infty} m_N(J) = m_J = \sigma(+0) - \sigma(-0).$$

Define

$$M_o = \sigma(+0) - \sigma(-0).$$

The non-decreasing function $\sigma(\theta)$ can be represented as:

$$(10) \qquad \sigma(\theta) = \sigma_o(\theta) + M_o \Delta(\theta),$$

where $\sigma_o(\theta)$ is an non-decreasing function continuous in $\theta = 0$, and $\Delta(\theta)$ is a non--decreasing function having only one discontinuity at $\theta = 0$ with the jump equal with 1.

From (10) it follows that

$$J = J_o + M_o Q \ (= J_o + m_J Q),$$

where $J_o$ is the positive Toeplitz matrix generated by the non-decreasing function $\sigma_o$. Thus the conditions of Theorem 2 are fulfilled for $\lambda = 0$ by any positive Toeplitz matrix. We proved:

**THEOREM 3.** *For the infinite positive Toeplitz matrix* $J$, *generated by the non-decreasing function* $\sigma$, *we have:*

$$\lim_{N\to\infty} d_N(J) = M_o^{-1}, \quad M_o = \sigma(+0) - \sigma(-0).$$

Another proof of Theorem 3 was previously obtained in [3].

By Hamburger Theorem, an infinite Hankel matrix $J = \left(J_{j+k}\right)^{\infty}_{j,k=0}$ is positive if and only if the terms of the sequence $(J_k)^{\infty}_o$ are represented as

$$J_k = \int_{-\infty}^{\infty} t^k d\sigma(t),$$

where $\sigma(t)$ is a non-decreasing function having an infinite set of points of growth. For a non-decreasing function $\sigma(t)$, the sequence $(J_k)^{\infty}_o$ and

$$m_N(J) = (1/N^2) \sum_{j,k=0}^{N-1} J_{j+k} = (1/N^2) \int_{-\infty}^{\infty} ((1 - t^N)/(1 - t))^2 d\sigma(t)$$

are bounded if and only if all the points of discontinuity of the function $\sigma(t)$ belong to the interval $[-1,1]$. We will assume in what follows this last condition is fulfilled. Denote $M_o = \sigma(1) - \sigma(1-0)$. In this case

$$\lim_{N\to\infty} m_N(J) = m_J = M_o.$$

The non-decreasing function $\sigma(t)$ can be represented as

$$(11) \qquad\qquad\qquad \sigma(t) = \sigma_o(t) + M_o \cdot \Delta(t - 1),$$

where $\sigma_o(t)$ is a non-decreasing function with an infinite number of points of growth in the interval $[-1,1)$, and $\Delta(t)$ was defined above. According to (11), for the Hankel matrix $J$ generated by the function $\sigma(t)$, we have:

$$J = J_o + M_o \cdot Q,$$

where $J_o$ is the positive Hankel matrix generated by the non-decreasing function $\sigma_o$. From Theorem 2 we infer

**THEOREM 4.** *For the infinite positive Hankel matrix* $J$ *generated by the non--decreasing function* $\sigma(t)$ *on the interval* $[-1,1]$, *we have:*

$$\lim_{N\to\infty} d_N(J) = M_o^{-1}, \qquad M_o = \sigma(1) - \sigma(1-0)$$

**4.**

For the applications to some problems of statistical mechanics it is important to find conditions under which for a given infinite positive matrix J the limit $\chi_J$ of the sequence

$$\chi_N(J) = (1/N) \sum_{j,k=1}^{N} (J^{-1})_{j,k} = (1/N)d_N(J)$$

exists and is finite; also it is useful to compute (in this case) $\chi_J$. Define

$$v_N(J) = (1/N) \sum_{j,k=1}^{N} J_{j,k} = N \cdot m_N(J), \qquad v_J = \lim_{N\to\infty} v_N(J).$$

Rewriting the inequality (3) in the form

$$v_N^{-1}(J) \leq \chi_N(J)$$

we obtain that the sequence $\chi_N(J)$ is bounded if $v_J > 0$, and if the sequence $\chi_N(J)$ converges, then

$$\chi_J \geq v_J^{-1}.$$

For a positive Toeplitz matrix J, generated by a non-decreasing function $\sigma(\theta)$ on the interval $[-\pi,\pi]$, the numbers

$$v_N(J) = (1/N) \int_{-\pi}^{\pi} \sin^2(N\theta/2)/\sin^2(\theta/2)d\sigma(\theta)$$

coincides, up to the factor $2\pi$, with the Cesaro means of the Fourier series of the non-negative measure $d\sigma(\theta)$ in the point $\theta = 0$. We decompose now the measure $d\sigma(\theta)$ as the sum of its absolutely continuous part $(1/2\pi)f(\theta)d\theta$ and its singular part $d\sigma_s(\theta)$:

$$d\sigma(\theta) = (1/2\pi)f(\theta)d\theta + d\sigma_s(\theta),$$

where $f \in L^1(-\pi,\pi)$, $f(\theta) \geq 0$, $d\sigma_s(\theta) \geq 0$. Accordingly, we shall represent the positive Toeplitz matrix J as the sum

$$J = J_a + J_s$$

where $J_a$ and $J_s$ are the non-negative Toeplitz matrices generated by the measures

$(1/2\pi)f(\theta)d\theta$ and $d\sigma_s(\theta)$, respectively.

**LEMMA 1.** *If for the measure* $d\sigma(\theta)$ *we have*

(i) *the sequence* $v_N(J_s)$ *is convergent and*

$$\lim_{N\to\infty} v_N(J_s) = 0 \; ; \tag{13}$$

(ii) *the function* $f(\theta)$ *is continuous at* $\theta = 0$, *then*

$$\varliminf_{N\to\infty} \chi_N(J) \geq (1/f(0)). \tag{14}$$

Indeed, the Cesaro means of the Fourier series of the function $f(\theta)$ converge to the values of this function in the point where it is continuous. Therefore, the sequence $v_N(J_a)$, associated to $d\sigma(\theta)$, is also convergent. By (13) we have

$$\lim_{N\to\infty} v_N(J) = \lim_{N\to\infty} v_N(J_a) + \lim_{N\to\infty} v_N(J_s) = f(0), \tag{15}$$

and, according to (12), we obtain (14).

**LEMMA 2.** *If the density* $f(\theta)$ *of the absolutely continuous part of the measure* $d\sigma(\theta)$ *satisfies the condition* $f^{-1} \in L^1(-\pi,\pi)$, *then*

$$\chi_N(J) \leq (1/2\pi) \int_{-\pi}^{\pi} (\sin^2(N\theta/2)/\sin^2(\theta/2))(1/f(\theta))d\theta. \tag{16}$$

PROOF. Since $J = J_a + J_s \geq J_a$, then $J^{-1} < J_a^{-1}$, and consequently

$$\chi_N(J) \leq \chi_N(J_a). \tag{17}$$

Let us suppose first that the density $f(\theta)$ satisfies the more restrictive condition

$$0 < c < f(\theta) < C < \infty \quad (a.e.). \tag{18}$$

Under these conditions, the operator $\hat{J}_f$ of multiplication by the function f in $L^2(-\pi,\pi)$ is a continuous, positive and with continuous inverse operator, such that

$$\hat{J}_f^{-1} = \hat{J}_{f-1}.$$

Let us consider in $L^2(-\pi,\pi)$, the subspace $Q_N = \{\phi ; \phi = \sum_{k=0}^{N-1} \xi_k e^{ik\theta}\}$ of trigonometric polynomials, and let $P_N$ be the orthogonal projection from $L^2(-\pi,\pi)$ onto

$Q_N$. Define $\hat{J}_{f,N} = P_N \hat{J}_f | Q_N$. The matrix of the operator $\hat{J}_{f,N}$ in the orthogonal basis $\{\xi^k = e^{ik\theta}, k = 0, 1, \ldots, N - 1\}$ coincides with the matrix of $J_{a,N}$. Therefore, the matrix of the inverse operator $\hat{J}_{f,N}^{-1}$ coincides with the matrix of $\hat{J}_{a,N}^{-1}$. Taking

$$\hat{e}_N = (1/\sqrt{N}) \sum_{k=0}^{N-1} \xi^k = (1/\sqrt{N})((1 - \xi^N)/(1 - \xi)).$$

we have that

$$(\hat{J}_{f,N}^{-1}\hat{e}_N, \hat{e}_N) = (1/N) \sum_{s,k=0}^{N-1} (\hat{J}_{f,N}^{-1}\xi^k, \sigma^s) = (1/N)(J_{a,N}^{-1}e_N, e_N) = \chi_N(J_a).$$

Putting in the general inequality (1) $A = \hat{J}_f$, $P_1 = P_N$, $P_2 = I - P_N$, we obtain

$$\hat{J}_{f,N}^{-1}P_N \leq \hat{J}_f^{-1} = \hat{J}_{f^{-1}}.$$

Therefore

$$\chi_N(J_a) = (\hat{J}_{f,N}^{-1}\hat{e}_N, \hat{e}_N) \leq \hat{J}_{f^{-1}}\hat{e}_N, \hat{e}_N) = (1/2\pi N) \int_{-\pi}^{\pi} (\sin^2(N\theta/2)/\sin^2(\theta/2)f^{-1}(\theta)d\theta.$$

By (17) it remains to get rid of the supplementary condition (18). For this, let us remark that the functions

$$f_n(\theta) = \begin{cases} n^{-1}, & f(\theta) < n \\ f(\theta), & n - 1 \leq f(\theta) < n, \\ n, & f(\theta) > n \end{cases}$$

satisfy the condition (18) and converge to $f$ as $n \longrightarrow \infty$; moreover $f_n^{-1}$ converge to $f^{-1}$ in $L^1(-\pi,\pi)$. Taking into account the continuous dependence of the entries of the matrix $J$ with respect to $f$, and the same for the right hand side of (16) with respect to $f^{-1}$, we obtain (by passing to the limit) that (16) holds when $f^{-1} \in L^1(-\pi,\pi)$.

If the density $f$ is continuous at $\theta = 0$ and $f(0) > 0$, then the function $f^{-1}(\theta)$ is also continuous at $\theta = 0$. If, in addition, $f^{-1} \in L^1(-\pi,\pi)$, then

$$\lim_{N\to\infty} (1/2\pi N) \int_{-\pi}^{\pi} (\sin^2(N\theta/2)/\sin^2(\theta/2))(1/f(\theta))d\theta = 1/f(0).$$

Consequently, in these conditions, the relation (16) implies that

$$(19) \qquad\qquad\qquad \overline{\lim_{N\to\infty}} \chi_N(J) \leq 1/f(\theta).$$

Comparing the inequality (19) with the results of Lemma 1, we obtain:

**THEOREM 5.** *If the non-decreasing, bounded function $\sigma(\theta)$ satisfies the conditions:*

(i) *the density $f(\theta)$ of the absolutely continuous part $(1/2\pi)f(\theta)d\theta$ of the measure $d\sigma(\theta)$ is continuous at $\theta = 0$ and*

$$\lim_{N\to\infty} (1/N) \int_{-\pi}^{\pi} (\sin^2(N\theta/2)/\sin^2(\theta/2))d\sigma(\theta) = f(0) \; ;$$

(ii) $f^{-1} \in L^{-1}(-\pi,\pi)$,

*then the sequence $\chi_N(J)$, associated to the Toeplitz matrix $J$ generated by the function $\sigma(\theta)$, converges and*

(20)
$$\lim_{N\to\infty} \chi_N(J) = (1/f(0)).$$

**REMARK 1.** Condition (ii) of Theorem 5 can be weakened. For example, if $f(\theta)$ is a non-negative trigonometric polynomial of degree $\ell$, the this condition can be even dropped. In this case, using only the inequality (1) one can show that

(21)
$$\chi_N(J) \le (1 + (\ell + 1)/N)(1/f(0)),$$

and hence the inequality (19) holds. But Condition (ii) in Theorem 5 was imposed to f only for proving (19).

**5.**

We complete now the results obtained above by using the connection between the already considered problems and some results concerning the theory of orthogonal polynomials on the unit circle.

Let $J$ be the positive Toeplitz matrix generated by a non-decreasing function $\sigma(\theta)$. Let us consider the system of linear equations

(22)
$$\sum_{q=0}^{N-1} J_{p-q}X_q^{(N)} = 1, \qquad p = 0,\ldots,N-1.$$

We associate to the solution of the system (22) the trigonometric polynomial

$$Q_N(\zeta) = \sum_{q=0}^{N-1} X_q^{(N)}\zeta^q, \quad \zeta = e^{i\theta}, \quad -\pi \le \theta \le \pi.$$

It is obvious that the numbers $d_N(J)$ have the representation

(23)
$$d_N(J) = \sum_{q=0}^{N-1} X_q^{(N)} = Q_N(1).$$

Multiply now the $p^{th}$ eqation of the system (22) by $\zeta^p$ and add all the obtained equalities. Noting that the numbers $J_p$ are the trigonometric moments of the measure $d\sigma(\theta)$, we obtain that the system (22) can be written as

(24)
$$\int_{-\pi}^{\pi} [(1 - \zeta\overline{\zeta'})^N/(1 - \zeta\overline{\zeta'})]Q(\zeta')d\sigma(\theta) = (1 - \zeta^N)/(1 - \zeta), \quad \zeta = e^{i\theta}, \quad \zeta' = e^{i\theta'}.$$

If we replace in the left hand side of the equation (24) the non-decreasing function $\sigma(\theta)$ by any other function $\tilde\sigma(\theta)$ which satisfies the conditions

(25)
$$\int_{-\pi}^{\pi} \zeta^p d\sigma(\theta) = \int_{-\pi}^{\pi} \zeta^p d\tilde\sigma(\theta) = J_p, \quad p = 0,\ldots,N - 1,$$

and the polynomials $Q_N(\zeta)$ remain as above, then the equality (24) remains valid. It is known that among the non-decreasing functions satisfying the conditions (25) there exists a simple function $\sigma_o(\theta)$ which has a non-zero jump $h_o$ at the point $\theta = 0$ and, besides zero, it has at most $N - 1$ points of discontinuity $\theta_1,\ldots,\theta_{N-1}$ with corresponding jumps $h_1,\ldots,h_{N-1}$ [4]. Let us put this function in the place of $\sigma(\theta)$ in (24). In this case (24) can be written as

(26)
$$[(1 - \zeta^N)/(1 - \zeta)]Q_N(1)h_o + \sum_{s=1}^{N-1} [(1 - (\zeta\overline{\zeta}_s)^N)/(1 - \zeta\overline{\zeta}_s)]h_s Q_N(\zeta_s) =$$
$$= (1 - \zeta^N)/(1 - \zeta), \quad \zeta_s = e^{i\theta s}.$$

From (26) it follows that the polynomial $Q_N(\zeta)$ is uniquely defined by the conditions

(27)
$$Q_N(1) = 1/h_o, \quad Q_N(\zeta_s) = 0, \quad s = 1,\ldots,N - 1.$$

Let us form, using the moments $J_o,\ldots,J_{N-1}$, the system of polynomials

(28)
$$\phi_n(\zeta) = (1/\sqrt{\Delta_{n-1}\Delta_n})\det\begin{bmatrix} J_o & J_1 & & J_n \\ J_{-1} & J_o & & J_{n-1} \\ \cdot & & & \\ \cdot & & & \\ \cdot & & & \\ J_{-n+1} & J_{-n+2} & & J_1 \\ 1 & \zeta & & \zeta^n \end{bmatrix} \qquad \Delta_n = \det\begin{bmatrix} J_o & J_1 & & J_n \\ J_{-1} & J_o & & J_{n-1} \\ \cdot & & & \\ \cdot & & & \\ J_{-n} & J_{-n+1} & & J_o \end{bmatrix}$$

for n = 0, 1, ... ,N - 1.

The polynomials $\{\phi_\ell(\zeta)\}_0^{N-1}$ form an orthnormal system both in $L^2_{\sigma_0}(-\pi,\pi)$ and in $L^2_{\sigma_0}(-\pi,\pi)$. By (27) we infer

$$\int_{-\pi}^{\pi} Q_N(\zeta)\overline{\phi}_\ell(\zeta)d\sigma_0(\zeta) = h_0 Q_N(1)\overline{\phi}_\ell(1) + \sum_{s=1}^{N} h_s Q_N(\zeta_s)\overline{\phi}_\ell(\zeta_s) = \overline{\phi}_\ell(1)$$

so

$$Q_N(\zeta) = \sum_{\ell=0}^{N-1} \overline{\phi}_\ell(1)\phi_\ell(\zeta).$$

Therefore

(29)     $$d_N(J) = Q_N(1) = \sum_{\ell=0}^{N-1} |\phi_\ell(1)|^2, \qquad \chi_N(J) = (1/N) \sum_{\ell=0}^{N-1} |\phi_\ell(1)|^2,$$

and, consequently:

**COROLLARY 1** (of Theorem 3). *If $\{\phi_\ell(\zeta)\}_0^\infty$ is the system of orthogonal polynomials associated to the non-negative measure $d\sigma(\theta)$, $-\pi \leq \theta \leq \pi$, then*

(30)     $$\lim_{N\to\infty} \sum_{\ell=0}^{N-1} |\phi_\ell(1)|^2 = (1/M_0), \qquad M_0 = \sigma(+0) - \sigma(-0).$$

**COROLLARY 2** (of Theorem 5). *If the non-negative measure $d\sigma(\theta)$ and the density $f(\theta)$ of its absolutely continuous part satisfy the conditions* (i) *and* (ii) *of Theorem 5, then*

(31)     $$\lim_{N\to\infty} (1/N) \sum_{\ell=0}^{N-1} |\phi_\ell(1)|^2 = (1/f(0)).$$

Formula (30) has been established by Ia. L. Geronimus as early as 1948, see [5]. The limit relation in (31), under different assumptions on the function $f(\theta)$, was established by many authors, who worked in the theory of orthonormal polynomials. The most general result, which asserts that (31) is valid if

1) the function $\sigma(\theta)$ is absolutely continuous on the interval $[-\delta,\delta]$, $0 < \delta \leq \pi$, and the density $f(\theta) \geq m > 0$ is continuous:

2)

(32)     $$\int_{-\pi}^{\pi} \ln f(\theta)d\theta > -\infty,$$

was also obtained by Ia. L. Geronimus in [5]. Comparing this result with Theorem 5 it is clear that in the case when the density $f(\theta)$ is continuous and strictly positive in some neighborhood of $\theta = 0$, then the condition (ii) in Theorem 5 can be replaced by the weaker one (32).

**CONJECTURE.** If the function $\sigma(\theta)$ verifies only the first condition of Theorem 5, then the limit relations (20) and (31) are valid.

Let $\sigma(t)$ be a non-decreasing, bounded function on the interval $[-1,1]$ and let $J = (J_{j+k})_0^\infty$ be the positive Hankel matrix generated by the moments of the distribution $d\sigma(t)$. Consider also the canonical system $\{P_\ell(t)\}_0^\infty$ of orthogonal polynomials (with respect to the measure $d\sigma(t)$) on the interval $[-1,1]$.

By similar arguments as those used above, we obtain the formula

$$d_N(J) = \sum_{\ell=0}^{N-1} P_\ell^2(1).$$

**COROLLARY 3** (of Theorem 4). *If $\{P_\ell(t)\}_0^\infty$ is the system of orthogonal polynomials related to the measure $d\sigma(t)$ on the inteval $[-1,1]$, then*

$$\lim_{N\to\infty} \sum_{\ell=0}^{N-1} P_\ell^2(1) = (1/M_1), \qquad M_1 = \sigma(1) - \sigma(1-0).$$

The following is also true:

**THEOREM 6.** *If the positive Hankel matrix $J = (J_{j+k})_0^\infty$, constructed with the moments of the distribution $d\sigma(t)$ on the interval $[-1,1]$, is the matrix of a positive continuous Hilbert space operators, then*

$$\lim_{N\to\infty} \sum_{\ell=0}^{N-1} P_\ell^2(1) = +\infty.$$

The proof of Theorem 6 can be obtained using the relations which connect the orthogonal polynomials on the interval $[-1,1]$ and those on the unit circle, H. Widom continuity criterion for an operator generated by a positive Hankel matrix, and the constructions for Toeplitz matrices presented above. Due to its length, we omit this proof.

### 6.

Consider in each space $\mathbf{C}^N$ the vectors $e_N(\psi)$, $0 \leq \psi < 2\pi$, with the coordinates

$$[e_N(\psi)]_k = (1/\sqrt{N})e^{-ik\psi}, \qquad k = 0, 1, \ldots, N - 1.$$

For a positive Toeplitz matrix, besides the numbers $\chi_N(J)$, let us consider the quadratic forms

$$\chi_N(J;\psi) = (J_N^{-1} e_N(\psi), e_N(\psi)) = (1/N) \sum_{j,k=1}^{N} (J_N^{-1})_{j,k} e^{-i(k-j)}.$$

For a fixed $N$ and for each $\psi_0 \in [0,2\pi)$, the vectors $e_N(\psi_0 + (2\pi\ell/N))$, $\ell = 0,\ldots,N-1$, form an orthonormal basis in $\mathbf{C}^N$. Therefore

$$(33) \qquad S_N(J) = (1/N)\mathbf{Tr}\, J_N^{-1} = (1/N) \sum_{\ell=0}^{N-1} \psi_N(J; \psi_0 + (2\pi\ell/N));$$

the sum in (33) is not depending on $\psi_0$.

Suppose that the density $f(\theta)$ of the absolutely continuous part of the measure which generates the Toeplitz matrix $J$ is such that $f^{-1} \in L^1(-\pi,\pi)$. The same arguments used to obtain (16) and (12) lead to the inequalities

$$\chi_N(J;\psi) \leq (1/2\pi) \int_{-\pi}^{\pi} [\sin^2(N(\theta - \psi)/2)/\sin^2((\theta - \psi)/2)](1/f(\theta))d\theta,$$

$$(34)$$

$$\chi_N(J;\psi) \geq F_N^{-1}(\psi),$$

where $2\pi F_N(\psi)$ is the Cesaro mean $(c,1)$ of the measure $d\sigma(\theta)$. Noting that because

$$\sum_{\ell=0}^{N-1} \sin^2(N/2)(\theta - \psi_0 - (2\pi\ell/N))/\sin^2(\theta - \psi_0 - (2\pi\ell/N)/2) \equiv N.$$

and that, for any continuous periodic function $g$

$$(1/2\pi) \int_{-\pi}^{\pi} \sum_{\ell=0}^{N-1} g(\theta - \psi_0 - (2\pi\ell/N))d\theta = (N/2\pi) \int_{-\pi}^{\pi} g(\theta)d\theta,$$

from (33) and (34) we infer that

$$(35) \qquad (1/2\pi) \int_{-\pi}^{\pi} (1/F_N(\theta))d\theta \leq S_N(J) \leq (1/2\pi) \int_{-\pi}^{\pi} (1/f(\theta))d\theta.$$

As it is known (see [6]), for almost all $\theta \in [-\pi,\pi]$ we have

$$\lim_{N\to\infty} F_N(\theta) = 2\pi\sigma'(\theta) = f(\theta).$$

Therefore, from (35) and Fatou Lemma, it follows that

$$\lim_{N\to\infty} S_N(J) = (1/2\pi) \int_{-\pi}^{\pi} (1/f(\theta))d\theta.$$

If $1/f \notin L^1(-\pi,\pi)$, then replacing $J$ by $J_\lambda = J + \lambda I$, $\lambda > 0$, generated by the measure $d\sigma(\theta) + (\lambda/2\pi)d\theta$, we obtain that $S_N(J) > S_N(J_\lambda)$ and, by the first inequality in (35), we have

$$\lim_{N \to \infty} S_N(J) \geq \lim_{N \to \infty} S_N(J_\lambda) = (1/2\pi) \int_{-\pi}^{\pi} 1/(f(\theta) + \lambda)d\theta.$$

But in this case

$$(1/2\pi) \int_{-\pi}^{\pi} 1/(f(\theta) + \lambda)d\theta \xrightarrow[\lambda \to 0]{} \infty.$$

So, we proved

**THEOREM 7.** *For the positive Toeplitz matrix* $J$, *generated by the measure* $d\sigma(\theta)$, *we have*

$$(36) \qquad \lim_{N \to \infty} (1/N)\mathbf{Tr}\, J_N^{-1} = (1/4\pi^2) \int_{-\pi}^{\pi} d\theta/\sigma'(\theta).$$

**COROLLARY 3** (Szegö). *For the positive Toeplitz matrix* $J$, *generated by the measure* $d\sigma(\theta)$, *we have*

$$\lim_{N \to \infty} (1/N)\mathbf{ln}\,\det J_N = (1/2\pi) \int_{-\pi}^{\pi} \mathbf{ln}[2\pi\sigma'(\theta)]d\theta.$$

PROOF. Let us note that for any $x > 0$

$$\mathbf{ln}\,x = \int_0^{\infty} [1/(1 + u) - 1/(u + x)]du.$$

Therefore

$$(1/N)\mathbf{ln}\,\det J_N = (1/N)\mathbf{Tr}\,\mathbf{ln}\,J_N = \int_0^{\infty} [1/(1 + u) - S_N(J_u)]du,$$

where $J_u = J + uI$. By Theorem 7, for a sufficiently small $\lambda > 0$, we get

$$\overline{\lim_{N \to \infty}} (1/N)\mathbf{ln}\,\det J_N \leq \overline{\lim_{N \to \infty}} (1/N)\mathbf{ln}\,\det J_{\lambda,N} =$$

$$= \lim_{N \to \infty} \int_0^{\infty} [1/(1 + u) - S_N(J_{u+\lambda})]du = \int_0^{\infty} [1/(1 + u) - (1/2\pi) \int_{-\pi}^{\pi} d\theta/(f(\theta) + \lambda + u)]du =$$

$$= (1/2\pi) \int_{-\pi}^{\pi} \mathbf{ln}[f(\theta) + \lambda]d\theta.$$

On the other hand, from (35) we infer

$$(1/N)\ln \det J_n = \int_0^\infty [1/(1 + u) - S_N(J_n)]du = \int_0^\infty [1/(1 + u) - S_N(J_{1+u})]du -$$

$$- \int_0^1 S_N(J_u)du \geq \int_0^\infty [1/(1 + u) - (1/2\pi) \int_{-\pi}^\pi 1/(1 + u + f(\theta))d\theta]du -$$

$$- (1/2\pi)\int_0^1 du \int_{-\pi}^\pi d\theta/(f(\theta) + u) = (1/2\pi) \int_{-\pi}^\pi \ln f(\theta)d\theta.$$

Therefore

$$D_J = \lim_{N\to\infty} (1/N)\ln \det J_N = (1/2\pi) \int_{-\pi}^\pi \ln f(\theta)d\theta.$$

7.

Let us consider now the generalizations of Theorems 3, 5 and 7 for multi--dimensional analogues of positive Toeplitz matrices.

Let

$$\mathbf{T}^m = \{\theta : \theta = (\theta_1, \ldots, \theta_m), -\pi \leq \theta_s \leq \pi, s = 1, 2, \ldots, m\}$$

$$\mathbf{Z}^m = \{k ; k = (k_1, \ldots, k_m), k_s = 0, \pm 1, \ldots ; s = 1, 2, \ldots, m\}.$$

Embed canonically $\mathbf{Z}^m$ in the m-dimensional Euclidean space $\mathbf{R}^m$. Let $J(k)$ be a complex valued function on $\mathbf{Z}^m$. For a bounded domain $G \subseteq \mathbf{R}^m$ which satisfies the condition

(37)
$$G' = G \cap \mathbf{Z}^m \neq \{\emptyset\}$$

we consider in the linear space $L_G$ of complex-valued functions on G', the operator

$$(\hat{J}_G f)(k) = \sum_{\ell \in G'} J(k - \ell)f(\ell), \quad f \in L_G.$$

The function $J(k)$ will be called positive definite if for every bounded domain G for which G' is non-void and for every non-zero function $f \in L_G$ ($\sum_{k \in G'} |f(k)|^2 \neq 0$), we have

$$(\hat{J}_G f, f) = \sum_{k,\ell \in G'} J(k - \ell)f(\ell)\overline{f}(k) > 0.$$

By S. Bochner Theorem, the fact that $J(k)$ is positive definite is equivalent to the existence of the representation

(38)
$$J(k) = \int_{\mathbf{T}^m} e^{i\langle k,\theta \rangle} d\sigma(\theta), \quad \langle k,\theta \rangle = k_1\theta_1 + \ldots + k_m\theta_m$$

where $d\sigma(\theta)$ is a non-negative measure on $\mathbf{T}^m$.

For the positive definite function $J(k)$ and for any bounded domain $G$ which satisfies (37), the operator $\mathfrak{J}_G$ is invertible. Let us associate to each such operator the number

$$d_G(J) = \sum_{k,\ell \in G} \mathfrak{J}_G^{-1}(\ell,k) = (\mathfrak{J}_G^{-1} e_G, e_G),$$

where $e_G \in L_G$ is the function identically equal with the unit of $G'$. If $\tilde{G}$ is another bounded domain with $G \subset \tilde{G}$, then, by the inequality (2) we obtain that

$$(39) \qquad\qquad\qquad d_G(\mathfrak{J}) \leq d_{\tilde{G}}(J).$$

Let $d\sigma(\theta)$ be the measure from the representation (38), let $\{\delta_n\}$ be a descending sequence of intervals (parallelepipeds) in $\mathbf{T}^m$ which converges to the point $\theta_0 = (0, 0, , \ldots, 0)$ as $n \to \infty$, and

$$M_0 = \lim_{n \to \infty} \sigma(\delta_m) \geq 0.$$

**THEOREM 8.** *Let $G_1 \subset G_2 \subset \ldots$ be a sequence of bounded domains which satisfy the condition*

$$(40) \qquad\qquad \bigcup_{p=1}^{\infty} G_p = \mathbf{R}_T^m = \{x : x_1 > -1, \ldots, x_m > -1\}.$$

*Then*

$$(41) \qquad\qquad \lim_{n \to \infty} d_{G_n}(\mathfrak{J}) = (1/M_0).$$

PROOF. By our assumption, for each positive integer $N$ one finds a number $P(N)$ such that for all $P \geq P(N)$ the cube

$$V_N = \{x : 0 \leq x_1 < N, \ldots, 0 \leq x_m < N\} \subset \mathbf{R}^m$$

is included in $G$. On the other hand, using the boundedness of $G$ and the condition (40), for a sufficiently large $N$ we have $G_p' \subset V_N'$. Taking into account (39) we see that it is sufficient to prove (41) for the cubes $V_N$, $N = 1, 2, \ldots$ . But in this case the proof is the same as the proof of Theorem 3 excepting the minor difference that now

$$m_N(\mathfrak{J}) = (1/N^m) \int_{\mathbf{T}^m} \left( \prod_{s=1}^{m} \sin^2(N\theta_s/2)/\sin^2(\theta_s/2) \right) d\sigma(\theta).$$

Let $\nu(G)$ be the number of the elements of the set G'. Considering $\nu(G) \geq 1$, let us put

$$\chi_G(\hat{J}) = (1/\nu(G))d_G(\hat{J}).$$

Let us denote by $\mathbf{Z}_+^m$ the set of all multi-indices $\alpha = (N_s^{(\alpha)})_{s=1}^m$ of positive integers. We associate to each multi-index $\alpha \in \mathbf{Z}_+^m$ the parallelepiped $V_\alpha = \{x : 0 \leq x_s \leq N_s^\alpha, 1 \leq s \leq m\}$ and the number $\nu(\alpha)$ equal to the cardinal of $V_\alpha \cap \mathbf{Z}^m$.

**THEOREM 9.** *Let* $G_1 \subset G_2 \subset \ldots$ *be an increasing sequence of bounded domains with* $\bigcup_p G_p = \mathbf{R}_+^m$ *for which there exists an increasing system of parallelepipeds* $V_{\alpha_1} \subset V_{\alpha_2} \subset \ldots$, *such that* $V_{\alpha_p} \subset G_p \subset V_{\alpha_{p+1}}$ *and*

$$(42) \qquad\qquad \lim_{p \to \infty} (\nu(\alpha_p)/\nu(\alpha_{p+1})) = 1.$$

*Let* $d\sigma(\theta)$ *be a measure on* $\mathbf{T}^m$ *and* $J(k)$, $k \in \mathbf{Z}$, *the positive definite function generated by it. Suppose that the density* $f(\theta)$ *of the absolutely continuous part of* $d\sigma(\theta)$ *satisfies the conditions:*

1) $f(\theta)$ *is continuous at* $\theta_o = (0, \ldots, 0)$ *and*

$$\lim_{\min N_s \to \infty} \int_{\mathbf{T}^m} \prod_{s=1}^m (\sin^2(N_s\theta_s/2)/N_s\sin^2(\theta_s/2))d\sigma(\theta) = f(\theta_o)$$

2) $1/f \in L^1(\mathbf{T}^m)$.

*Then we have*

$$(43) \qquad\qquad \lim_{p \to \infty} \chi_{G_p}(\hat{J}) = 1/f(\theta_o).$$

PROOF. From the hypothesis

$$\nu(\alpha_p) \leq \nu(G_p) \leq \nu(\alpha_{p+1})$$

and by (42)

$$\lim_{p \to \infty} (\nu(\alpha_p)/\nu(G_p)) = \lim_{p \to \infty} (\nu(\alpha_{p+1})/\nu(G_p)) = 1.$$

According to (39), we obtain

$$(\nu(\alpha_p)/\nu(G_p))\chi_{\nu(\alpha_p)}(\hat{J}) \leq \chi_{G_p}(\hat{J}) \leq (\nu(\alpha_{p+1})/\nu(G_p))\chi_{\nu(\alpha_{p+1})}(\hat{J}).$$

Therefore it is sufficient to prove (43) for the increasing sequence of parallelepipeds

$V_{\alpha_1} \subset V_{\alpha_2} \subset \ldots, \ \bigcup_p V_{\alpha_p} = \mathbf{R}_+^m$. But this can be obtained as a consequence of a trivial generalization of the arguments in the proof of Theorem 5.

Some obvious modification of the construction presented in the proof of Theorem 7 enables us to obtain the conclusions of Theorem 7 and of its Corollary 3 in the case of multi-dimensional Toeplitz matrices. More precisely, we have:

**THEOREM 10.** *Let* $G_1 \subset G_2 \subset \ldots$ *be a sequence of domains as in Theorem 9 and suppose that the density* $f(\theta)$ *of the absolutely continuous part of the measure* $d\sigma(\theta)$ *on* $\mathbf{T}^m$ *(which generates a positive definite function* $J(k)$, $k \in \mathbf{Z}^m$*) satisfies the condition* $f^{-1} \in L^1(\mathbf{T}^m)$. *Then*

$$\lim_{p \to \infty} (1/\nu(G_p))\mathbf{Tr}\ \hat{J}_{Gp}^{-1} = (1/(2\pi)^m) \int_{\mathbf{T}^m} (1/f(\theta))d\theta \ ;$$

and (the multi-dimensional variant of Szego Theorem):

**COROLLARY 4.** [7]. *For the same sequence of domains as in Theorem 9 and for the multi-dimensional Toeplitz matrix generated by a non-negative measure on* $\mathbf{T}^m$ *with the density of its absolutely continuous part* $f(\theta)$, *we have:*

$$\lim_{p \to \infty} (1/\nu(G_p))\mathbf{ln\ det}\ \hat{J}_{G_p} = (1/(2\pi)^m) \int_{\mathbf{T}^m} \mathbf{ln}\ f(\theta)d\theta.$$

**8.**

Let $J(x)$ be a continuous hermitian-positive function on $\mathbf{R}^m$. By S. Bochner Theorem the function $J(x)$ is the Fourier transform of a bounded measure $d\sigma(h)$ on $\mathbf{R}^m$:

$$J(x) = \int_{\mathbf{R}^m} e^{i\langle k,x\rangle} d\sigma(k).$$

For an arbitrary bounded domain $G \subset \mathbf{R}^m$, denote by $\hat{J}_G$ the non-negative nuclear operator on $L^2(G)$ which corresponds to the quadratic form

$$\langle \hat{J}_G g, h\rangle = \int_G \int_G dx\, dx'\, J(x - x')g(x')\overline{h}(x).$$

Let $I_G$ be the identity operator on $L^2(G)$ and let $Q_G(x,x')$ be the non-negative nuclear operator $\hat{Q}$ obtained from $\hat{J}_G$ by

$$I_G = \hat{Q}_G = (I_G + \hat{J}_G)^{-1}.$$

Let $e_G(x)$ be the function on G identically equal with the unit of G. From the positivity of the operator $\hat{Q}_G$ we infer that

$$r_G(J) = (\hat{Q}_G e_G, e_G) = \iint_{GG} dx\, dx'\, Q_G(x, x') \leq V_G,$$

where $V_G$ is the volume (measure) of the domain G. Trivial generalizations of the constructions made in the proofs of Theorems 2, 5, 7-10, lead to the following result:

**THEOREM 11.** *Let* $G_1 \subset G_2 \subset \dots$ *be a sequence of domains as in Theorem 9.*

*a) If the measure* $d\sigma(k)$ *which generates the continuous hermitian-positive form* $J(x)$ *has a finite "mass"* $M_0$ *in* $k_0 = (0, \dots, 0)$, *then*

$$\lim_{p \to \infty} [V_{G_p} - r_{G_p}(J)] = 1/M_0 ;$$

*b) if* $M_0 = 0$, *then the density* $f(k)$ *of the absolutely continuous part* $1/(2\pi)^m f(k) d\sigma(k)$ *of the measure* $d\sigma(k)$ *is continuous in* $k_0$ *and the Fejér means of the Fourier integral of the measure* $d\sigma(k)$ *converge in* $k_0$ *to* $f(k_0)$, *then*

$$\lim_{p \to \infty} (1/V_G) r_G(J) = f(k_0)/(1 + f(k_0));$$

c)

$$\lim_{p \to \infty} (1/V_{G_p})\mathrm{Tr}\, \hat{Q}_{G_p} = (1/(2\pi)^m) \int_{R^m} (f(k)/(1 + f(k)))dk$$

d)

$$\lim_{p \to \infty} (1/V_{G_p})\det [I_{G_p} + \hat{Q}_{G_p}] = (1/(2\pi)^m) \int_{R^m} \ln[1 + f(k)]dk.$$

## REFERENCES


1. **Vladimirov, V.S. ; Volovich, I.V.** : The diophantic moment problem, orthogonal polynomials and some models of statistical physics, in *Lecture Notes in Math.*, Springer Verlag, 1984, v. **1043**, pp. 289-292.

2. **Vladimirov, V.S. ; Volovich, I.V.** : On a model of statistical physics (Russian), *Teoret. Mat. Fiz.* **54**:1(1983), 8-22.

3. **Sahnovich, A.L. ; Spitkovskii, I.M.** : On block-Toeplitz matrices and properties of the Gauss model on the half-line related to them (Russian), *Teoret. Mat. Fiz.* **63**:1(1985), 154-160.

4.      **Krein, M.G. ; Nudelman, A.A.** : *The Markov problem of moments and extremal problems* (Russian), Nauka, Moscow, 1973.

5.      **Geronimus, Ia.L.** : *Orthogonal polynomials on the circle and on an interval* (Russian), Fizmathgiz, Moscow, 1959.

6.      **Zygmund, A.** : *Trigonometrical series, vol. I* (Russian), Mir, Moscow, 1965, pp. 174-176.

7.      **Linnik, Yu.V.** : Multi-dimensional analogous of Szegö limit theorem (Russian), *Izv. Akad. Nauk. SSSR* 9(1975), 1323-1332.

**V.M. Adamjan**
Department of Physics,
State University of Odessa,
USSR.

# MULTICYCLIC SYSTEMS OF COMMUTING OPERATORS

**Ernst Albrecht** and **Peter Maria Wirtz**

## 0. INTRODUCTION

In this note we establish multidimensional variants of some of the analytic structure results of D.A. Herrero [6], [8], [9], [10] (see also Chapter 11 of the monograph [2]). Roughly speaking, Herrero proved among other statements that n–multicyclic operators on an infinite dimensional (complex) Banach space $X$ with the additional property that the set

$$\sigma_p^n(T^*) := \{z \in \mathbf{C} \; ; \dim \ker(z - T^*) = n\}$$

has non–empty interior behave "almost" like the operator of multiplication with the independent variable on the Banach space $A(D)^n$ for some open disc D contained in the spectrum $\mathbf{sp}(T,X)$ of T. Here, $T^*$ is the transposed of T and $A(D)$ denotes the Banach algebra of all those continuous functions on $\bar{D}$ which are analytic on D. Of course, in the case of N–tuples of commuting linear operators on $X$, several new phenomena occur.

The following section contains the necessary definitions and some basic general facts concerning n–multicyclic systems of commuting operators. In the second part we then investigate the analytic structure of n–multicyclic N–tuples $T = (T_1, \ldots, T_N)$ under the additional condition that the set

$$\sigma_p^n(T^*) := \{z \in \mathbf{C}^N \; ; \dim \bigcap_{j=1}^{N} \ker(z_j - T_j^*) = n\}$$

has non–empty interior.

## 1. GENERAL FACTS CONCERNING MULTICYCLICITY

Let $B$ be a subalgebra of the Banach algebra $L(X)$ of all (bounded) linear operators on an infinite dimensional (complex) Banach space $X$. As usual the *multiplicity* $\mu(B)$ of $B$ is defined as

$$\mu(B) := \mathbf{inf}\{\mathbf{card}(\Gamma) \; ; \Gamma \subseteq X \text{ and } X = \bigvee\{Ax \; ; A \in B, x \in \Gamma\}\}$$

where $\bigvee\{\ldots\}$ denotes the closed linear span of $\{\ldots\}$ in $X$. If $\mu(B) = m < \infty$ and if

$y_1, \ldots, y_m \in X$ satisfy $\bigvee \{ A y_j ; A \in B, \; 1 \leq j \leq m \} = X$, then $y = (y_1, \ldots, y_m)$ is called a *multicyclic* m-*tuple for B*. The set of all multicyclic m-tuples for $B$ will be denoted by $C_m(B)$.

Let now $T = (T_\lambda)_{\lambda \in \Lambda}$ be a family of commuting operators in $L(X)$. The compatibility properties of the Taylor spectrum [12], [14] and of the Taylor analytic functional calculus [13], [14] allow a natural extension of the Taylor theory to the case of infinite systems. See for example [11], [1] for the definition of the joint Taylor spectrum $\mathbf{sp}(T, X)$ in the case that $\mathbf{card}(\Lambda) = \infty$. A function f which is defined in an open neighborhood of $\mathbf{sp}(T, X)$ in $\mathbf{C}^\Lambda$ will be called analytic, if there exists a finite subset M of $\Lambda$ such that $f = g \circ \pi_M^\Lambda$ for some function g which is analytic in a neighborhood of the spectrum $\mathbf{sp}(T_M, X)$ of $T_M := (T_\mu)_{\mu \in M}$. Here, $\pi_M^\Lambda : \mathbf{C}^\Lambda \to \mathbf{C}^M$ denotes the natural projection given by $\pi_M^\Lambda(z) := (z_\mu)_{\mu \in M}$ for $z = (z_\lambda)_{\lambda \in \Lambda} \in \mathbf{C}^\Lambda$. We then may define $f(T) := g(T_M)$ where $g \longmapsto g(T_M)$ is the Taylor analytic functional calculus for $T_M$. We are interested in the following subalgebras of $L(X)$:

$\mathbf{C}[T]$, the algebra of all polynomials in $T = (T_\lambda)_{\lambda \in \Lambda}$,

$\mathbf{C}(T)$, the algebra of all rational functions of T (i.e. the set of all operators of the form $p(T)q(T)^{-1}$, where p, q are polynomials, and where q has no zeros on $\mathbf{sp}(T, X)$),

$\mathbf{Hol}(T)$, the range of the Taylor analytic functional calculus for T.

T is called m-*multicyclic* (resp. *rationally* m-*multicyclic*, resp. *analytically* m-*multicyclic*) if $\mu(\mathbf{C}[T]) = m$ (resp. $\mu(\mathbf{C}(T)) = m$, resp. $\mu(\mathbf{Hol}(T)) = m$). Of course, we have $\mu(\mathbf{Hol}(T)) \leq \mu(\mathbf{C}(T)) \leq \mu(\mathbf{C}[T])$.

We shall need multi-index notation. To this purpose, we define

$$\mathbf{N}_o(\Lambda) := \{ \alpha = (\alpha_\lambda)_{\lambda \in \Lambda} \in \mathbf{N}_o^\Lambda ; \alpha_\lambda \neq 0 \text{ for at most finitely many } \lambda \in \Lambda \} .$$

For a multi-index $\alpha \in \mathbf{N}_o(\Lambda)$ we write

$$|\alpha| := \sum_{\lambda \in \Lambda} \alpha_\lambda, \quad \alpha! := \prod_{\lambda \in \Lambda} \alpha_\lambda!, \quad T^\alpha := \prod_{\lambda \in \Lambda} T^{\alpha_\lambda}, \quad \partial^\alpha := \prod_{\lambda \in \Lambda} \frac{\partial^{\alpha_\lambda}}{\partial z_\lambda^{\alpha_\lambda}}, \quad w^\alpha := \prod_{\lambda \in \Lambda} w_\lambda^{\alpha_\lambda},$$

where $w = (w_\lambda)_{\lambda \in \Lambda} \in \mathbf{C}^\Lambda$. Notice, that the sum and the products are finite. If $q \in \mathbf{C}[Z_\lambda ; \lambda \in \Lambda]$ is an analytic polynomial, we have for the Taylor expansion of q around $z \in \mathbf{C}^\Lambda$ at $w \in \mathbf{C}^\Lambda$,

$$(1) \qquad q(w) = \sum_{|\alpha| \geq 0} \alpha!^{-1} (\partial^\alpha q)(z)(w - z)^\alpha$$

where the sum is finite, since $\partial^{\alpha}q \neq 0$ only for finitely many $\alpha \in \mathbf{N}_o(\Lambda)$.

We also introduce the notations $T^* := (T_\lambda^*)_{\lambda \in \Lambda}$,

$$\ker(T) := \bigcap_{\lambda \in \Lambda} \ker(T_\lambda),$$

and, for $\beta \in \mathbf{N}^N$, $T^{[\beta]} := (T_\lambda^{\beta_\lambda})_{\lambda \in \Lambda}$. Thus, we have especially,

$$\ker(T^{[\beta]}) = \bigcap_{\lambda \in \Lambda} \ker(T_\lambda^{\beta_\lambda}).$$

As in the case of single operators [6], Proposition 1(i), (see also [2], Proposition 11.1), we have a lower estimate for $\mu(\mathbf{C}(T))$:

**1.1. PROPOSITION.** *Let* $T = (T_\lambda)_{\lambda \in \Lambda}$ *be a system of commuting operators in* $L(X)$. *Then,*

$$\mu(\mathbf{C}[T]) \geq \mu(\mathbf{C}(T)) \geq \sup\{\dim \ker(z - T^*) \,;\, z \in \mathbf{C}^\Lambda\}.$$

PROOF. If $z \in \mathbf{C}^\Lambda \setminus \mathbf{sp}(T,X)$, then $z_M = (z_\mu)_{\mu \in M} \in \mathbf{C}^M \setminus \mathbf{sp}(T_M,X)$ for some finite subset M of $\Lambda$. It follows that

$$\{0\} = \bigcap_{\mu \in M} \ker(z_\mu - T_\mu^*) \supseteq \ker(z - T^*)$$

and hence $\dim \ker(z - T^*) = 0 \leq \mu(\mathbf{C}(T))$.

Fix now $z \in \mathbf{sp}(T,X)$ and consider $R := \bigvee_{\lambda \in \Lambda} \mathbf{ran}(z_\lambda - T_\lambda) = {}^\perp\ker(z - T^*)$. Thus, $R \in \mathbf{Lat}(\mathbf{C}(T))$ and $\dim X/R = \nu := \dim \ker(z - T^*)$. For fixed $y_1,\ldots,y_m \in X$, we define $M := \bigvee\{R,y_1,\ldots,y_m\} \in \mathbf{Lat}(\mathbf{C}[T])$ and $Y := \bigvee\{Ay_j \,;\, A \in \mathbf{C}(T), 1 \leq j \leq m\} \in \mathbf{Lat}(\mathbf{C}(T))$. It follows that $\dim M/R \leq m < \infty$.

We show now, that $Y \subseteq M$. To prove this, consider $y = \sum_{j=1}^m A_j y_j$ , where $A_j = p_j(T)q_j(T)^{-1} \in \mathbf{C}(T)$ with analytic polynomials $p_j, q_j$, $q_j \neq 0$ on $\mathbf{sp}(T,X)$. Write $x_j := q_j(T)^{-1}y_j$. From the Taylor expansion (1) of $q_j$ we obtain by means of the analytic functional calculus

$$(2) \qquad y_j = q_j(T)x_j = q_j(z)x_j + \sum_{|\alpha|>0} \alpha!^{-1}(\partial^\alpha q_j)(z)(T - z)^\alpha x_j .$$

Since $z \in \mathbf{sp}(T,X)$, we have $q_j(z) \neq 0$. Therefore, (2) implies $x_j = q_j(T)^{-1}y_j \in M$ for $j = 1,\ldots,m$. M being invariant for $\mathbf{C}[T]$, we conclude that $y \in M$.

Now, if $m = \mu(\mathbf{C}(T)) < \infty$ and if $y_1,\ldots,y_m$ are chosen such that $Y = X$, we obtain

$M = X$ and hence, $\nu = \dim X/R = \dim M/R \leq m$.                                   $\square$

Already now we can see some fundamental difference between the single operator theory and the several variable case.

**1.2. PROPOSITION.** (a) *If* $T \in L(X)$ *is a single operator, then* $\mu(\mathbf{Hol}(T)) = \mu(\mathbf{C}(T))$.

(b) *For* $N \geq 2$ *and* $n \in \mathbf{N} \cup \{\infty\}$ *there exist* $N$-*tuples* $T = (T_1, \ldots, T_N)$ *of*
*commuting bounded linear operators on some Banach space* $X$ *such that*

$$\mu(\mathbf{Hol}(T)) = 1 \leq n \leq \sup_{z \in \mathbf{C}^N} \dim \ker(z - T^*) \leq \mu(\mathbf{C}(T)).$$

PROOF. (a) By the Runge theorem the set of rational functions with poles off
$\mathbf{sp}(T,X)$ is dense in the algebra $O(\mathbf{sp}(T,X))$ of all (equivalence classes of $\mathbf{C}$-valued)
functions f which are holomorphic in some neighborhood $U_f$ of $\mathbf{sp}(T,X)$. $O(\mathbf{sp}(T,X))$ is
endowed with its natural inductive limit topology. Since the analytic functional calculus
is continuous with respect to this topology, we see that $\mathbf{C}(T)$ is dense in $\mathbf{Hol}(T)$.
Therefore, $\mu(\mathbf{Hol}(T)) = \mu(\mathbf{C}(T))$.

(b) For a compact set $K \subseteq \mathbf{C}^N$ let $A(K)$ be the closure of $O(K)$ in $C(K)$ (with
respect to the sup-norm). Thus, the $N$-tuple $T = (T_1, \ldots, T_N)$ in $L(A(K))$ defined by
$(T_j f)(z) := z_j f(z)$ for $f \in A(K)$, $z \in K$, $1 \leq j \leq N$, is analytically cyclic. For $w \in K$, let $\Delta_w$ be
the set of all those multiplicative linear functionals $\phi$ on $A(K)$ satisfying $(\phi(\pi_1), \ldots$
$\ldots, \phi(\pi_N)) = w$, where $\pi_j(z) := z_j$ for $z \in K$. We show that for each $n \in \mathbf{N} \cup \{\infty\}$ there
exists a compact set $K_n \subset \mathbf{C}^N$ such that

$$(3) \qquad\qquad \sup_{w \in K_n} \dim \bigvee \Delta_w \geq n .$$

Since $\bigvee \Delta_w \subseteq \ker(w - T^*)$, we obtain from Proposition 1.1 that $\mu(\mathbf{C}(T)) \geq n$.

The construction of $K_n$ is based on Hartogs phenomena. We follow a construc-
tion in [5], p. 43. For $z = (z_1, \ldots, z_N) \in \mathbf{C}^N$ we write $z' := (z_2, \ldots, z_N)$. If $w \in \mathbf{C}^N$ and
$r > 0$ are given, we denote by $D(w,r)$ the closed polydisc with centre w and radius r. We
introduce the Hartogs figure

$$H(w,r) := \left\{ z \in D(w,r) ; \; |z_1 - w_1| > r/4 \Longrightarrow \| z' - w' \|_\infty \geq 3r/4 \right\} .$$

For $0 \leq \alpha < 2\pi$, $k \in \mathbf{N}$, and $0 < r \leq 1$, we define the following compact sets:

$$A(\alpha,r,k) := \bigcup_{j=1}^{k} H(((1 - 4^{-j}r)e^{i\alpha},0),4^{-j}r)$$

$$\Gamma(\alpha,r,k) := \{((1 + \tfrac{r}{2}(1 - e^{it}))e^{i\alpha}, ((1 - t)13r + 15t\tfrac{r}{4})4^{-k-2}) ; 0 \le t \le 1\}$$

$$V(\alpha,r,k) := \{z \in \mathbf{C}^N; \mathrm{dist}(z,\Gamma(\alpha,r,k)) \le 4^{-k-3}r\}$$

$$\Gamma(\alpha,r,0) := \{((1 + \tfrac{r}{2}(1 - e^{it}))e^{i\alpha}, ((1 - t)13r + 15t\tfrac{r}{4})4^{-2}) ; 0 \le t \le 1\}$$

$$V(\alpha,r,0) := \{z \in \mathbf{C}^N ; \mathrm{dist}(z,\Gamma(\alpha,r,0)) \le 4^{-3}r\} \ .$$

Now, if $n \in \mathbf{N}$, we consider

$$K_n := H(0,1) \cup A(0,1,n - 1) \cup \bigcup_{j=0}^{\infty} V(0,1,j) \ .$$

Notice, that every function $f$ which is analytic in the interior of $H((1 - 4^{-j},0),4^{-j})$ has a unique analytic extension $E_j(f)$ to the open polydisc $U((1 - 4^{-j},0),4^{-j})$ with centre $(1 - 4^{-j},0)$ and radius $4^{-j}$ $(0 \le j < n)$. Thus, if $w \in U((1 - 4^{1-n},0), 4^{1-n})$, then $\phi_j$ with

$$\phi_j(f) := E_j(f \,|\, H((1 - 4^{-j},0),4^{-j}))(w)$$

is a multiplicative linear functional on $A(K_n)$ $(j = 0,1,\ldots,n - 1)$. To prove that $\phi_0,\ldots,\phi_{n-1}$ are linearly independent, note that the function $z \longmapsto (z_1 - 3/2)^{1/n}$ has a well defined branch $g$ which is analytic in a neighborhood of $K_n$. $g$ can be chosen such that

$$\phi_j(g) = \rho^{1/n}e^{i\alpha/n}\xi_j \qquad (0 \le j < n)$$

where $w_1 - 3/2 = \rho e^{i\alpha}$ $(0 \le \alpha < 2\pi)$ and $\xi_j := e^{(2j\pi/n)i}$ $(0 \le j < n)$. It follows that

$$\left| \det((\phi_j(g^k))_{j,k=0,\ldots,n-1}) \right| = \rho^{(n-1)/2} \prod_{o \le j < k < n} |\xi_k - \xi_j| \ne 0 \ .$$

This shows that $\dim \bigvee\{\phi_0 \ldots,\phi_{n-1}\} = n$ and we obtain (3).

In the case $n = \infty$, we choose a sequence $(r_k)_{k=1}^{\infty}$ in $(0,1)$ and a strictly increasing sequence $(\alpha_k)_{k=1}^{\infty}$ in $[0,2\pi]$ such that the polydiscs

$$D((1 - \tfrac{1}{4}r_k)e^{i\alpha_k}, \tfrac{1}{4}r_k) \qquad (k \in \mathbf{N})$$

are pairwise disjoint. We then define

$$K_{\infty} := H(0,1) \cup \bigcup_{k=1}^{\infty} (A(\alpha_k,r_k,k) \cup \bigcup_{j=0}^{k-1} V(\alpha_k,r_k,j)) \ .$$

Now, if $w \in \mathbf{int}\, H(((1 - 4^{-k} r_k)e^{i\alpha_k}, 0), 4^{-k} r_k)$, then as in the proof for $n < \infty$, we see that $\dim \bigvee \Delta_w \geq k + 1$. This shows that $\sup_{w \in K_\infty} \dim \bigvee \Delta_w = \infty$ and the proof is complete. $\qquad\square$

For a single operator $T \in L(X)$ it is well known (cf. [7], Lemma 4.3) that rationally multicyclic operators are already multicyclic. More precisely, one has

$$\mu(\mathbf{C}[T]) \leq \mu(\mathbf{C}(T)) + 1 .$$

We do not know if this statement has a natural generalization to the several variable case.

**1.3. PROBLEM.** *Is the implication*

$$\mu(\mathbf{C}(T)) < \infty \;\Rightarrow\; \mu(\mathbf{C}[T]) < \infty$$

*valid also for $N$-tuples of commuting operators?*

If the spectrum is a product of compact sets in $\mathbf{C}$, then the answer is positive.

**1.4. LEMMA.** *If $T = (T_1, \ldots, T_N)$ is a system of $N$ commuting operators in $L(X)$ such that $\mathbf{sp}(T,X) = \prod_{j=1}^{N} \mathbf{sp}(T_j, X)$, then*

$$\mu(\mathbf{C}[T]) \leq \mu(\mathbf{C}(T)) + N$$

*and $\mu(\mathbf{C}(T)) = \mu(\mathbf{Hol}(T))$.*

PROOF. The statement is nontrivial only for $\mu(\mathbf{C}(T)) = m < \infty$. Assume that $X = \bigvee\{Ax_j \,;\, A \in \mathbf{C}(T), 1 \leq j \leq m\}$. By the proof of Lemma 4.3 in [7], there exists $x_{m+1} \in X$ such that for all $z_1 \in \mathbf{C} \setminus \mathbf{sp}(T_1, X)$ and for $1 \leq k \leq m$ we have

$$(z_1 - T_1)^{-1} x_k \in M_1 := \bigvee \{p(T_1)x_j \,;\, p \in \mathbf{C}[Z], 1 \leq j \leq m + 1\} .$$

By the same argument, we find $x_{m+2} \in X$ such that for all $z_2 \in \mathbf{C} \setminus \mathbf{sp}(T_2, X)$ and for $1 \leq k \leq m + 1$ we have

$$(z_2 - T_2)^{-1} x_k \in M_2 := \bigvee \{p(T_2)x_j \,;\, p \in \mathbf{C}[Z], 1 \leq j \leq m + 2\} .$$

It follows that for $z_1 \in \mathbf{C} \setminus \mathbf{sp}(T_1, X)$, $z_2 \in \mathbf{C} \setminus \mathbf{sp}(T_2, X)$ we have

$$(z_2 - T_2)^{-1}(z_1 - T_1)^{-1} x_k \in N_1 := \bigvee \{p(T_1, T_2)x_j \,;\, p \in \mathbf{C}[Z_1, Z_2], 1 \leq j \leq m + 2\} .$$

Continuing this procedure, we find $x_{m+1}, \ldots, x_{m+N} \in X$ such that for $z_r \in \mathbf{C} \setminus \mathbf{sp}(T_r, X)$ $(1 \leq r \leq N)$ and $1 \leq k \leq m$ we have

$$( \prod_{r=1}^{N} (z_r - T_r)^{-1})x_k \in N_N := \vee \{p(T)x_j ; p \in \mathbf{C}[Z_1, \ldots, Z_N], 1 \leq j \leq m + N\}$$

If $f$ is any complex valued function which is analytic in a neighborhood U of $\mathbf{sp}(T,X) =$ $= \mathbf{sp}(T_1,X) \times \ldots \times \mathbf{sp}(T_N,X)$, we find neighborhoods $U_j$ of $\mathbf{sp}(T_j,X)$ and admissible contours $\Gamma_j$ surrounding $\mathbf{sp}(T_j,X)$ in $U_j$ $(1 \leq j \leq N)$ satisfying $U_1 \times \ldots \times U_N \subseteq U$. Then we have for the analytic functional calculus:

$$(4) \qquad f(T) = (2\pi i)^{-N} \int_{\Gamma_1} \ldots \int_{\Gamma_N} ( \prod_{j=1}^{N} (z_j - T_j)^{-1})f(z)dz_1 \ldots dz_N .$$

It follows that $f(T)x_j \in N_N$ for $j = 1, \ldots, N$ and hence

$$X = \vee \{Ax_j ; A \in \mathbf{C}(T), 1 \leq j \leq N\} \subseteq N_N \subseteq X .$$

Thus, $T = (T_1, \ldots, T_N)$ is multicyclic with $\mu(T) \leq m + N$. Moreover, in this special situation, we see from (4) that $\mu(\mathbf{C}(T)) = \mu(\mathbf{Hol}(T))$.                                          $\square$

If $K \subseteq \mathbf{C}^N$ is compact, then $\mathbf{Rat}(K)$ denotes the set of rational functions with singularities off K. Let us also note the following difference between the one and the several variable case:

**1.5. LEMMA.** *Let T be a rationally cyclic operator on an infinite dimensional Banach space X and let $x \in X$ be a cyclic vector for $\mathbf{C}(T)$.*

*(a) The mapping $f \longmapsto f(T)x$ from $\mathbf{Rat}(\mathbf{sp}(T,X))$ to X is injective.*

*(b) For $N \geq 2$, the system $S := (T, \ldots, T) \in L(X)^N$ is rationally cyclic with cyclic vector x for $\mathbf{C}(S)$ but the mapping $f \longmapsto f(T)x$ from $\mathbf{Rat}(\mathbf{sp}(T,X))$ to X is not injective.*

PROOF. (a) Assume that there is some $0 \neq f \in \mathbf{Rat}(\mathbf{sp}(T,X))$ with $f(T)x = 0$. Then, for every $g \in \mathbf{Rat}(\mathbf{sp}(T,X))$, we have $f(T)g(T)x = g(T)f(T)x = 0$. Since $\{g(T)x; g \in \mathbf{Rat}(\mathbf{sp}(T,X))\}$ is dense in $X$, we obtain $f(T) = 0$. By the spectral mapping theorem this implies $\mathbf{sp}(T,X) = \mathbf{sp}(T,X) \cap f^{-1}(0)$. Hence, the spectrum of T is a finite set; $\mathbf{sp}(T,X) = \{\zeta_1, \ldots, \zeta_k\}$. Consider the polynomial

$$p(z) := \prod_{j=1}^{k} (\zeta_j - z)^{m_j}, \quad z \in \mathbf{C},$$

where $m_j$ is the order of the zero $\zeta_j$ of $f$. If follows that $p = fg$, where $g := p/f \in \mathbf{Rat}(\mathbf{sp}(T,X))$. Therefore, $p(T) = g(T)f(T) = 0$ and we also must have $p(T^*) = 0$. A straightforward computation shows that this implies $\mathbf{dim\,ker}(\zeta_j - T^*) = \infty$ for some $j \in \{1,\ldots,k\}$. By Proposition 1.1 we obtain $\mu(T) = \infty$, which is a contradiction.

(b) We have $f(S)x = 0$ for $f \in \mathbf{C}[Z_1,\ldots,Z_N] \subset \mathbf{Rat}(\mathbf{sp}(T,X))$ with $f(z) := z_1 - z_2$ for all $z = (z_1,\ldots,z_N) \in \mathbf{C}^N$.                                                                    $\square$

**1.6. REMARK.** Of course, if $T = (T_1,\ldots,T_N)$ is a rational cyclic system of commuting operators on $X$ with cyclic vector $x$ for $\mathbf{C}(T)$ and if $\mathbf{int}(\mathbf{sp}(T,X)) \neq \emptyset$, then the mapping $f \mapsto f(T)x$ from $\mathbf{Rat}(\mathbf{sp}(T,X))$ to $X$ is injective. Indeed, if $f(T)x = 0$ for some $f \in \mathbf{Rat}(\mathbf{sp}(T,X))$, then, as in the single operator case, we obtain $f(T) = 0$ and thus $\mathbf{sp}(T,X) = \mathbf{sp}(T,X) \cap f^{-1}(0)$. Since $f$ is rational, we have $\mathbf{int}(f^{-1}(0)) \neq \emptyset$ only if $f \equiv 0$.

The following two elementary facts will be needed later.

**1.7. LEMMA.** *Let* $T = (T_1,\ldots,T_N)$ *be a commutative system in* $L(X)$. *For all* $z \in \mathbf{C}^N$, $h \in O(\mathbf{sp}(T^*,X^*)) = O(\mathbf{sp}(T,X))$ *and* $\phi \in \mathbf{ker}(z - T^*)$ *we have,* $h(T)^*\phi = h(z)\phi = h(T^*)\phi$.

PROOF. We write $m_j$ for the operator of multiplication with the complex number $z_j$ on $\mathbf{C}$ resp. on $\mathbf{ker}(z - T^*)$ and denote by $\iota : \mathbf{ker}(z - T^*) \to X^*$ the canonical inclusion mapping. Because of the obvious intertwining relations

$$m_j\phi = \phi \circ T_j \text{ and } m_j\iota = \iota T_j^* \quad (1 \leq j \leq N).$$

The statement now follows from the compatibility properties of the analytic functional calculus (see for example [14], Corollary III 9.10).                                             $\square$

**1.8. LEMMA.** *Let* $T = (T_1,\ldots,T_N)$ *be a commutative system in* $L(X)$ *and suppose that* $B$ *is a subalgebra of* $\mathbf{Hol}(T)$ *such that* $\mu(B) = m < \infty$. *Then, for every* $z \in \sigma_p^{\geq m}(T^*) := \{z \in \mathbf{C}^N ; \mathbf{dim\,ker}(z - T^*) \geq m\}$ *and every multicyclic* $m$-*tuple* $y = (y_1,\ldots,y_m)$ *for* $B$, *there are* $\phi_1,\ldots,\phi_m \in \mathbf{ker}(z - T^*)$ *such that*

(5)                                     $$\mathbf{det}\big(<\phi_j,y_k>\big)_{1\leq j,k\leq m} \neq 0.$$

PROOF. Choose linear independent vectors $\phi_1,\ldots,\phi_m \in \mathbf{ker}(z - T^*)$ and assume that $\mathbf{det}\big(<\phi_j,y_k>\big)_{1\leq j,k\leq m} = 0$. Then there is some $0 \neq \alpha = (\alpha_1,\ldots\alpha_m) \in \mathbf{C}^m$ such that, with $\phi := \sum_{j=1}^{m} \alpha_j\phi_j$, we have $<\phi,y_k> = 0$ for $k = 1,\ldots,m$. If $A$ is an arbitrary operator in $B$ then $A$ is of the form $A = f(T)$ for some $f \in O(\mathbf{sp}(T,X))$. Hence, by Lemma 1.7, we obtain $<\phi,Ay_k> = <f(T)^*\phi,y_k> = f(z)<\phi,y_k> = 0$ for $k = 1,\ldots,m$. Since $y = (y_1,\ldots,y_m)$ is a multicyclic m-tuple for $B$ this implies $\phi = 0$ in contradiction to the linear independence of $\phi_1, \ldots, \phi_m$. Therefore, (5) must be true. $\qquad\square$

## 2. ANALYTIC STRUCTURE

We shall need the fact that every weak*-analytic function with values in a dual Banach space is already analytic with respect to the norm topology. The following slightly more general result which is due to A. Grothendieck ([4], Théorème 1, Remarques 1.5) is of interest also in other situations. For the sake of completeness, we include the proof.

**2.1. LEMMA.** *Let* $\langle E,F \rangle$ *be a dual system of Banach spaces satisfying the following two conditions:*

(i) $\| x \| = \sup\{ | \langle u,x \rangle | ; u \in E, \| u \| \leq 1\}$ *for all* $x \in F$.

(ii) *The* $\sigma(F,E)$-*closed convex hull of every* $\sigma(F,E)$-*compact set in F is* $\sigma(F,E)$-*compact.*

*Let* $f : \Omega \to F$ *be an F-valued function on an open set* $\Omega \subseteq \mathbf{C}^N$ *which is analytic with respect to the topology* $\sigma(F,E)$ *on F. Then f is analytic with respect to the norm topology on F.*

PROOF. Fix an arbitrary point $w \in \Omega$. We have to show, that $f$ has a power series representation around $w$ which converges uniformly in a neighborhood of $w$ with respect to the norm topology. To this purpose, fix $r > 0$ such that $D(w,r) \subseteq \Omega$ and let $K \subset U(w,r)$ be a compact neighborhood of $w$. We write $\partial_o D$ for the distinguished boundary of $D(w,r)$, i.e. $\partial_o D = \{ z \in \mathbf{C}^N ; | z_j - w_j | = r$ for $j = 1,\ldots,N\}$. Notice that for $\zeta \in \partial_o D$, $z \in K$, we have with $\eta := (1,\ldots,1)$,

$$(6) \qquad \prod_{j=1}^{N} (\zeta_j - z_j)^{-1} = \sum_{\alpha \in \mathbf{N}_o^N} \frac{(z-w)^\alpha}{(\zeta-w)^{\alpha+\eta}}$$

with uniform convergence on $\partial_o D \times K$. Recall, that for every $\sigma(F,E)$-continuous function $g : \partial_o D \to F$, the integral

$$\int_{\partial_o D} g(\zeta)d\zeta_1 \cdots d\zeta_N$$

exists in the sense of [3], Proposition 1.1, and satisfies

$$\left\| \int_{\partial_o D} g(\zeta)d\zeta_1 \cdots d\zeta_N \right\| \leq (2\pi r)^{-N} \sup_{\zeta \in \partial_o D} \| g(\zeta) \|$$

and

$$\forall u \in E : \langle u, \int_{\partial_o D} g(\zeta)d\zeta_1 \cdots d\zeta_N \rangle = \int_{\partial_o D} \langle u, g(\zeta) \rangle d\zeta_1 \cdots d\zeta_N .$$

Using this and the fact that f is $\sigma(F,E)$-analytic, we obtain from (6),

$$f(z) = (2\pi i)^{-N} \int_{\partial_o D} \left( \prod_{j=1}^{N} (\zeta_j - z_j)^{-1} \right) f(\zeta)d\zeta_1 \cdots d\zeta_N =$$

$$= (2\pi i)^{-N} \sum_{\alpha \in \mathbf{N}_o^N} \int_{\partial_o D} \frac{1}{(\zeta-w)^{\alpha+\eta}} f(\zeta)d\zeta_1 \cdots d\zeta_N (z - w)^{\alpha}$$

with uniform convergence on K with respect to the norm on F.                    $\square$

Let now $X$ be a Banach space. The conditions (i) and (ii) are then satisfied for
the dual systems $\langle X, X^* \rangle$ and $\langle X^*, X \rangle$. Thus, Lemma 2.1 implies the well known fact
that $X$-valued functions which are analytic with respect to the weak topology are
already analytic in the norm topology. We also obtain:

**2.2. COROLLARY.** *Let* $f : \Omega \to X^*$ *be an* $X^*$-*valued function on an open set*
$\Omega \subseteq \mathbf{C}^N$ *which is analytic in the weak*$^*$-*topology. Then, f is analytic with respect to the*
*norm topology.*

**2.3. COROLLARY.** *Let* $K \subseteq \mathbf{C}^N$ *be a compact set and denote by* $A(K)$ *the*
*closure of* $O(K)$ *in* $C(K)$ *with respect to the sup-norm* $\| \cdot \|_K$. *Then, for every closed*
*subspace F of* $A(K)$, *the function* $E : K \to F^*$, *defined by*

$$\langle f, E(z) \rangle := f(z) \quad for \ f \in F, \ z \in K,$$

*is weak*$^*$-*continuous on K and analytic on* $\mathrm{int}(K)$ *with respect to the norm topology.*

Let K be a given compact set in $\mathbf{C}^N$ and let $F$ be a commutative algebra of
continuous functions (or germs of functions on some compact set S) containing K in
their domains of definition. We also assume that $F | K := \{f | K; f \in F\}$ contains all holo-
morphic polynomials restricted to K. The closure $\overline{F | K}$ of $F | K$ with respect to $\| \cdot \|_K$ is
then a commutative Banach algebra. The Shilov boundary of the uniform algebra $\overline{F | K}$
will be denoted by $\partial_F K$. Recall, that $\partial_F K$ is the smallest subset B of K such that
$\| f \|_K = \| f \|_B$ for all $f \in \overline{F | K}$. We write $\Delta(F,K)$ for the character space of $\overline{F | K}$ and
define the $F$-hull $\hat{K}_F$ of K by

$$K_{\hat{F}} := \{(\chi(\pi_1), \ldots, \chi(\pi_N)) \, ; \, \chi \in \Delta(F, K)\}$$

where $\pi_1, \ldots, \pi_N$ are the coordinate functions on K ($\pi_j(z) = z_j$ for $1 \leq j \leq N$ and $z = (z_1, \ldots, z_N) \in K$). $K_{\hat{F}}$ is precisely the joint spectrum $\sigma(\pi, \overline{F \mid K})$ of $\pi = (\pi_1, \ldots, \pi_N)$ in $\overline{F \mid K}$ and coincides with the Taylor spectrum $\mathbf{sp}(M, \overline{F \mid K})$ of the system $M = (M_1, \ldots, M_N)$ of the operators of multiplication with $\pi_1, \ldots, \pi_N$ on $\overline{F \mid K}$. Let us mention some cases of special interest:

$F = \mathbf{C}[Z_1, \ldots, Z_N]$, the set of all analytic polynomials on $\mathbf{C}^N$. $K_{\hat{F}}$ is then the *polynomial convex hull* $K_{\hat{p}}$ of K. In this case, we write $P(K)$ instead of $\overline{F \mid K}$. If $N = 1$, then $K_{\hat{p}}$ is the union of K together with all bounded components of $\mathbf{C} \setminus K$ and $\partial_{\mathbf{C}[Z]} K$ coincides with the topological boundary of $K_{\hat{p}}$. In several complex variables, $\partial_{\mathbf{C}[Z_1, \ldots, Z_N]} K$ is in general a proper subset of the topological boundary of $K_{\hat{p}}$.

$F = \mathbf{Rat}(S)$, the set of all (germs of) rational functions with poles off S, where S is a compact set containing K. In the case $S = K$, we call $K_{\hat{F}}$ the *rational convex hull* of K and write $K_{\hat{F}} = K_{\hat{r}}$. In this special case, $\overline{F \mid K}$ will be denoted by $R(K)$.

$F = O(U)$, where U is a holomorphically convex domain containing K. In this case, $K_{\hat{O}(U)}$ is the $O(U)$-convex hull in the usual sense.

$F = O(S)$, where S is a compact set containing K. We shall write $A(K)$ instead of $\overline{O(K) \mid K}$.

Recall from [12] that the *spectral convex* hull $K_{\hat{s}}$ of a compact set K in $\mathbf{C}^N$ is by definition the set of all $w \in \mathbf{C}^N$ such that there do *not* exist functions $u_1, \ldots, u_N$ which are analytic in a neighborhood of K and satisfy $\sum_{j=1}^{N} (z_j - w_j) u_j(z) \equiv 1$ in a neighborhood of K.

**2.4. LEMMA.** $K_{\hat{s}} = K_{\hat{O}(K)}$ *for every compact set* $K \subseteq \mathbf{C}^N$.

PROOF. Since $K_{\hat{O}(K)} = \sigma(\pi, A(K))$, we obviously have $K_{\hat{O}(K)} \subseteq K_{\hat{s}}$. Conversely, if $w \in \mathbf{C}^N \setminus K_{\hat{O}(K)}$, then there are $u_1, \ldots, u_N \in A(K)$ such that $\sum_{j=1}^{N} (z_j - w_j) u_j(z) \equiv 1$ on K. These functions $u_1, \ldots, u_N$ can be approximated uniformly on K by functions which are analytic in (individual) neighborhoods of K. Hence, there are $g_1, \ldots, g_N$, analytic in a neighborhood U of K, such that the function

$$h(z) := \sum_{j=1}^{N} (z_j - w_j)g_j(z)$$

has no zero in U. With $f_j(z) := g_j(z)/h(z)$ $(z \in U)$, we have $\sum_{j=1}^{N} (z_j - w_j)f_j(z) \equiv 1$ on U. It follows that $K_s^\wedge \subseteq K_{O(K)}^\wedge$.                                                                        $\square$

We can now prove our first analytic structure result. In the single operator case, it contains Propositions 3 and 3a in [6] (cf. also [2], Proposition 11.3 and Proposition 11.10).

**2.5. PROPOSITION.** *Let $T = (T_1, \ldots, T_N)$ be a finite system of commuting operators on a Banach space $X$ and suppose that $\mu(B) = m < \infty$ for $B = \mathbf{C}[T]$, resp. $B = \mathbf{C}(T)$. Fix a multicyclic $m$-tuple $y = (y_1, \ldots, y_m) \in X^m$ for $B$, let $K \subseteq \mathbf{sp}(T,X)$ be a compact set and suppose that, with $F = \mathbf{C}[Z_1, \ldots, Z_N]$, resp. $F = \mathbf{Rat}(\mathbf{sp}(T,X))$, there exists a function $\phi : \partial_F K \rightarrow (X^*)^m$, $\phi(z) = (\phi_1(z), \ldots, \phi_m(z))$, and constants $\delta > 0$, $C > 0$, such that for all $z \in \partial_F K$, the following three conditions are fulfilled:*

    (i) $\phi_1(z), \ldots, \phi_m(z) \in \mathbf{ker}(z - T^*)$,

    (ii) $\| \phi(z) \| \leq C$, *and*

    (iii) $| \mathbf{det}\, H(z) | \geq \delta$, *where*

$$H(z) := \left( h_{j,k}(z) \right)_{j,k=1}^{m} := \left( \langle y_k, \phi_j(z) \rangle \right)_{j,k=1}^{m}.$$

*We also suppose that $\mathbf{int}(K_F^\wedge) \neq \varnothing$.*

*Then there exists a function $\Phi = (\Phi_1, \ldots, \Phi_m) : K_F^\wedge \rightarrow (X^*)^m$, such that the following statements hold:*

    (a) $\Phi(z) = \phi(z)H(z)^{-1}$ *for all $z \in \partial_F K$ ($\Phi(z)$, $\phi(z)$ are considered as row vectors in $(X^*)^m$, the elements of $X^m$ as column vectors).*

    (b) $\Phi$ *is weak$^*$-continuous on $K_F^\wedge$ and analytic on $\mathbf{int}(K_F^\wedge)$ with respect to the norm topology.*

    (c) *Let $A$ be the Banach algebra $\overline{F \mid K_F^\wedge}$. Then $J : X \rightarrow A^m$, defined by $(Jx)(z) := \left( \langle x, \Phi_1(z) \rangle, \ldots, \langle x, \Phi_m(z) \rangle \right)$ for $x \in X$, $z \in K_F^\wedge$, is a continuous linear operator with dense range satisfying*

$$(7) \qquad\qquad\qquad JT_j = M_j J \qquad for\ j = 1, \ldots, N.$$

*Here, $M_j \in L(A^m)$ is again the operator of multiplication with $\pi_j$ (i.e. $(M_j f)(z) = (z_j f_1(z), \ldots, z_j f_m(z))$ for $f = (f_1, \ldots, f_m) \in A^m$, $z \in K_F^\wedge$).*

(d) $K_F^\wedge \subseteq \sigma_p^m(T^*)$. *For all* $z \in K_F^\wedge$ *the vectors* $\Phi_1(z),\ldots,\Phi_m(z)$ *form a basis for* $\ker(z - T^*)$. *Moreover, for* $\eta := (1,\ldots,1) \leq \beta \in \mathbf{N}_0^N$, *a holomorphic basis* $\{\Phi_{j,\alpha}(\,\cdot\,);\ 1 \leq j \leq m,\ \alpha \leq \beta - \eta\}$ *for* $\ker(\,\cdot\,-T^*)^{[\beta]}$ *is given on* $\mathrm{int}(K_F^\wedge)$ *by*

$$\langle x,\Phi_{j,\alpha}(z)\rangle := (\partial^\alpha \langle x,\Phi_j(\,\cdot\,)\rangle)(z) \qquad \text{for } x \in X,\ z \in \mathrm{int}(K_F^\wedge)\,.$$

(e) *If* $\Psi \neq \varnothing$ *is a connected component of* $\mathrm{int}(K_F^\wedge)$ *and if* $\varnothing \neq U \subseteq \Psi$ *is open, then, for* $x \in X$, *the following statements are equivalent:*

($\alpha$) $\forall\, z \in U : x \in {}^\perp\ker(z - T^*)\,.$

($\beta$) $\exists\, z \in U\ \forall\, \beta \in \mathbf{N}^N : x \in {}^\perp\ker(z - T^*)^{[\beta]}.$

($\gamma$) $\forall\, z \in \Psi : x \in {}^\perp\ker(z - T^*).$

(f) $\ker J = \displaystyle\bigcap_{z \in \partial_F K} {}^\perp\ker(z - T^*) = \bigcap_{k=1}^{\infty} \bigcap_{\beta \in \mathbf{N}^N} {}^\perp\ker(w_k - T^*)^{[\beta]}$, *where* $(w_k)_{k=1}^\infty$ *is a sequence in* $\mathrm{int}(K_F^\wedge)$ *such that each connected component of* $\mathrm{int}(K_F^\wedge)$ *contains at least one* $w_k$.

PROOF. We write $B(\partial_F K)$ for the Banach algebra of all bounded complex valued functions on $\partial_F K$, endowed with the sup-norm $\|\cdot\|_{\partial_F K}$. For a bounded function $g$ on $\partial_F K$, we denote by $M(g) \in L(B(\partial_F K)^m)$ the operator of multiplication with $g$, i.e. $M(g)f = (gf_1,\ldots,gf_m)$ for $f = (f_1,\ldots,f_m) \in B(\partial_F K)^m$. Consider the linear operator $L : X \to B(\partial_F K)^m$ defined by

$$(Lx)(z) := \big(\langle x,\phi_1(z)\rangle,\ldots,\langle x,\phi_m(z)\rangle\big) \qquad \text{for } x \in X,\ z \in \partial_F K\,.$$

Then $\|L\| \leq C\sqrt{m}$ (if we endow $B(\partial_F K)^m$ with the norm given by

$$\|f\|_{\partial_F K}^2 = \sum_{j=1}^m \|f_j\|_{\partial_F K}^2 \qquad \text{for } x \in X,\ z \in \partial K_F^\wedge\,.$$

For all $x \in X$ and $z \in \partial_F K$ we have by Lemma 1.7,

$$(Lf(T)x)(z) = \big(\langle f(T)x,\phi_j(z)\rangle\big)_{j=1}^m = \big(\langle x,f(T^*)\phi_j(z)\rangle\big)_{j=1}^m = f(z)(Lx)(z)$$

and hence,

(8) $$Lf(T^*) = M(f)L\,.$$

Moreover, for an arbitrary (column vector) $f = (f_1,\ldots,f_m) \in F^m$ we obtain with $x :=$

$$:= \sum_{j=1}^{m} f_j(T)y_j \, ,$$

$$\forall \, z \in \partial_F K : (Lx)(z) = \sum_{j=1}^{m} (Lf_j(T)y_j)(z) = \sum_{j=1}^{m} f_j(z)(\langle y_j, \phi_i(z)\rangle)_{i=1}^{m} = H(z)f(z) \, .$$

Therefore,

$$(9) \qquad\qquad H(\cdot)^{-1}Lx = f \,\big|\, \partial_F K \, .$$

It follows that

$$(10) \qquad \|f\|_{\partial_F K} = \|H(z)^{-1}Lx\|_{\partial_F K} \leq \sup_{z \in \partial_F K} \|H(z)^{-1}\| \cdot \|L\| \cdot \|x\| \leq$$

$$\leq \sqrt{m} \cdot m! \cdot C^m \cdot \delta^{-1} \|x\| \, .$$

Since y is a multicyclic m-tuple for $B$, we see from (9) and (10),

$$(11) \qquad\qquad (F \,|\, \partial_F K)^m \subseteq H(\cdot)^{-1}\mathbf{ran}(L) \subseteq \overline{(F \,|\, \partial_F K)}^m \, .$$

By the definition of $\partial_F K$, the restriction mapping

$$R : A^m \longrightarrow \overline{(F \,|\, \partial_F K)}^m \, , \qquad f \mapsto f \,|\, \partial_F K$$

is an isometric isomorphism.

For $j = 1,\ldots,m$ consider the function $E_j : \hat{K}_F \longrightarrow (A^m)^* = (A^*)^m$ given by

$$\langle f, E_j(z)\rangle := f_j(z) \qquad \text{for} \ \ f = (f_1,\ldots,f_m) \in A^m, \ \ z \in \hat{K}_F \, .$$

By Corollary 2.3, $E_j$ is weak*-continuous on $\hat{K}_F$ and (norm-) analytic on $\mathbf{int}(\hat{K}_F)$. Hence, if we define $\Phi_j := (R^{-1}H(\cdot)^{-1}L)^* E_j$ , we see that $\Phi = (\Phi_1,\ldots,\Phi_m)$ fulfils (b). Moreover, it follows from the definition of $\Phi$, that the operator J in (c) coincides with $R^{-1}H(\cdot)^{-1}L$. From (11) and (8) we see that J has dense range and satisfies the intertwining condition (7). This completes the proof of (c).

If we denote the entries of $H(\cdot)^{-1}$ by $\tilde{h}_{i,j}(\cdot)$, $1 \leq i,j \leq m$, we have for all $x \in X$, $z \in \partial_F K$, $1 \leq j \leq m$,

$$\langle x, \Phi_j(z)\rangle = \langle x, J^* E_j(z)\rangle =$$

$$= \langle R^{-1}H(\cdot)^{-1}(\langle x,\phi_1(\cdot)\rangle,\ldots,\langle x,\phi_m(\cdot)\rangle), E_j(z)\rangle =$$

$$= \langle x, \sum_{i=1}^{m} \tilde{h}_{i,j}(z)\phi_i(z)\rangle \, .$$

This implies (a).

For the proof of (d), we first notice that for all $x \in X$, $w \in K_F^\wedge$, $1 \leq k \leq N$ we have

$$\left(\langle x,(w_k - T_k^*)\Phi_j(w)\rangle\right)_{j=1}^m = \left(\langle(w_k - T_k)x,\Phi_j(w)\rangle\right)_{j=1}^m = (J(w_k - T_k)x)(w) =$$

$$= ((w_k - M_k)Jx)(w) = 0$$

and hence $\Phi_1(w),\ldots,\Phi_m(w) \in \ker(w - T^*)$.

Now we show that for all $\beta \in \mathbf{N}^N$, $w \in \operatorname{int}(K_F^\wedge)$, $\zeta \in K_F^\wedge$, we have

(12) $$\mathsf{V}\{\Phi_{j,\alpha}(w); 1 \leq j \leq m, \alpha \leq \beta - \eta\} \subseteq \ker(w - T^*)^{[\beta]},$$

(13) $$\dim \mathsf{V}\{\Phi_{j,\alpha}(w); 1 \leq j \leq m, \alpha \leq \beta - \eta\} = m \prod_{j=1}^N \beta_j \leq \dim \ker(w - T^*)^{[\beta]},$$

and

(14) $$\dim \ker(\zeta - T^*) \geq m.$$

Fix an arbitrary $x \in X$, $w \in \operatorname{int}(K_F^\wedge)$, $\zeta \in K_F^\wedge$. For $1 \leq k \leq N$, $\alpha \leq \beta - \eta$, we have with $\tilde{\alpha} :=$

$$:= (a_1,\ldots,\alpha_{k-1},0,\alpha_{k+1},\ldots,\alpha_N),$$

$$\left(\langle x,(w_k - T_k^*)^{\beta_k}\Phi_{j,\alpha}(w)\rangle\right)_{j=1}^m = \left(\langle(w_k - T_k)^{\beta_k}x,\Phi_{j,\alpha}(w)\rangle\right)_{j=1}^m =$$

$$= \partial^\alpha\left(\langle(w_k - T_k)x,\Phi_j(z)\rangle\right)_{j=1}^m \big|_{z=w} =$$

$$= \partial^\alpha\left(J((w_k - T_k)^{\beta_k}x)\right)\big|_{z=w} =$$

$$= \partial^\alpha\left((w_k - z_k)^{\beta_k}Jx\right)\big|_{z=w} = \frac{\partial^{\alpha_k}}{\partial z_k^{\alpha_k}}\left((w_k - z_k)^{\beta_k}\partial^{\tilde{\alpha}} Jx\right)\big|_{z=w} = 0,$$

where we used (7) and the fact that $\alpha_k < \beta_k$. This proves (12). Recall that $\Phi_j(\cdot) \equiv J^* E_j(\cdot)$ on $K_F^\wedge$ and hence $\Phi_{j,\alpha}(\cdot) \equiv J^* \partial^\alpha E_j(\cdot)$ on $\operatorname{int}(K_F^\wedge)$. Since $J^*$ is injective and since $\{(\partial^\alpha E_j)(w); \alpha \leq \beta - \eta, 1 \leq j \leq m\}$ resp. $\{\Phi_1(\zeta),\ldots,\Phi_m(\zeta)\}$ is easily seen to be linearly independent, we obtain (13) and (14). By Proposition 1.1 it follows, that for all $\zeta \in K_F^\wedge$, the vectors $\Phi_1(\zeta),\ldots,\Phi_m(\zeta)$ form a basis of $\ker(\zeta - T^*)$. To complete the proof of (d), it now suffices to show that for all $\beta \in \mathbf{N}^N$, $z \in \operatorname{int}(K_F^\wedge)$,

(15)
$$\dim \ker(w - T^*)^{[\beta]} \leq m \prod_{j=1}^{N} \beta_j.$$

We proceed by induction. For $\beta = \eta$ the result has already been proved. Let now $\beta$ be a multi-index in $\mathbf{N}^N$ such that (15) holds for all $\gamma \in \mathbf{N}^N$ with $\gamma \leq \beta$ and let $\tilde{\beta} := \beta + \delta_j$, where $\delta_j = (\delta_{j,1}, \ldots, \delta_{j,N})$ and $\delta_{j,k}$ is the Kronecker symbol. We write $K := \ker(z - T^*)^{[\beta]}$ and $\tilde{K} := \ker(z - T^*)^{[\tilde{\beta}]}$. Of course, $K \subseteq \tilde{K}$. Assume now,

(16)
$$\dim \tilde{K}/K > d := m \prod_{j \neq k} \beta_j$$

and fix $x_1^*, \ldots, x_{d+1}^* \in \tilde{K}$ such that their equivalence classes $[x_1^*], \ldots, [x_{d+1}^*]$ in $\tilde{K}/K$ are linearly independent. Then the vectors

$$(z_k - T_k^*)^{\beta_k} x_1^*, \ldots, (z_k - T_k^*)^{\beta_k} x_{d+1}^*$$

are linearly independent in $X^*$. Indeed, if $c_1, \ldots, c_{d+1} \in \mathbf{C}$ satisfy $\sum_{i=1}^{d+1} c_i (z_k - T_k^*)^{\beta_k} x_i^* = 0$ then it is easily seen, that $\sum_{i=1}^{d+1} c_i [x_i^*] = 0$ in $\tilde{K}/K$ and hence $c_1 = c_2 = \ldots \ldots = c_{d+1} = 0$. Notice that because of $x_i^* \in \tilde{K}$, we have $(z_k - T_k^*) x_i^* \in \ker(z - T^*)^{[\gamma]}$, where $\eta \leq \gamma := \tilde{\beta} - \beta_k \delta_k \leq \beta$. Therefore,

$$\dim \ker(z - T^*)^{[\gamma]} \leq d = m \prod_{j=1}^{N} \gamma_j$$

which contradicts our induction hypothesis. Hence, the assumption (16) is false and we must have $\dim \tilde{K}/K \leq d = m \prod_{j \neq k} \beta_j$. It follows that

$$\dim \tilde{K} \leq \dim K + d \leq m \prod_{j=1}^{N} \beta_j + m \prod_{j \neq k} \beta_j = m \prod_{j=1}^{N} \tilde{\beta}_j.$$

This completes the proof of (15) and of (d).

For the proof of (e) and (f) notice that by (c) and (d) the following three statements are equivalent for $x \in X$, $z \in \hat{K}_F$:

$$x \in {}^{\perp}\ker(z - T^*),$$

$$\langle x, \Phi_j(z) \rangle = 0 \quad \text{for } j = 1, \ldots, m,$$

$$(Jx)(z) = 0.$$

Hence, the equivalences in (e) are straightforward consequences of the identity theorem for analytic functions. Moreover, $x \in \mathbf{ker}(J)$ if and only if $\langle x, \phi_j(z) \rangle \equiv 0$ on $K_F^{\wedge}$ for $j = 1, \ldots, m$, which is equivalent to $\langle x, \phi_j(z) \rangle \equiv 0$ on $\partial_F K$ for $j = 1, \ldots, m$. This proves the first equation in (f). The second one is again a consequence of the identity theorem.   $\square$

In the case $B = \mathbf{Hol}(T)$, $F = O(\mathbf{sp}(T,X))$, Proposition 1.1 is no longer applicable and Proposition 1.2 shows that the statements (d)-(f) of Proposition 2.5 are no longer valid in this situation. However, with the remaining parts of the proof of Proposition 2.5 we still obtain:

**2.5'. PROPOSITION.** *Let* $T = (T_1, \ldots, T_N)$ *be an analytically m-cyclic system of commuting operators in* $L(X)$ *and let* $y = (y_1, \ldots, y_m)$ *be a multicyclic m-tuple for* $\mathbf{Hol}(T)$. *Let* $K \subseteq \mathbf{sp}(T,X)$ *be a compact set and suppose that* $\mathrm{int}(K_O^{\wedge}) \neq \emptyset$, *where* $O := O(\mathbf{sp}(T,X))$. *Suppose also that there exist a function* $\phi = (\phi_1, \ldots, \phi_m) : \partial_O K \to (X^*)^m$ *and constants* $\delta > 0$, $C > 0$ *such that for all* $z \in \partial_O K$ *the conditions* (i)-(iii) *in Proposition 2.5 are satisfied. Then there exists a function* $\Phi = (\Phi_1, \ldots, \Phi_m) : \partial_O K \to (X^*)^m$ *such that the statements* (a)-(c) *in Proposition 2.5 hold (with F replaced by O). Moreover, we have*

(d') $K_O^{\wedge} \subseteq \sigma_p^{\geq m}(T^*) := \{ z \in \mathbf{C}^N ; \ \dim \ker(z - T^*) \geq m \}$ *and for all* $z \in \mathrm{int}(K_O^{\wedge})$,

$$\dim \ker(z - T^*)^{[\beta]} \geq m \prod_{j=1}^{N} \beta_j \ .$$

Since the conditions (i)-(iii) in Proposition 2.5 are somewhat technical, we describe now some situations in which they are fulfilled. $T = (T_1, \ldots, T_N)$ will always be a system of commuting operators in $L(X)$. As in 2.5 and 2.5', $B$ will denote one of the algebras $\mathbf{C}[T]$, $\mathbf{C}(T)$ resp. $\mathbf{Hol}(T)$, and $F$ will be the corresponding algebra of functions $\mathbf{C}[Z]$, $\mathbf{Rat}(\mathbf{sp}(T,X))$ resp. $O(\mathbf{sp}(T,X))$.

**2.6. REMARK.** Suppose, that $\mu(B) = m < \infty$. Let $K \subseteq \mathbf{C}^N$ be a compact set such that $\partial_F K \subseteq \sigma_p^m(T^*)$. If $\phi = (\phi_1, \ldots, \phi_m) : \partial_F K \to (X^*)^m$ is weak$^*$-continuous and satisfies

(i') $\forall z \in \partial_F K : \phi_1(z), \ldots, \phi_m(z)$ are linearly independent elements of $\mathbf{ker}(z - T^*)$,

then $\phi$ already fulfils (ii) and (iii) in Proposition 2.5. This can be proved precisely in the same way as in the single operator case (cf. [6], [2]).

The following variant of a result of J.L. Taylor [12] follows directly from the proof of the original result (cf. also [14], Lemma II.10.5).

**2.7. LEMMA.** ([12]). *Let* $X$, $Y$, $Z$ *be Banach spaces and let* $\Omega \subseteq \mathbf{C}^N$ *be an open set. Suppose that* $\alpha \in O(\Omega, L(X,Y))$, $\beta \in O(\Omega, L(Y,Z))$ *are analytic operator valued functions such that* $\beta(z)\alpha(z) \equiv 0$ *on* $\Omega$ *and* $\mathbf{ran}(\alpha(w)) = \mathbf{ker}(\beta(w))$ *for some* $w \in \Omega$. *Then, for every* $g \in O(\Omega, Y)$ *with* $\beta(z)g(z) \equiv 0$ *on* $\Omega$, *there are a neighborhood* $U$ *of* $w$ *and a function* $f \in O(U,X)$ *such that* $\alpha(z)f(z) \equiv g(z)$ *on* $U$. *Moreover, if* $g(w) = 0$ *and* $x_o \in \mathbf{ker}(\alpha(w))$, *then* $f$ *can be chosen such that* $f(w) = x_o$.

**2.8. COROLLARY.** *If, in the situation of Lemma 2.7, we have in addition* $\mathbf{ran}(\alpha(z)) = \mathbf{ker}(\beta(z))$ *and* $\dim \mathbf{ker}(\alpha(z)) = m < \infty$ *for all* $z \in \Omega$, *then* $\mathbf{ker}(\alpha(\,\cdot\,))$ *has locally a holomorphic basis, i.e. for every* $w \in \Omega$ *there exist a neighborhood* $U$ *of* $w$ *and functions* $f_1, \ldots, f_m \in O(U,X)$ *such that for every* $z \in U$ *the vectors* $f_1(z), \ldots, f_m(z)$ *form a basis for* $\mathbf{ker}(\alpha(z))$.

PROOF. Fix an arbitrary $w \in \Omega$ and let $x_1, \ldots, x_m$ be a basis for $\mathbf{ker}(\alpha(w))$. By Lemma 2.7, there exist a neighborhood $V$ of $w$ and functions $f_1, \ldots, f_m \in O(V,X)$ such that $\alpha(z)f_j(z) \equiv 0$ on $V$ and $f_j(w) = x_j$ for $j = 1, \ldots, m$. By Auerbach's Lemma there are functionals $x_1^*, \ldots, x_m^* \in X^*$ with $\langle x_j^*, x_k \rangle = \delta_{j,k}$ $(1 \le j,k \le m)$. It follows that $\det\left(\langle x_j^*, f_k \rangle\right)_{j,k=1}^m \ne 0$ in a neighborhood $U$ of $w$. Hence, for every $z \in U$, the vectors $f_1(z), \ldots, f_m(z)$ are linearly independent and thus form a basis for $\mathbf{ker}(\alpha(z))$. $\qquad\Box$

We shall consider the following subset of $\sigma_p^m(T)$,

$$\mathbf{sp}_o^m(T) := \left\{ z \in \mathbf{C}^N \,;\, \dim \mathbf{ker}(z - T) = m \text{ and } H_p(z - T) = 0 \text{ for } p \ge 1 \right\} .$$

By a perturbation theorem of F.-H. Vasilescu (cf. [14], Theorem III. 5.8) this set is open in $\mathbf{C}^N$.

**2.9. COROLLARY.** *Let* $T = (T_1, \ldots, T_N)$ *be a commutative system in* $L(X)$ *and suppose that* $\mu(B) = m < \infty$. *Let* $y = (y_1, \ldots, y_m)$ *be a multicyclic* $m$-*tuple for* $B$ *and suppose that* $K$ *is a compact subset of* $\mathbf{sp}_o^m(T^*)$. *Then there exist a function* $\phi = (\phi_1, \ldots$ $\ldots, \phi_m) : K \longrightarrow (X^*)^m$ *and constants* $\delta > 0$, $C > 0$, *such that conditions* (i)–(iii) *in 2.5 are fulfilled on* $K$.

PROOF. We apply Corollary 2.8 with $\Omega = \mathbf{sp}_o^m(T^*)$, $\alpha(z) = \alpha_o(z - T^*)$, and $\beta(z) = \alpha_1(z - T^*)$. Since we have locally a holomorphic basis for $\mathbf{ker}(z - T^*) = \mathbf{ker}\,\alpha_o(z - T^*)$ on $K$, it is easy to construct a piecewise continuous function $\phi = (\phi_1, \ldots, \phi_m)$ :

$: K \rightarrow (X^*)^m$ fulfilling the requirements of the statement.                    $\square$

It would be interesting to know if in the situation of Corollary 2.9, the set $\mathbf{sp}_o^m(T^*)$ is $F$-convex, i.e. if for every compact set $K \subseteq \mathbf{sp}_o^m(T^*)$ we have $\hat{K}_F \subseteq \mathbf{sp}_o^m(T^*)$. Notice that by Proposition 2.5 (resp. Proposition 2.5'), we already have $\hat{K}_F \subseteq \sigma_p^m(T^*)$ (resp. $\hat{K}_F \subseteq \sigma_p^{\geq m}(T^*)$ in the case $F = O(\mathbf{sp}(T,X)))$. In the single operator case the answer to this question is positive (cf. [9], Theorems 2 and $2^a$).

Because of Propositions 2.5 and 2.5', the following example may be considered as a kind of standard example. It also shows that the set of multicyclic m-tuples may be rather small.

**2.10. EXAMPLE.** *Let* $S$ *be a compact set in* $\mathbf{C}^N$ *and let* $F$ *be a subalgebra of* $O(S)$ *containing* $\mathbf{C}[Z]$. *Suppose that* $K \subseteq S$ *is a compact set with* $\mathbf{int}(K) \neq \emptyset$. *Write again* $A := \overline{F|K}$ *and* $M = (M_1, \ldots, M_N) \in L(A^m)$ *for the system of operators of multiplication with the coordinate functions* $\pi_1, \ldots, \pi_N$. *For* $h \in F$, *denote by* $M(h) \in L(A^m)$ *the operator of multiplication with* h. *Then* $\mu(M(F)) = m$ *and* $e = (e_1, \ldots, e_m) \in (A^m)^m =: M_m(A)$ *with* $e_{j,k}(z) \equiv \delta_{j,k}$ *on* $K$ $(1 \leq j, k \leq m)$ *is a multicyclic m-tuple for* $M(F)$. *For the set* $C_m(M(F))$ *of all multicyclic m-tuples for* $M(F)$ *we have*

$$\mathbf{int}(M_m(A) \setminus C_m(M(F))) \neq \emptyset .$$

PROOF. Fix $w \in \mathbf{int}(K)$ and $r > 0$ such that $D := D(w,r) \subseteq \mathbf{int}(K)$. For $1 \leq j, k \leq m$ we define $f_{j,k} \in A$ by $f_{j,k}(z) := \delta_{j,k} \prod_{i=1}^{N} (z_i - w_i)$ $(z \in K)$. We shall show that $f = (f_1, \ldots$ $\ldots, f_m) \in \mathbf{int}(M_m(A) \setminus C_m(M(F)))$, where $f_j := (f_{j,1}, \ldots, f_{j,m})$. Consider

$$F(z) := \mathbf{det}(f(z)) = \prod_{i=1}^{N} (z_i - w_i)^m .$$

By the Cauchy integral formula,

$$(2\pi i)^{-N} \int_{\partial_o D} (\prod_{j=1}^{N} (\zeta_j - z_j)^{m-1}) F(z)^{-1} dz_1 \ldots dz_N = 1 .$$

Hence, there exists some $\delta > 0$ such that for all $g \in M_m(A)$ with

$$\| f - g \|_{\partial_F K} := (\sum_{j,k=1}^{m} \| f_{j,k} - g_{j,k} \|_{\partial_F K}^2)^{\frac{1}{2}} < \delta$$

we have, with $G(z) := \det(g(z))$,

$$(2\pi i)^{-N} \int_{\partial_o D} \left( \prod_{j=1}^{N} (\zeta_j - z_j)^{m-1} \right) G(z)^{-1} dz_1 \ldots dz_N \neq 0 .$$

Therefore, $G(z)^{-1}$ cannot exist for all $z \in D$, i.e. $G$ must have zeros in D. Assume now that $g \in C_m(M(F))$ and $\|f - g\|_{\partial_F K} < \delta$. Then, for every $\epsilon > 0$, there is some $h = (h_{j,k})_{j,k=1}^{m} \in M_m(F)$ such that

$$\left\| \sum_{k=0}^{m} h_{j,k}(M) g_k - e_j \right\|_{A^m} < \epsilon .$$

For $\epsilon$ sufficiently small, this implies that $G(z)\det(h(z)) = \det(h(z)g(z)) \neq 0$ for all $z \in K_{\hat{F}}$ in contradiction to the fact that $G(z)$ has some zeros in D. This shows that $C_m(M(F)) \cap \cap U_\delta(f) = \emptyset$. $\qquad\qquad\qquad\qquad\qquad\qquad\qquad\qquad\qquad\qquad\qquad\qquad\qquad\qquad\square$

For the single operator version of the following theorem we refer to [2], [10].

**2.11. THEOREM.** *Let* $T = (T_1, \ldots, T_N)$ *be a commutative system in* $L(X)$ *with* $\mu(B) = m < \infty$, *where* $B$ *is* **C**[T], **C**(T) *or* **Hol**(T). *Fix a multicyclic m-tuple for B and let F be the algebra* **C**[Z], *resp.* **Rat**(sp(T,X)), *resp.* $O(sp(T,X))$. *Then the following three statements are equivalent:*

(a) *There is some compact set K with* $\mathbf{int}(K_{\hat{F}}) \neq \emptyset$ *and with* $\partial_F K \subseteq \sigma_p^m(T^*)$ *(resp.* $\partial_F K \subseteq \sigma_p^{\geq m}(T^*)$ *in the case* $B = \mathbf{Hol}(T)$*) and there are a function* $\phi = (\phi_1, \ldots, \phi_m)$: $\partial_F K \to (X^*)^m$ *and constants* $C > 0$, $\delta > 0$, *such that (i)-(iii) in Proposition 2.5 are satisfied.*

(b) *There are some compact set K with* $\mathbf{int}(K_{\hat{F}}) \neq \emptyset$ *and a continuous linear operator* $J : X \to A^m$ *(where* $A := \overline{F|K}$*) such that* $\mathbf{ran}(J)$ *is dense in* $A^m$ *and,*

$$JT_j = M_j J \quad \text{for } 1 \leq j \leq N$$

*where again* $M_j$ *is the operator of multiplication with* $\pi_j$.

(c) $\mathbf{int}(\sigma_p^m(T^*)) \neq \emptyset$ *(resp.* $\mathbf{int}(\sigma_p^{\geq m}(T^*)) \neq \emptyset$*).*

*Moreover, if T satisfies one (and hence all) of these three conditions, then*

(d) *The set* $C_m(B)$ *of all multicyclic m-tuples for B is not dense in* $X^m$.

PROOF. The implication (a) $\Rightarrow$ (b) follows from Proposition 2.5 resp.

Proposition 2.5'. If (b) holds, then, for $w \in \mathbf{int}(K_F^\wedge)$, the functionals $E_j(w) \in (A^m)^*$, given by $E_j(w)f = f_j(w)$ for $f = (f_1, \ldots, f_m) \in A^m$, are obviously linearly independent elements of $\mathbf{ker}(z - M^*)$. Since J is injective, this implies $\dim \mathbf{ker}(z - T^*) \geq m$ and hence $\emptyset \neq \mathbf{int}(K_F^\wedge) \subseteq \sigma_p^{\geq m}(T^*)$. Because of Proposition 1.1, we actually have $\dim \mathbf{ker}(z - T^*) = m$ in the cases $B = \mathbf{C}[T]$, resp. $B = \mathbf{C}(T)$. This shows that in these cases we have $\emptyset \neq \mathbf{int}(K_F^\wedge) \subseteq \sigma_p^m(T^*)$.

Suppose now that $\mathbf{int}(\sigma_p^m(T^*)) \neq \emptyset$ and fix an arbitrary closed polydisc $D = D(\zeta, r)$ contained in $\sigma_p^m(T^*)$. We define

$$D_k := \left\{ z \in D \;\middle|\; \begin{array}{l} \exists \phi = (\phi_1, \ldots, \phi_m) \in (\mathbf{ker}(z - T^*))^m \text{ such that } \|\phi\| \leq 1 \text{ and} \\[2mm] |\mathbf{det}(\langle y_i, \phi_j \rangle)_{1 \leq i, j \leq m}| \geq 1/k \end{array} \right\}$$

and show first that $D_k$ is a closed subset of D. Indeed, if $(z_n)_{n=1}^\infty$ is a sequence in $D_k$ converging to some $w \in D$, then there exist $\phi^{(n)} = (\phi_1^{(n)}, \ldots, \phi_m^{(n)})$ in the unit ball of $(\mathbf{ker}(z_n - T^*))^m$, such that

$$|\mathbf{det}(\langle y_i, \phi_j^{(n)} \rangle)_{1 \leq i, j \leq m}| \geq 1/k .$$

Since X must be a separable Banach space, the unit ball of $(X^*)^m = (X^m)^*$ is metrizable and compact with respect to the weak$^*$-topology. Hence, by passing to a subsequence, we may assume that $(\phi^{(n)})_{n=1}^\infty$ converges to some $\phi = (\phi_1, \ldots, \phi_m) \in (\mathbf{ker}(w - T^*))^m$ in the weak$^*$-topology. By continuity, we still have

$$|\mathbf{det}(\langle y_i, \phi_j \rangle)_{1 \leq i, j \leq m}| \geq 1/k .$$

This shows that $w \in D_k$. Because of Lemma 1.4, D is the union of the closed sets $D_k$, $k \in \mathbf{N}$. Thus, by the Baire category theorem, $\mathbf{int}(D_k) \neq \emptyset$ for suitable k. If we choose now $K := D_k$, then, for every $z \in K$, we can find $\phi(z) = (\phi_1(z), \ldots, \phi_m(z)) \in (\mathbf{ker}(z - T^*))^m$ such that (i)-(iii) in 2.5 are satisfied with $C = 1$ and $\delta = 1/k$.

Suppose now that (a) is satisfied and let $J : X \longrightarrow A^m$ be the mapping given by Proposition 2.5, resp. Proposition 2.5'. Define $J_m : X^m \longrightarrow (A^m)^m = M_m(A)$ by $J_m x :=$ $:= (J_m x_1, \ldots, J_m x_m)$ for $x = (x_1, \ldots, x_m) \in X^m$. Since J has dense range, so does $J_m$. Let $\delta$ and $f = (f_1, \ldots, f_m)$ be as in the proof of 2.10 (with $S := \mathbf{sp}(T, X)$). Then $V := J_m^{-1}(U_\delta(f))$ is a nonempty open set in $X^m$. Assume there exists some $x = (x_1, \ldots, x_m) \in V$ which is a multicyclic m-tuple for B, i.e. $\bigvee\{Bx_j ; 1 \leq j \leq m\} = X$. Since J has dense range, this implies $\bigvee\{JBx_j ; 1 \leq j \leq m\} = A^m$. Because of $Jh(T) = M(h)J$ for all $h \in F$ (cf. Proposition 2.5, resp.

2.5'), we obtain $V\{M(h)Jx_j\,;\, h \in F,\, 1 \leq j \leq m\} = A^m$. Thus $J_m x \in C_m(M(F))$ in contradiction to $C_m(M(F)) \cap U_\delta(f) = \emptyset$ (by the proof of 2.10). Hence our assumption was false, we must have $V \cap C_m(B) = \emptyset$. This shows that $C_m(B)$ is not dense in $X^m$. $\quad\square$

In the case $N = 1$, statement (d) of Theorem 2.11 is actually equivalent to (a) – (c) (see [10]). The following elementary example shows that this is no longer true in several variables.

**2.12. EXAMPLE.** Let $S \in L(X)$ be a cyclic operator with $\mathrm{int}(X \setminus C_1(S)) \neq \emptyset$ (for example the unilateral shift on $\ell^2(\mathbf{N}_0)$) and consider the pair $T := (S,1)$. Then $T$ is cyclic and has the same cyclic vectors as $S$. Especially, $\mathrm{int}(X \setminus C_1(T)) \neq \emptyset$ so that $T$ fulfils (d) in 2.11. However, $\mathrm{int}(\mathrm{sp}(T,X)) = \mathrm{int}(\mathrm{sp}(S,X) \times \{1\}) = \emptyset$. Therefore, condition (c) in 2.11 is not satisfied.

**REFERENCES**


1.      **Albrecht, E.** : Spectral decompositions for systems of decomposable operators, *Proc. Roy. Irish Acad. Sect. A* **81**(1981), 81–98.

2.      **Apostol, C. ; Fialkow, L.A. ; Herrero, D.A. ; Voiculescu, D.** : *Approximation of Hilbert space operators. II*, Research Notes in Mathematics, **102**, Pitman Advanced Publ. Program, Boston-London-Melbourne, 1984.

3.      **Arveson, W.** : On groups of automorphisms of operator algebras, *J. Funct. Anal.* **15**(1974), 217–243.

4.      **Grothendieck, A.** : Sur certains espaces de fonctions holomorphes. I, *J. Reine Angew. Math.* **192**(1953), 35–64.

5.      **Gunning, R.C. ; Rossi, H.** : *Analytic functions of several complex variables*, Prentice-Hall, Englewood Cliffs, N.J., 1965.

6.      **Herrero, D.A.** : On multicyclic operators, *Integral Equations Operator Theory* **1** (1978), 57–102.

7.      **Herrero, D.A.** : *Approximation of Hilbert space operators. I*, Research Notes in Mathematics, **72**, Pitman Advanced Publ. Program, Boston-London-Melbourne, 1982.

8.      **Herrero, D.A.** : On multicyclic operators. II, *Proc. London Math. Soc. (3)* **48** (1984), 247–282.

9.      **Herrero, D.A.** : The Fredholm structure of n–multicyclic operators, *Indiana Univ. Math. J.* **36**(1987), 549–566.

10.     **Herrero, D.A. ; Rodman, L.** : The multicyclic n–tuples of n–multicyclic operators, *Indiana Univ. Math. J.* **34**(1985), 619–629.

11.     **Słodkowski, Z. ; Żelasko, W.** : On joint spectra of commuting families of operators, *Studia Math.* **50**(1974), 127–148.

12.    **Taylor, J.L.** : A joint spectrum for several commuting operators, *J. Funct. Anal.* **6**(1970), 172-191.

13.    **Taylor, J.L.** : The analytic functional calculus for several commuting operators, *Acta Math.* **125**(1970), 1-38.

14.    **Vasilescu, F.-H.** : *Analytic functional calculus and spectral decomposition*, D. Reidel Publ. Comp., Dordrecht and Editura Academiei, Bucureşti, 1982.

**Ernst Albrecht**
Fachbereich Mathematik
Universität des Saarlandes
D-6600 Saarbrücken
Fed. Rep. Germany.

**Peter Maria Wirtz**
Institut für Medizinische
Statistik und Dokumentation
Universität Erlangen
Waldstrasse 6, D-8520 Erlangen
Fed. Rep. Germany.

Operator Theory:
Advances and Applications, Vol. 43
© 1990 Birkhäuser Verlag Basel



# REGULAR J-INNER MATRIX-FUNCTIONS AND
# RELATED CONTINUATION PROBLEMS

D.Z. Arov

## § 1. STATEMENT OF THE PROBLEM AND BASIC RESULTS

In this paper, especially in Sections 1-3, we give an account of the author's results concerning $J$-inner matrix-functions and related continuation problems stated in [1] and whose proofs can be found in [2]. They are supplemented, in Section 4,5, with some results about the solutions of the continuation problem with maximal jump of the spectral function, which are obtained using a generalization of Carathéodory Theorem on holomorphic contractive scalar functions in the case of matrix-valued functions. For the significance of $J$-inner matrix-functions in various problems of Mathematical Analysis and their applications, see [3–6], [25,26].

**1.** We introduce now the basic notation and definitions. Consider

(i) $D = \{z : |z| < 1\} (\subset \mathbf{C})$, $\ \Pi = \{z : \mathbf{Re}\ z > 0\}$, $\ E = D$ or $\Pi$;

(ii) $H^{\infty}_{p \times q}$ = the space of holomorphic bounded matrix-functions (in short **m.f.**) $h(z)$ on E, of order $p \times q$, with

$$\| h \|_{\infty} = \mathbf{sup}\{ \| h(z) \| : z \in E\} < \infty;$$

(iii) $H^{\infty}_1 = H^{\infty}_{1 \times 1}$;

(iv) $\mathbf{B}_{p \times q} = \{h : h \in H^{\infty}_{p \times q}, \| h \|_{\infty} \le 1\}$, $\ \mathbf{B}_p = \mathbf{B}_{p \times p}$, $\ \mathbf{B} = \bigcup_{p,q=1}^{\infty} \mathbf{B}_{p \times q}$;

(v) $\mathbf{J}$ = a self-adjoint unitary matrix, different from $\pm I$, that is, unitarily equivalent to the matrix

$$\mathbf{j} = \begin{bmatrix} I_n & 0 \\ 0 & -I_m \end{bmatrix}$$

for some positive integers n and m (to avoid ambiguity, we shall consider $m \le n$);

(vi) $\mathbf{B}_J$ = the class of **m.f.** meromorphic on E and $\mathbf{J}$-contractive, i.e. $W^{*}(z)\mathbf{J}W(z) \le \mathbf{J}$, $z \in E$;

(vii) $P_{\pm} = \frac{1}{2}(I \pm J)$.

The Potapov-Ginsburg transform

(1)
$$\tilde{S} = [P_- + P_+ W][P_+ + P_- W]^{-1}$$

maps **m.f.** W from $B_J$ into **m.f.** $\tilde{S}$ from $B_{n+m}$. By the same transform, any **m.f.** W from $B_J$ can be represented in terms of the corresponding **m.f.** S from $B_{n+m}$. According to this, any **m.f.** W from $B_J$, can be written as $h_2^{-1}h_1$, where $h_1 \in H^\infty_{p \times p}$, $h_2 \in H^\infty_1$ (p = n + m). Consequently, for every W from $B_J$ the boundary values

$$W(\zeta_0) = \lim_{z \to \zeta_0} W(z), \qquad \zeta_0 \in \partial E$$

exist a.e. on $\partial E$ (lim denotes here the limit when z tends to $\zeta_0$ along the normal to $\partial E$ at the point $\zeta_0$). $\underset{z \to \zeta_0}{}$

**DEFINITION.** An **m.f.** W in $B_J$ is called **J**-inner if it has **J**-unitary values a.e. on $\partial E$, i.e.

$$W^*(z) J W(z) = J, \qquad z \in \partial E \text{ a.e.}$$

The class of these W is denoted by U(**J**). By (1) they are related to inner **m.f.** $\tilde{S} \in B$, with $\tilde{S}^*(z)\tilde{S}(z) = I$ a.e. on $\partial E$.

An **m.f.** W in U(**J**) has a pseudo-continuation, defined on the exterior $E_e$ of E, which can be obtained by the symmetry principle:

$$W(z) = J[W^*(z^\sim)]^{-1} J, \qquad z \in E_e,$$

where $z^\sim = 1/\bar{z}$ for E = **D** and $z^\sim = -\bar{z}$ for E = II.

For **J** = **j**, any **m.f.** W (and $\tilde{S}$) can be cannonically decomposed into blocks: $W = [W_{jk}]_1^2$ ($\tilde{S} = [S_{jk}]_1^2$) with the diagonal blocks $W_{11}$ and $W_{22}$ ($S_{11}$ and $S_{22}$) of order n and order m, respectively. Accordingly, formula (1) can be written

(2)
$$\begin{bmatrix} S_{11}(z) & S_{12}(z) \\ \\ S_{21}(z) & S_{22}(z) \end{bmatrix} = \begin{bmatrix} [W_{11}^*(z^\sim)]^{-1} & W_{12}(z)W_{22}^{-1}(z) \\ \\ -W_{22}^{-1}(z)W_{21}(z) & W_{22}^{-1}(z) \end{bmatrix}.$$

Here we take into account that for the **j**-inner **m.f.** $W = [W_{jk}]_1^2$, we have

$$(S_{11}(z)=)W_{11}(z) - W_{12}(z)W_{22}^{-1}(z)W_{21}(z) = [W_{11}^*(z^\sim)]^{-1}.$$

An **m.f.** h(z) from $H^\infty_{p \times p}$ with **det** h(z) $\not\equiv$ 0 is outer if

$$\ln|\det h(z_0)| \;=\; \int_{\partial E} \ln|\det h(z)|\,d\mu,$$

where $d\mu = |dz|/\pi(1+|z|^2)$, and $z_0 = 0$ for $E = D$ and $z_0 = 1$ for $E = \Pi$.

Let us denote:

$$D_{p\times q} = \{h : h = h_2^{-1}h_1,\; h_1 \in H^{\infty}_{p\times q},\; h_2 \in H^{\infty}_1,\; h_2 \text{ outer}\},$$

and $D_p = D_{p\times p}$. A holomorphic **m.f.** h on E is called *outer*, if it is represented as $h = h_2^{-1}h_1$, where $h_1 \in H^{\infty}_{p\times p}$ and $h_2 \in H^{\infty}_1$ can be chosen to be outer. For h with $\det h(z) \not\equiv 0$ this holds iff $h^{\pm 1} \in D_p$. An arbitrary **m.f.** h from $D_p$ with $\det h(z) \not\equiv 0$ has an (essentially unique) representation in the form $h = b\phi$, where $b\;(\in \mathbf{B}_p)$ and $\phi\;(\in D_p)$ are inner, resp. outer **m.f.** From (2), it follows that for an $W = [W_{jk}]_1^2$ in U(j) we have:

$$[W_{11}^{*}(\tilde{z})]^{-1} \in \mathbf{B}_n, \qquad W_{22}^{-1} \in \mathbf{B}_m.$$

Therefore $W_{11}$ and $W_{22}$ have the essentially unique representations

$$(3) \qquad\qquad W_{11}(z) = b_1(z)p_-(z), \qquad W_{22}(z) = b_2^{-1}(z)p_+(z),$$

where $b_1\;(\in \mathbf{B}_n)$ and $b_2\;(\in \mathbf{B}_m)$ are inner, and $p_-^{*}(\tilde{z})$ and $p_+(z)$ are outer **m.f.** ($p_-(z)$ and $p_+(z)$ have a pseudo-continuations on $E_e$). From (2) it can be also seen that for the **j**-inner **m.f.** $W = [W_{jk}]_1^2$, we have:

$$W \in D_{n+m} \Longleftrightarrow b_2 = \mathbf{const}\; ; \qquad W^{-1} \in D_{n+m} \Longleftrightarrow b_1 = \mathbf{const}$$

$$(W^{-1} \in D_{n+m} \Longleftrightarrow W(\bar{\tilde{z}}) \in D_{n+m}).$$

**DEFINITION.** A **J**-inner **m.f.** W is called *singular* if it is an outer **m.f.**, i.e. if $W^{\pm 1} \in D_p$.

Since for an **m.f.** W in U(**J**) we have $|\det W(z)| = 1$ a.e. on $\partial E$, we can replace the condition $W^{\pm 1} \in D_p$ with the condition $W \in D_p$, $\det W(z) = \mathbf{const.} = c$ with $|c| = 1$.

**DEFINITION.** A **J**-inner **m.f.** W is called *regular* if it has no non–constant singular right divisors in the class of **J**-inner **m.f.**

An arbitrary **J**-inner **m.f.** W has an essentially unique representation $W = W_r W_s$, where $W_r$ and $W_s$ are regular, resp. singular **J**-inner **m.f.** [3].

Let us denote by U(n,m), $U_r(n,m)$ and $U_s(n,m)$ the classes of **j**-inner **m.f.**, resp. regular **m.f.** in U(n,m), and singular **m.f.** in U(n,m).

If $W = [W_{jk}]_1^2 \in U(n,m)$ and $W_{11}$, $W_{22}$ are represented as in (3), then

$$W \in U_s(n,m) \Longleftrightarrow b_1 = \text{const}, \ b_2 = \text{const},$$

and therefore $W \in U_s(n,m)$ iff $W_{11}(\bar{z}^{\backsim})$ and $W_{22}(z)$ are outer **m.f.** (iff $S_{11}(z)$ and $S_{22}(z)$ are outer **m.f.**).

An **m.f.** $W = [W_{jk}]_1^2$ in $U(n,m)$ is singular ($W \in U_s(n,m)$) iff the function

$$\phi(z) := \det W_{11}(\bar{z}^{\backsim}) \det W_{22}(z)$$

is outer, that is iff

$$\int_{\partial E} \ln|\phi(z)| \, d\mu = \ln|\phi(z_o)|.$$

In [1] and [3] it was obtained a criterium for $W$ to be in $U_r(n,m)$.

**2.** There exists a two-ways connection between the class of **m.f.** $U_r(n,m)$ and the following generalized bitangent Schur - Nevanlinna - Pick problem, in the case when it is completely indetermined.

**The SNP(n,m) problem.** *Let* $S_o$ *($\in \mathbf{B}_{n \times m}$), $b_1$ ($\in \mathbf{B}_n$) and $b_2$ ($\in \mathbf{B}_m$) be given* **m.f.** *such that* $b_1$ *and* $b_2$ *are inner. Find the* **m.f.** *S such that*

$$(4) \qquad\qquad b_1^{-1}(S - S_o)b_2^{-1} \in H_{n \times m}^{\infty}, \quad S \in \mathbf{B}_{n \times m}.$$

Let us denote by $F_{S_o;b_1,b_2}$ the set of all solutions $S$ of this problem. Let $z_*$ be a fixed point in $E$, with $\det b_1(z_*) \neq 0$, $\det b_2(z_*) \neq 0$. Then $K(z_*) :=$
$:= \{S(z_*) : S \in F_{S_o;b_1,b_2}\}$ is a matrix sphere:

$$K(z_*) = \{S(z_*) \, ; \, S(z_*) = S_c(z_*) + R_\Lambda(z_*)VR_\Pi(z_*), \ V^*V \leq I\},$$

where $S_c(z_*)$ is the center of the sphere, $R_\Lambda(z_*)$ $(\geq 0)$ and $R_\Pi(z_*)$ $(\geq 0)$ matrices of order n, resp. m are the left and the right semi-radii of the sphere, respectively, and V is an arbitrary contractive matrix of order $n \times m$. The problem (4) is called *completely indetermined* if $R_\Lambda(z_*) > 0$ and $R_\Pi(z_*) > 0$ (in fact, it is sufficient that $R_\Lambda(z_*) > 0$ or $R_\Pi(z_*) > 0$). This definition does not depend on the choice of the point $z_*$.

For $W = (W_{jk})_1^2$, we denote:

$$F_W(E) = [W_{11}E + W_{12}][W_{21}E + W_{22}]^{-1},$$

$$F_W(\mathbf{B}_{n,m}) = \{S : S = F_W(E), \ E \in \mathbf{B}_{n \times m}\}.$$

The two-ways connection between the **m.f.** W in $U_r(n,m)$ and the completely indetermined problem (4) is established by the following theorem, announced in [4] and [5], and proved in [3].

**THEOREM A** 1. *Let* $W = [W_{jk}]_1^2 \in U(n,m)$, $S_o = F_W(0)$, *and* $b_1$ *and* $b_2$ *be the inner* **m.f.** *which appear in representation* (3). *Then:*

a) *the problem* (4) *with these* $S_o$, $b_1$ *and* $b_2$ *is completely indetermined;*

b) $F_W(\mathbf{B}_{n \times m}) \subset F_{S_o;b_1,b_2}$;

c) $F_W(\mathbf{B}_{n \times m}) = F_{S_o;b_1,b_2}$ *iff* $W \in U_r(n,m)$.

2.*For an arbitrary completely indetermined problem* (4) *there exists* $W \in U_r(n,m)$, *which has in the representation* (3) *the same* $b_1$ *and* $b_2$ *as in problem* (4) *and such that* $F_W(\mathbf{B}_{n \times m}) = F_{S_o;b_1,b_2}$. *This* **m.f.** *W is defined by the probem* (4) *up to a constant j-unitary right factor.*

**REMARK 1.** For an arbitrary **m.f.** $W = [W_{jk}]_1^2$ in $U(n,m)$ there exists a unique constant **j**-unitary matrix $U$ such that the **m.f.** $\hat{W} = [\hat{W}_{jk}]_1^2 := WU \ (\in U(n,m))$ is normalized by the conditions

$$\beta_-(z_o^\sim) > 0, \quad \beta_+(z_o) > 0, \quad \hat{S}_{21}(z_o) = 0$$

and has, in the representation (3), the same inner **m.f.** $b_1$ and $b_2$ as the representation of the **m.f.** W.

The set of the matrices in $U_r(n,m)$ normalized by these conditions is denoted by $U_r^o(n,m)$.

The **m.f.** W $(\in U_r(n,m))$ which appears in the second part of Theorem A can be considered normalized; such a **m.f.** is also uniquely defined by problem (4).

**REMARK 2.** If $W \in U(n,m)$, $\tilde{W} \in \mathbf{B}_j$ and $F_{\tilde{W}}(\mathbf{B}_{n \times m}) = F_W(\mathbf{B}_{n \times m})$, then by Simakow Theorem [7], we have

$$(5) \qquad \tilde{W}(z) = \rho(z)W(z)\,U, \quad \rho(z) = \overset{o}{b}_2(z)/\overset{o}{b}_1(z),$$

where $U$ is a constant **j**-unitary matrix, $\overset{o}{b}_1$ and $\overset{o}{b}_2$ are scalar inner functions which are divisors of the inner **m.f.** $b_1$ and $b_2$, respectively, appearing in the representation (3) for $W = [W_{jk}]_1^2$ $(\overset{o}{b}_1{}^{-1}b_1 \in \mathbf{B}, \overset{o}{b}_2{}^{-1}b_2 \in \mathbf{B})$. Therefore we clearly have that $\tilde{W} \in U(n,m)$. Conversely, if $W \in U(n,m)$ and the **m.f.** $\tilde{W}$ is defined by formula (5), then $\tilde{W} \in U(n,m)$ and $F_{\tilde{W}}(\mathbf{B}_{n \times m}) = F_W(\mathbf{B}_{n \times m})$. Moreover, if $W \in U_r(n,m)$ then $\tilde{W} \in U_r(n,m)$ also.

**DEFINITION.** An **m.f.** W from $\mathbf{B}_j$ such that $F_{S_o;b_1,b_2} = F_W(\mathbf{B}_{n\times m})$ is called a *resolvent matrix* for the completely indetermined problem (4).

According to Theorem A and Remarks 1 and 2, we have:

1) For any completely indetermined problem (4) there exists a unique resolvent matrix $W = [W_{jk}]_1^2$ in $\overset{\circ}{U}_r(n,m)$ having in the representation (3) the same $b_1$ and $b_2$ as in the considered problem (4);

2) all the resolvent matrices $\widetilde{W}$ for this problem are expressed with this **m.f.** W by formula (5) and are regular **j**-inner **m.f.**;

3) an arbitrary regular **j**-inner **m.f.** W is a resolvent matrix for the completely indetermined problem (4) in which $b_1$ and $b_2$ are taken from the representation (3) and $S_o = W_{12}W_{22}^{-1}$ $(= F_W(0))$.

**3.** For W in $U(n,m)$ we denote by $H_W$ the set of the points in the extended complex plane, in which the **m.f.** W is holomorphic (regarding the pseudo-continuation in $E_e$). For any ordered pair $\{b_1,b_2\}$ of inner **m.f.** $b_1$ $(\epsilon\, \mathbf{B}_n)$ and $b_2$ $(\epsilon\, \mathbf{B}_m)$ we consider: $H_{b_1}^-$ = the set of the points in $E_e$ in which $b_1$ is holomorphic; $H_{b_2}^+$ = the set of the points in $E$ in which $b_2^{-1}$ is holomorphic; $H_{b_1,b_2}^{\circ}$ = the set of points of $\partial E$, in which the following function

$$b_{b_1,b_2}(z) := \mathbf{det}\, b_1(z)\cdot \mathbf{det}\, b_2(z);$$

is holomorphic;

$$H_{b_1,b_2} := H_{b_1}^- \cup H_{b_2}^+ \cup H_{b_1,b_2}^{\circ}.$$

From (2) it follows that if $W = [W_{jk}]_1^2 \in U(n,m)$ and if $b_1$ and $b_2$ are given by (3), then $H_W \cap E = H_{b_2}^+$ and $H_W \cap E_e = H_{b_1}^-$. Now, it is easy to prove that $H_W \cap \partial E \subset H_{b_1,b_2}^{\circ}$. However, the equalities $H_W \cap \partial E = H_{b_1,b_2}^{\circ}$ and $H_W = H_{b_1,b_2}$ are not always true. Nevertheless, for $W \in U_r(n,m)$ it will be shown that $H_W = H_{b_1,b_2}$. So, we have

**THEOREM 1.** *Let* $W = [W_{jk}]_1^2 \in U(n,m)$ *and let* $b_1$ *and* $b_2$ *be those given by* (3). *Then:*

a) $H_W \cap E = H_{b_2}^+$, $\quad H_W \cap E_e = H_{b_1}^-$, $\quad H_W \cap \partial E \subset H_{b_1,b_2}^{\circ}$.

b) *if, in addition,* $W \in U_r(n,m)$, *then* $H_W \cap \partial E = H_{b_1,b_2}^{\circ}$ *and* $H_W = H_{b_1,b_2}$.

**COROLLARY.** *Let* $E = \Pi$, $W = [W_{jk}]_1^2 \in U(n,m)$, $b_1$ *and* $b_2$ *be given by* (3). *Then:*

a) *If* W *is an entire* **m.f.**, *then* $b_1$ *and* $b_2$ *are entire* **m.f.**

b) *If* $b_1$ *and* $b_2$ *are entire* **m.f.** *and* $W \in U_r(n,m)$ *then* W *is an entire* **m.f.**

Theorem 1, together with Theorem A and the remarks following it, enable us to conclude that

**THEOREM 2.** *If the problem (4) is completely indetermined, then there exists a resolvent matrix* W *for it such that* $H_W = H_{b_1,b_2}$, *where* $b_1$ *and* $b_2$ *are inner* **m.f.** *from problem (4) and for any resolvent matrix* $\widetilde{W}$ *for this problem we have* $H_{\widetilde{W}} \cap \partial E = H^\circ_{b_1,b_2}$.

Using (5) we get

**COROLLARY.** *If* $E = \Pi$ *and in a completely indetermined problem (4)* $b_1$ *and* $b_2$ *are entire* **m.f.**, *then any resolvent matrix of this problem is an entire* **m.f.**

This is in fact a generalization of M.G. Krein's result which asserts that the resolvent matrix of the indetermined continuation problem for the helicoidal function on the interval [-a,a] is an entire **m.f.** [8], [9]. This problem is equivalent to the particular case of (4) when $b_1 = I$, $b_2(z) = e^{-az}I$, $E = \Pi$.

From Theorem 1a) and (5) we get

**THEOREM 3.** *If* $W \in U(n,m)$, $\widetilde{W} \in B_j$ *and* $F_{\widetilde{W}}(\mathbf{B}_{n \times m}) = F_W(\mathbf{B}_{n \times m})$ *then* $\widetilde{W} \in U(n,m)$ *and* $H_{\widetilde{W}} \cap \partial E = H_W \cap \partial E$; *if, moreover,* $W \in U_r(n,m)$, *then* $\widetilde{W} \in U_r(n,m)$.

**4.** Problem (4), with the changing $f = b_1^{-1}Sb_2^{-1}$ and with a changing of argument if $E = \Pi$, leads to the following Nehari matrix problem.

Let $L^\infty_{p \times q}$ denote the space of measurable **m.f.** $f(z)$ of order $p \times q$ on $\partial D$ with the norm $\|f\|_\infty = \text{ess sup}\{\|f(z)\| : z \in \partial D\} < \infty$; denote by $\gamma_k(f)$ the Fourier coefficients of the **m.f.** f:

$$\gamma_k(f) = \int_{\partial D} z^k f(z)d\mu, \quad k = 0, \pm 1, \ldots$$

**Problem N(n,m).** *For a given sequence* $\{\gamma_k\}_1^\infty$ *of matrices of order* $n \times m$, *find the* **m.f.** f *in* $L^\infty_{n \times m}$ *such that* $\gamma_k(f) = \gamma_k$ *for* $k \geq 1$ *and with* $\|f\|_\infty \leq 1$, *that is*

(6)
$$f \sim \sum_1^\infty \gamma_k z^{-k} + \ldots \quad (z \in \partial D), \quad \|f\|_\infty \leq 1.$$

It is known [10] that the study of this problem is related to the operator $\boldsymbol{\Gamma}$ defined by the Hankel block-matrix $[\gamma_{j+k-1}]_1^\infty$ and acting from the Hilbert space

$$\ell_m^2 := \{\xi : \xi = \{\xi_k\}_1^\infty \ , \ \|\xi\|^2 = \sum_1^\infty \|\xi_k\|^2 < \infty, \ \xi_k \in \mathbf{C}^m\}$$

into the space $\ell_n^2$ by the formula

$$\eta = \boldsymbol{\Gamma}\xi, \quad \xi = \{\xi_k\}_1^\infty \ , \quad \eta = \{\eta_j\}_1^\infty \ , \quad \eta_j = \sum_{k=1}^\infty \gamma_{j+k-1}\xi_k.$$

The condition for the existence of the solution to the problem (6) is $\|\boldsymbol{\Gamma}\| \leq 1$; moreover, $\|\boldsymbol{\Gamma}\| < 1$ iff there exists a solution to (6) with $\|f\|_\infty < 1$. For $\|\boldsymbol{\Gamma}\| < 1$, in [10] there were obtained algebraic formulae for **m.f.** $p_\pm(z)$ and $q_+(z)$ in terms of $\boldsymbol{\Gamma}$, such that the solution of (6) can be written as

(7) $\qquad\qquad f(z) = [p_-(z)E(z) + q_-(z)][q_+(z)E(z) + p_+(z)]^{-1}, \quad E \in \mathbf{B}_{n\times m}$

where

$$U(z) := \begin{bmatrix} p_-(z) & q_-(z) \\ q_+(z) & p_+(z) \end{bmatrix}$$

has the properties: $U(z)$ takes **j**-unitary values a.e. on $\partial D$. $p_-^*(z^\sim)$ and $p_+(z)$ are outer **m.f.**, and $\chi := -p_+^{-1}q_+ \in \mathbf{B}_{m\times n}$. The **m.f.** $U(z)$ verifying these properties are called in [3] $\gamma$-derived matrices; the class of them is denoted by $M(n,m)$. Formula (7) with $U \in M(n,m)$ describes the solution f for (6) in a more general (than $\|\boldsymbol{\Gamma}\| < 1$) case of complete indetermination, namely when $N \subset (I - \boldsymbol{\Gamma}^*\boldsymbol{\Gamma})^{\frac{1}{2}}\ell_m^2$, where $N := \{\xi : \xi = \{\xi_k\}_1^\infty \ , \ \xi_k \in \mathbf{C}^m, \ \xi_k = 0 \text{ for } k > 1\}$.

The **m.f.** $U(z)$ from $M(n,m)$ which are used in (7) for obtaining all the solutions of the completely indetermined problem N(n,m) form the class $M_r(n,m)$ of those regular $\gamma$-derived matrices, which, as elements of $M(n,m)$, have no right non-constant divisors in $U_s(n,m)$ [3]. The **m.f.** $U \ (\in M_r(n,m))$ in (7) can be normalized by the conditions:

$$p_-(\infty) > 0, \quad p_+(0) > 0, \quad q_+(0) = 0;$$

such a normalized $U(z)$ in $M_r(n,m)$ is uniquely defined by the completely indetermined problem. The algebraic formulae mentioned above for $\|\boldsymbol{\Gamma}\| < 1$ give exactly such a **m.f.** $U(z)$.

Problem (4) is completely indetermined iff the corresponding problem (6) with $\gamma_k = \gamma_k(b_1^{-1}s_0b_2^{-1})$ is completely indetermined. For it, we have

$$F_{S_0;b_1,b_2} = F_W(\mathbf{B}_{n\times m}), \qquad W = \begin{bmatrix} b_1 & 0 \\ 0 & b_2^{-1} \end{bmatrix} U \in U_r^\circ(n,m).$$

Thus, if among the solutions S of problem (4) there exists one with $\|S_\infty\| < 1$, then the results concerning problem (6) obtained in [10] for $\|\mathbf{\Gamma}\| < 1$, give simple algebriac formulae for finding a resolvent matrix W for problem (4) which has in the representation (3) the same $b_1$ and $b_2$ as in problem (4) and which is normalized. If problem (6) is completely indetermined, but $\|\mathbf{\Gamma}\| = 1$, then the **m.f.** $p_\pm$ and $q_\pm$ in (7) can be no longer obtained directly by algebraic formulae as for $\|\mathbf{\Gamma}\| < 1$. In [11] it was shown that they can still be obtained by a regularization of problem (6): consider first the **m.f.** $p_\pm^{(\rho)}$ and $q_\pm^{(\rho)}$ corresponding to $\mathbf{\Gamma}_\rho := \rho\mathbf{\Gamma}$ $(0 < \rho < 1)$ and then take the limit as $\rho \uparrow 1$:

$$p_-(z^\sim) = \lim_{\rho \uparrow 1} p_-^{(\rho)}(z^\sim), \qquad q_-(z^\sim) = \lim_{\rho \uparrow 1} q_-^{(\rho)}(z^\sim),$$

$$q_+(z) = \lim_{\rho \uparrow 1} q_+^{(\rho)}(z), \qquad p_+(z) = \lim_{\rho \uparrow 1} p_+^{(\rho)}(z).$$

For problem (4), it will be obtained a stronger result:

**THEOREM 4.** *Let problem (4) be completely indetermined,* $0 < \rho < 1$, *and* $F_{\rho S_0;b_1,b_2} = F_{W_\rho}(\mathbf{B}_{n\times m})$, *where* $W_\rho \in U_r^\circ(n,m)$, $W_\rho$ *having in (3) the same* $b_1$ *and* $b_2$ *as in problem (4). Then* $F_{S_0;b_1,b_2} = F_W(\mathbf{B}_{n\times m})$, *where*

$$W(z) = \lim_{\rho \uparrow 1} W_\rho(z), \qquad z \in H_{b_1,b_2}, \; W \in U_r^\circ(n,m),$$

W *having in (3) the same* $b_1$ *and* $b_2$ *as in problem (4).*

In other classes of continuation problems, Katznelson obtained earlier, by a similar regularization method, the resolvent matrix for singular completely indetermined problems [12].

### § 2. PROOF OF THEOREM 1

1. Let $W = [W_{jk}]_1^2 \in U(n,m)$, and $b_1$, $b_2$ from (3); $H_w$, $H_{b_1}^-$, $H_{b_2}^+$, $H_{b_1,b_2}^\circ$, $H_{b_1,b_2}$ are the notations introduced in Section 1.3. We already showed that $H_w \cap E_e = H_{b_1}^-$, $H_W \cap E = H_{b_2}^+$. Let us show that $H_W \cap \partial E \subset H_{b_1,b_2}^\circ$. Consider $z_* \in H_w \cap \partial E$. Since

$$\det W_{11}(z) = \det p_-(z) \cdot \det b_1(z), \quad \det W_{22}(z) = \det p_+(z)/\det b_2(z),$$

$p_-^*(\tilde{z})$ and $p_+(z)$ are outer **m.f.**, and $[\det W_{11}(z)]^{\pm 1}$ and $\det[W_{22}(z)]^{\pm 1}$ are holomorphic functions in $z_*$, it follows that $\det b_1$ and $\det b_2$ have no zeros and poles in some neighbourhood of $z_*$ in E and $E_e$. Since $|\det W_{11}(z)| \geq 1$ and $|\det W_{22}(z)| \geq 1$ on $\partial$ E, then $\ln|\det W_{11}(z)|$ and $\ln|\det W_{22}(z)|$ are continuous functions in a neighbourhood of $z_*$. We obtain then that $\det b_1$ and $\det b_2$ are holomorphic in $z_*$, i.e. $z_* \in H^{\circ}_{b_1,b_2}$.

Thus, part (a) of Theorem 1 is proved. The proof of part (b) of this theorem is considerably more difficult. The rest of this paragraph is devoted to prove it, that is, to show that $H_W \cap \partial E = H^{\circ}_{b_1,b_2}$ for $W \in U_r(n,m)$.

**2.** Let $W \in U_r(n,m)$ and $b_{b_1,b_2} (= \det b_1 \cdot \det b_2)$ is a Blaschke product. Then W is a product of elementary Blaschke-Potapov factors with poles in E - the zeros of the **m.f.** $b_2$ - , and in $E_e$ - the poles of the **m.f.** $b_1$ - (see [3], Theorem 11). It is known [13] that a Blaschke - Potapov product converges uniformly on any compact in the complex plane which does not intersect the closure of the set of poles of the elementary factors. Therefore, the **m.f.** is holomorphic on such a compact. Consequently, in this case $H^{\circ}_{b_1,b_2} \subset H_W \cap \partial E.$

**3.** We will show that it is suffcicient to prove part (b) of Theorem 1 only in the case when $b_1 = b_0 I_n$, $b_2 = I_m$, where $b_0$ is a scalar inner function. Let $W = [W_{jk}]_1^2 \in$ $\in U_r(n,m)$, and let $b_1$, $b_2$ be as in (3), and

$$\tilde{W} := \begin{bmatrix} b_{b_1,b_2} b_1^{-1} & 0 \\ & \\ 0 & b_2 \end{bmatrix} W.$$

Then $\tilde{W} \in U_r(n,m)$ and in the representation (3) for $\tilde{W}$ we have that $\tilde{b}_1 = b_{b_1,b_2} I_n$, $\tilde{b}_2 = I_m$. Since $H^{\circ}_{b_1,b_2} = H^{\circ}_{\tilde{b}_1,\tilde{b}_2}$ and $H_{\tilde{W}} \cap \partial E \cap H^{\circ}_{b_1,b_2} \subset H_W \cap \partial E$, then if $H^{\circ}_{\tilde{b}_1,\tilde{b}_2} \subset H_{\tilde{W}} \cap \partial E$, we have $H^{\circ}_{b_1,b_2} \subset H_W \cap \partial E$. Thus, it is sufficient to prove (b) of Theorem 1 in case when $b_1 = b_0 I_n$, $b_2 = I_m$, where $b_0$ is a scalar function. As it was shown in § 2.2 this is true if $b_0$ is a Blaschke product. The case of an arbitrary $b_0$ will be reduced to this previous one.

**4.** We can suppose that $E = D$. Let $L^2_{p \times q}$ be the Hilbert space of measurable **m.f.** $h(z)$ on $\partial D$ of order $p \times q$ with the norm

$$\|h\|^2 = \int_{\partial D} \mathrm{Tr}[h^*(z)h(z)]d\mu < \infty;$$

$H^2_{p\times q} := \{h : h \in L^2_{p\times q}, \ \gamma_k(h) = 0 \ \text{ for } \ k \geq 1\}; \ K^2_{p\times q} = L^2_{p\times q} - H^2_{p\times q} \ (=\{h : h \in L^2_{p\times q}, \ \gamma_k(h) = 0 \text{ for } k \leq 0\}; \ H^2_p = H^2_{p\times 1}; \ K^2_p = K^2_{p\times 1}.$

**LEMMA 1.** *Let* $W = [W_{jk}]^2_1 \in U(n,m)$. *Then* $[W_{21}E + W_{22}]^{-1} \in H^2_{m\times m}$, $[W_{11} + W_{12}E^*]^{-1} \in zK^2_{n\times n}$, *for every* $E \in \mathbf{B}_{n\times m}$.

PROOF. We have

$$W = \begin{bmatrix} b_1 & 0 \\ & \\ 0 & b_2^{-1} \end{bmatrix} U, \qquad U \in M(n,m),$$

where $b_1$ and $b_2$ are from the representation (3). It remains to use Lemma 1 from [3] ([26]).

For a **m.f.** $E$ in $\mathbf{B}_{n\times m}$ we denote by $H_E$ the set of the points in $\partial E$ in which $E$ is holomorphic.

**LEMMA 2.** *Let* $W = [W_{jk}]^2_1 \in U(n,m)$, $S = F_W(E)$, $E \in \mathbf{B}_{n\times m}$. *Then* $H_W \cap H_E \subset H_S$.

PROOF. By Lemma 1, we have $[W_{21}E + W_{22}]^{-1} \in H^2_{m\times m}$. Therefore, if $z_* \in H_W \cap H_E$, then the **m.f.** $[W_{21}E + W_{22}]^{-1}$ is holomorphic in the point $z_*$. So $S \ (= F_W(E))$ is also holomorphic in $z_*$.

**DEFINITION.** If $S = F_W(E)$, where $W \in U(n,m)$, $E = $ **const**, $E^*E = I_m$, then we call $S$ a *canonical* **m.f.** for the family $F_W(\mathbf{B}_{n\times m})$.

**LEMMA 3.** *Let* $W \in U(n,m)$ *and* $z_* \in H_S$ ($\subset \partial E$) *for all canonical* **m.f.** $S$ *from* $F_W(\mathbf{B}_{n\times m})$. *Then either* $z_* \in H_W$ *or* $z_*$ *is a pole of the* **m.f.** $W$; *if in addition* $W \in U_r(n,m)$, *then* $z_* \in H_W$.

PROOF. Let $W = [W_{jk}]^2_1 \in U(n,m)$, $S_E = F_W(E)$, $E = $ **const**, $E^*E = I_m$. Then

$$E^*[W_{11} + W_{12}E^*]^{-1}S_E = [W_{21}E + W_{22}]^{-1}.$$

If $z_* \in H_{S_E}$, then, using Lemma 1 and a well-known sufficient condition for the existence of an analytic continuation across an arc of $\partial E$ [14], we get that the **m.f.** $[W_{21}E + W_{22}]^{-1}$ is holomorphic in $z_*$. Therefore $z_*$ is a regular point or a pole of the **m.f.** $[W_{21}E + W_{22}]$ and $[W_{11}E + W_{12}]$ ($= S_E[W_{21}E + W_{22}]$). Since this holds for any isometric matrix $E$, then $z_*$ is a regular point or a pole of the **m.f.** $W = [W_{jk}]_1^2$. If, moreover, $W \in U_r(n,m)$, then $z_*$ can not be a pole of the **m.f.** W. Indeed, if $z_*$ is a pole of the **m.f.** $W\ (\in U(n,m))$, then from W we can "take off" on the right an elementary Blaschke -- Potapov factor $W_o$ with a pole in $z_*$ [13]: $W = W_1 W_o$, $W_1 \in U(n,m)$. Since $W_o \in U_s(n,m)$, then $W \notin U_r(n,m)$.

**5. DEFINITION.** A solution f of the completely indetermined problem (6) is called *canonical* if it is obtained by the general formula (7) for $E = $ **const**, $E^* E = I_m$.

Such a solution is an isometry-valued **m.f.** on $\partial E$. In § 3.12 of [3], [26] the following information about canonical solutions was stated in a form which we will need here.

**LEMMA 4.** *Let* $f\ (\in L^\infty_{n\times m})$ *be an isometry-valued* **m.f.***, and* $\gamma_k = \gamma_k(f)$ *its Fourier coefficients for* $k \geq 1$. *The problem* (6) *is completely indetermined and f is a solution to this problem iff : for m = n, we have*

$$K_n^2 \cap f^* H_n^2 = \{0\}, \quad \textbf{dim}(zK_n^2 \cap f^* H_n^2) = n,$$

*and for m < n, there exists an inner* **m.f.** $b\ (\in B_{n\times(n-m)})$ *such that* $\tilde{f} = [f,b]$ *is a unitary--valued* **m.f.***, i.e.* $b^* f = 0$ *and*

$$K_n^2 \cap \tilde{f}^* H_n^2 = \{0\}, \quad \textbf{dim}(zK_n^2 \cap \tilde{f}^* H_n^2) = n.$$

**LEMMA 5.** *Let* f *be a canonical solution of the completely indetermined problem* (6), $\phi^{\pm 1} \in H_1^\infty$, $f_\phi := (\phi/\bar{\phi})f$. *Then the new problem* (6) *with* $\gamma_k = \gamma_k(f_\phi)$ $(k \geq 1)$ *is completely indetermined and* $f_\phi$ *is a canonical solution of it.*

PROOF. It is sufficient to use Lemma 4 and to take into account the fact that for $\phi^{\pm 1} \in H_1^\infty$, we have

$$h_- \in K_r^2 \Longleftrightarrow \bar{\phi}h_- \in K_n^2; \quad \overset{o}{h}_- \in zK_n^2 \Longleftrightarrow \bar{\phi}\overset{o}{h}_- \in zK_n^2; \quad h_+ \in H_n^2 \Longleftrightarrow \phi h_+ \in H_n^2.$$

**6. DEFINITION.** If W is a resolvent matrix of a completely indetermined problem (4), then the canonical **m.f.** of the family $F_W(B_{n\times m})$ are called the *canonical solutions* of the considered problem (4).

**REMARK 1.** From (5) it follows that this definition does not depend on the choice of the resolvent matrix of problem (4).

**REMARK 2.** An **m.f.** S is a canonical solution of the problem (4) iff $f = b_1^{-1} S b_2^{-1}$ is a canonical solution of the corresponding completely indetermined problem (6) with $\gamma_k = \gamma_k(b_1^{-1} S b_2^{-1})$.

**LEMMA 6.** *Let* $S_o$ *be a canonical solution of the completely indetermined problem (4) with* $b_1 = b_o I_n$, $b_2 = I_m$, *when* $b_o$ *is a scalar inner function. Let us consider*

$$b_\alpha = (b_o - \alpha)/(1 - \overline{\alpha} b_o), \qquad \alpha \in D,$$

*and the new problem (4) with the same* **m.f.** $S_o$, $b_2 = I_m$, *but* $b_1 = b_\alpha I_n$. *Then this problem is completely indetermined and* $S_o$ *is a canonical solution of it.*

PROOF. The **m.f.** $f_o := \overline{b}_o S_o$ is a canonical solution of the completely indetermined problem (6) with $\gamma_k = \gamma_k(f_o)$ $(k \geq 1)$. Since $f_\alpha := \overline{b}_\alpha S_o = (\phi_\alpha / \overline{\phi}_\alpha) f_o$, where $\phi_\alpha = 1 - \overline{\alpha} b_o$, $\phi_\alpha^{\pm 1} \in H_1^\infty$, then by Lemma 5 the new problem (6) with $\gamma_k = \gamma_k(f_\alpha)$ $(k \geq 1)$ is completely indetermined and $f_\alpha$ is a canonical solution of it. Therefore, problem (4) with **m.f.** $S_o$, $b_1 = b_\alpha I_n$, $b_2 = I_m$ is completely indetermined and $S_o$ is a canonical solution of it.

7. We end now the proof of Theorem 1. It remains to show that if $W = [W_{jk}]_1^2 \in$ $\in U_r(n,m)$ and in representation (3) we have $b_1 = b_o I_n$, $b_2 = I_m$, where $b_o$ is a scalar inner function, then $\overset{o}{H}_{b_1,b_2} \subset H_W \cap \partial E$. By Frostman Theorem [15] it follows that there exists $\alpha_o$ $(\in D)$ such that $b_{\alpha_o}$ is a Blaschke product. Let $S_o$ be a canonical **m.f.** of the family $F_W(\mathbf{B}_{n \times m})$. Since $W \in U_r(n,m)$, then $S_o$ is a canonical solution of the completely indetermined problem (4) with $b_1 = b_o I_n$, $b_2 = I_m$. By Lemma 6 the new problem (4) with the same **m.f.** $S_o$, $b_2 = I_m$, but with $b_1 = b_{\alpha_o} I_n$ is also completely indetermined and $S_o$ is a canonical solution of it. Let us consider a resolvent matrix $W_o = [W_{jk}^o]_1^2 \in U_r(n,m)$ of this problem, which has in (3) $b_1 = b_{\alpha_o} I_n$, $b_2 = I_m$. According to the facts already proved in § 2.2, in the points in which $b_{\alpha_o}$ is holomrphic on $\partial E$, the **m.f.** $W_o$ is also holomorphic. We obtain that $\overset{o}{H}_{b_1,b_2} \subset H_{W_o} \cap \partial E$. By Lemma 2, we have $H_{W_o} \cap \partial E = H_{S_o}$. Therefore $\overset{o}{H}_{b_1,b_2} \subset H_{S_o}$. Since $S_o$ is an arbitrary canonical **m.f.** of the family $F_W(\mathbf{B}_{n \times m})$, then by Lemma 3 we obtain that $\overset{o}{H}_{b_1,b_2} \subset H_W \cap \partial E$. Theorem 1 is now completely proved.

### § 3. PROOF OF THEOREM 4

**1.** Let problem (4) be completely indetermined, $F_{\rho S_o; b_1, b_2} = F_{W_\rho}(\mathbf{B}_{n \times m})$, $W_\rho \in \overset{\circ}{U}_r(n,m)$; $F_{S_o; b_1, b_2} = F_W(\mathbf{B}_{n \times m})$, $W \in \overset{\circ}{U}_r(n,m)$, $W$ and $W_\rho$ having in representation (3) the same $b_1$ and $b_2$ as in problem (4). Such $W$ and $W_\rho$ exist and are unique (Theorem A and Remark 1 following it). By Theorem 1:

$$H_{W_\rho} = H_{b_1, b_2}, \qquad H_W = H_{b_1, b_2}.$$

It remains to prove that

$$(8) \qquad W(z) = \lim_{\rho \uparrow 1} W_\rho(z), \qquad z \in H_{b_1, b_2}.$$

**2.** Let us consider the **m.f.** $\tilde{S}_\rho = [S_{jk}^{(\rho)}]_1^2$, obtained from $W_\rho = [W_{jk}^{(\rho)}]_1^2$ by the transform (2). Since $\tilde{S}_\rho \in \mathbf{B}_{n+m}$, then by any sequence $\rho_\ell \uparrow 1$ in E, there exists the limit

$$[S_{jk}(z)]_1^2 = \tilde{S}(z) = \lim_{\ell \to \infty} \tilde{S}_{\rho_\ell}(z) \ (\in \mathbf{B}_{n+m}).$$

Let us show that $\det S_{22}(z) \not\equiv 0$. We have

$$\det S_{22}(z) = \lim_{\ell \to \infty} \det S_{22}^{(\rho_\ell)}(z) = \det b_2(z) \lim_{\ell \to \infty} [\det p_+^{(\rho_\ell)}(z)]^{-1}.$$

Consider $R_\Lambda(0)$ $(>0)$ and $R_\Pi(0)$ $(>0)$ the left and the right semi-radii of the matrix sphere

$$K(0) := \{h(0) : h = b_1^{-1}(S_1 - S_o)b_2^{-1}, S \in F_{S_o; b_1, b_2}\}.$$

They are uniquely defined by $K(0)$ under the normalization $\det R_\Lambda(0) = \det R_\Pi(0)$. For the spheres

$$K_\rho(0) := \{h(0) : h = b_1^{-1}(S_1 - \rho S_o)b_2^{-1}, S_1 \in F_{\rho S_o; b_1, b_2}\}$$

their left and right semi-radii $R_\Lambda^{(\rho)}(0)$ and $R_\Pi^{(\rho)}(0)$, normalized by the condition $\det R_\Lambda^{(\rho)}(0) = \det R_\Pi^{(\rho)}(0)$, can be computed by the formulae [3]:

$$R_\Lambda^{(\rho)}(0) = [p_-^{(\rho)}(\infty)], \qquad R_\Pi^{(\rho)}(0) = [p_+^{(\rho)}(0)]^{-1},$$

where

$$p_-^{(\rho)}(z) = b_1^{-1}(z)W_{11}^{(\rho)}(z), \quad p_+^{(\rho)}(z) = b_2(z)W_{22}^{(p)}(z).$$

If $h(0) \in K(0)$, then $\rho h(0) \in K_\rho(0)$, that is $\rho K(0) \subset K_\rho(0)$. Therefore [16]:

$$\det[\rho^{\frac{1}{2}} R_\Lambda(0)] \cdot \det[\rho^{\frac{1}{2}} R_\Pi(0)] \leq \det R_\Lambda^{(\rho)}(0) \cdot \det R_\Pi^{(\rho)}(0).$$

Consequently

$$\rho^{(n+m)/2} \det[p_+(0)]^{-2} \leq [\det p_+^{(\rho)}(0)]^{-2}$$

$$\lim_{\ell \to \infty} [\det p_+^{(\rho_\ell)}(0)] \geq \det p_+(0); \quad \det S_{22}(z) \not\equiv 0.$$

**3.** Let us consider now the **m.f.** $\widetilde{W} = [\widetilde{W}_{jk}]_1^2$ related to a **m.f.** $\widetilde{S} = [S_{jk}]_1^2$ by the transformation (1). On the set of points where $W_\rho$ is holomorphic in E (which is equal to the set where $b_2^{-1}$ is holomorphic), we shall have

$$\widetilde{W}(z) = \lim_{\ell \to \infty} W_{\rho_\ell}(z) \in B_j, \quad \det \widetilde{W}(z) = \lim_{\ell \to \infty} \det W_{\rho_\ell}(z) = \det b_1(z)/\det b_2(z).$$

**4.** Let us show that $F_{S_o;b_1,b_2} \in F_{\widetilde{W}}(B_{n \times m})$. Obviously

$$F_{S_o;b_1,b_2} = \bigcap_{0 < \rho < 1} \rho^{-1} F_{\rho S_o;b_1,b_2} = \bigcap_{\ell=1}^{\infty} \rho_\ell^{-1} F_{\rho_\ell S_o;b_1,b_2},$$

where

$$\rho^{-1} F_{\rho S_o;b_1,b_2} := \{\rho^{-1} S_1 ; S_1 \in F_{\rho S_o;b_1,b_2}\}.$$

Therefore, if $S \in F_{S_o;b_1 b_2}$, then there exists a sequence $E_\ell \in B_{n \times m}$ such that

(9)
$$S = \rho_\ell^{-1} F_{W_{\rho_\ell}}(E_\ell).$$

There exists a subsequence $E_{\ell_k}$ of $E_\ell$, such that

$$E(z) := \lim_{k \to \infty} E_{\ell_k}(z) \in B_{n \times m}.$$

Taking the limit (as $\ell_k \to \infty$) in (9) we get that $S = F_{\widetilde{W}}(E)$, so $F_{S_o;b_1,b_2} \subset F_{\widetilde{W}}(B_{n \times m})$.

Let now $S = F_{\widetilde{W}}(E)$ with $E \in B_{n \times m}$. Consider $S_\rho = F_{W_\rho}(E)$. We have:

$$S(z) = \lim_{\ell \to \infty} S_{\rho_\ell}(z).$$

Since $S_\rho \in F_{\rho S_o;b_1,b_2}$, then

$$S_{\rho_\ell}(z) = \rho_\ell S_0(z) + b_1(z)h_{\rho_\ell}(z)b_2(z), \qquad \|h_{\rho_\ell}\|_\infty < 2, \qquad h_{\rho_\ell} \in H^\infty_{n\times m}.$$

There exists a subsequence $h_{\rho_{\ell_k}}(z)$ such that

$$h(z) := \lim_{k\to\infty} h_{\rho_{\ell_k}}(z) \in H^\infty_{n\times m} \qquad (\rho_{\ell_k} \uparrow 1).$$

Therefore

$$S(z) = S_0(z) + b_1(z)h(z)b_2(z), \quad h \in H^\infty_{n\times m},$$

that is $S \in F_{S_0;b_1,b_2}$. Thus $F_{S_0;b_1,b_2} = F_{\widetilde{W}}(\mathbf{B}_{n\times m})$.

**5.** For the **m.f.** $\widetilde{W}$ and $W$ we have $F_{\widetilde{W}}(\mathbf{B}_{n\times m}) = F_W(\mathbf{B}_{n\times m})$ and $\det \widetilde{W}(z) = \det W(z)\ (= \det b_1(z)/\det b_2(z))$. Therefore $\widetilde{W}(z) = W(z)V$, where $V$ is a constant **j**-unitary matrix. Since $W \in U^0_r(n,m)$ and $\widetilde{W} \in U^0_r(n,m)$, it follows that $V = I$.

**6.** We show now that

$$(10) \qquad\qquad \widetilde{S}(z) = \lim_{\rho\uparrow 1} \widetilde{S}_\rho(z), \quad z \in E.$$

Suppose that this is not true. Then there exist $z_* \in E$ and $\rho'_\ell \uparrow 1$ such that $\widetilde{S}_1(z) := \lim_{\ell\to\infty} \widetilde{S}_{\rho'_\ell}(z) \in \mathbf{B}_{n+m}$ and $\widetilde{S}_1(z_*) \neq \widetilde{S}(z_*)$. Similar to $\widetilde{S}(z)$, there exists a **m.f.** $\widetilde{W}_1 (\in U^0_r(n,m))$ associated to $\widetilde{S}_1(z)$, with $F_{S_0;b_1,b_2} = F_{\widetilde{W}_1}(\mathbf{B}_{n\times m})$, $\det \widetilde{W}_1 = \det b_1/\det b_2$, such that $\widetilde{W}_1(z) = W(z)\ (= W(z))$, that is $\widetilde{S}_1(z) = \widetilde{S}(z)$, $z \in E$. This is a contradiction. Consequently, we have relation (10). It follows from it that in the points where $W_\rho$ is holomorphic in E, i.e. where $b_1^{-1}$ is holomorphic in E, we have (8).

**7.** Since

$$W_\rho(z) = \mathbf{j}[W_\rho^*(\bar{z})]^{-1}\mathbf{j}, \qquad W(z) = \mathbf{j}[W^*(\bar{z})]^{-1}\mathbf{j}, \qquad z \in E_e$$

and the **m.f.** $W$ and $W_\rho$ are holomorphic on $E_e$ in the points where $b_1$ is holomorphic, then in all these points we have (8).

**8.** According to Theorem 1 for $W$ and $W_\rho$, we have

$$H_{W_\rho} \cap \partial E = H_W \cap \partial E = H^0_{b_1,b_2}$$

where $H^{o}_{b_1,b_2}$ is the set of points where $\det b_1 \cdot \det b_2$ is holomorphic on $\partial E$. Using this, and the fact that relation (8) holds in the points from $E$ and $E_e$ in a neighbourhood of the point $z_* \in H^{o}_{b_1,b_2}$, we obtain that it holds also in the point $z_*$.

## §.4. MAXIMAL JUMP OF SPECTRAL FUNCTION IN CONTINUATION PROBLEMS; CARATHÉODORY THEOREM FOR MATRIX FUNCTIONS IN THE CLASSES B AND $B_j$ (STATEMENT OF RESULTS)

**1.** Let $C_n$ and $P_n$ be the classes of **m.f.** of order n, holomorphic in $D$ and $\Pi$ respectively, and having there $\mathbf{Re}\, f(z) \geq 0$. These classes are represented by the Riesz-Herglotz and Nevanlinna formulae, respectively, as:

$$f(z) = i\gamma_f + (1/2\pi) \int_{-\pi}^{\pi} [(e^{i\mu} + z)/(e^{i\mu} - z)] d\sigma_f(\mu), \quad z \in D,\ f \in C_n,$$

where $\gamma_f^* = \gamma_f$, $d\sigma_f(\mu) \geq 0$, $\int_{-\pi}^{\pi} \| d\sigma_f(\mu) \| < \infty$;

$$f(z) = i\alpha_f + \beta_f z + (1/2\pi i) \int_{-\infty}^{\infty} [1/(\mu - iz) - \mu/(1 + \mu^2)] d\sigma_f(\mu), \quad z \in \Pi,\ f \in P_n,$$

where $\alpha_f^* = \alpha_f$, $\beta_f \geq 0$, $d\sigma_f(\mu) \geq 0$, $\int_{-\infty}^{\infty} (1 + \mu^2)^{-1} \| d\sigma_f(\mu) \| < \infty$.

The non-decreasing **m.f.** $\sigma_f(\mu)$ of order n which appear in these formulae are called *spectral*; by the normalization $\sigma_f(\mu - 0) = \sigma_f(\mu)$ they are determined by f up to a constant term by the Stieltjes formula.

Let $\zeta_0 = e^{i\mu}$ if $E = D$ and $\zeta_0 = i\mu$ if $E = \Pi$; put

$$m_f(\zeta_0) := \sigma_f(\mu+0) - \sigma_f(\mu), \quad \zeta_0 \in \partial E.$$

It is known [19] that

$$m_f(\zeta_0) = \lim_{z \to \zeta_0} |z - \zeta_0|\,\mathbf{Re}\, f(z) = \lim_{z \to \zeta_0} |z - \zeta_0|\, f(z),$$

$$((m_f(\infty) = )\ \beta_f = \lim_{x \uparrow +\infty} x^{-1}\mathbf{Re}\, f(x) = \lim_{x \uparrow +\infty} x^{-1} f(x), \quad f \in P_n).$$

For many problems in Harmonic Analysis, the description of the solutions in the completely indetermined case leads to the consideration of a family of **m.f.** $\{f_{p,g}\}$ in $C_n$ or in $P_n$, obtained by the formula

(11)   $f_{p,q}(z) = [a_{11}(z)p(z) + a_{12}(z)q(z)][a_{21}(z)p(z) + a_{22}(z)q(z)]^{-1}$,    $[p,q] \in O_{J_r}$.

where $O_{J_r}$ is the class of pairs $\xi(z) = [p(z),q(z)]$ of meromorphic **m.f.** on E, of order n, with **rang** $\xi(z_*) = n$ for some $z_* \in E$, and with $\xi(z)J_r\xi^*(z) \leq 0$ on E, while $A(z) = [a_{jk}(z)]_1^2 \in B_{J_r}$, with

$$J_r = \begin{bmatrix} 0 & -I_n \\ -I_n & 0 \end{bmatrix}$$

is a fixed **m.f.** which is the resolvent matrix of the problem.

The family of **m.f.** $\{f_{p,q}\}$ obtained by (11) is denoted by $F_A(O_{J_r})$. Let us remark that formula (11) is defined for any pair $\xi = (p,q)$ in $O_{J_r}$ iff

(12)                      $[a_{21}(z), a_{22}(z)]J_r[a_{21}(z), a_{22}(z)]^* < 0$,   $z \in E$.

**THEOREM 5.**  *Let*  $A(z) = [a_{jk}]_1^2 \in B_{J_r}$; *suppose that for A(z) there exists a* $J_r$-*unitary boundary value* $A(\zeta_o)$ *and that condition (12) is fulfilled. Then:*

  1) *there exists the limit*

$$m(\zeta_o) = \lim_{z \to \zeta_o} 2|z - \zeta_o| [a_{21}(z), a_{22}(z)](-J_r)[a_{21}(z), a_{22}(z)^*]^{-1};$$

  2) $m_{f_{p,q}}(\zeta_o) \leq m(\zeta_o)$    $\forall f_{p,q} \in F_A(O_{J_r})$;

  3) $m_{f_{p_o,q_o}}(\zeta_o) = m(\zeta_o)$ *for* $p_o(z) \equiv a_{22}^*(\zeta_o)$, $q_o(z) \equiv -a_{21}^*(\zeta_o)$.

This theorem is a generalization of the results about the solutions with a maximal jump of spectral function for the moments problem ([17]), for strings ([18]) and for the generalized moments problem ([19], [20]).

The limit theorem of Sahnovich [20] on maximal jump spectral functions for the monotone family $F_{A_x}(O_{J_r})$ has the following generalization useful in applications. Let X be an ordered set filtered on the right, which satisfies the condition that there exists a sequence $x_j$ ($\in$ X) such that $\forall x \in X$, $\exists j \in N$ with $x < x_j$.

**THEOREM 6.**  *Let* $A_x(z) \in B_{J_r}$, *and* $M_x = F_{A_x}(O_{J_r})$, *with* $M_{x'} \subset M_{x''}$ *for* $x' < x''$. *Consider* $M = \bigcap_{x \in X} M_x$. *Let the maximal jump* $m(\zeta_o;x)$ *of the spectral function* $\sigma_f(\mu)$ *be*

*defined for some* $\zeta_0 \in \partial E$, *when* f *runs* $M_x$, $\forall x \in X$. *Then there exists the limit*
$m(\zeta_0) = \lim\limits_{x \in X} m(\zeta_0;x)$ *and* $m(\zeta_0)$ *is the maximal jump of* $\sigma_f(\mu)$ *when* f *runs* M, *that is, there exists* $f^\circ$ *in* M *such that*

$$m_f(\zeta_0) \leq m_{f^\circ}(\zeta_0) = m(\zeta_0), \quad \forall f \in M.$$

**2.** Theorem 5 is obtained using the next generalization of Carathéodory Theorem on functions in $\mathbf{B}_1$ to the class $\mathbf{B}$ of **m.f.** .

**THEOREM 7.** *Let* $S(z) \in \mathbf{B}_{n \times m}$, $S^*(z)S(z) < I$ $(z \in E)$. *Then the limit*

(13) $$M(\zeta_0) = \lim\limits_{z \to \zeta_0} 2|z - z_0| [I - S^*(z)S(z)]^{-1}$$

*exists and is finite.*

*For any* **m.f.** $E(z)$ *in* $\mathbf{B}_{n \times m}$ *the limit*

(14) $$M_E(\zeta_0) = \lim\limits_{z \to \zeta_0} |z - \zeta_0| [I - E(z)S(z)]^{-1}$$

*exists and is finite. If, in addition,* $S_0 = \lim\limits_{z_n \to \zeta_0} S(z_n)$ *and* $E(z) \equiv S_0^*$, *then* $M_{S_0^*}(\zeta_0) = M(\zeta_0)$
*and*

(15) $$M(\zeta_0) = \lim\limits_{z \to \zeta_0} 2^{-1}|z - \zeta_0| [I - S^*(z)S_0]^{-1}[I - S^*(z)S(z)][I - S_0^*S(z)]^{-1},$$

(16) $$M_E(\zeta_0) = M^{\frac{1}{2}}(\zeta_0)V_E(\zeta_0)M^{\frac{1}{2}}(\zeta_0), \quad 0 \leq V_E(\zeta_0) \leq I, \quad \forall E \in \mathbf{B}_{m \times n},$$

(17) $$M_E(\zeta_0) \leq M_{S_0^*}(\zeta_0) = M(\zeta_0), \quad \forall E \in \mathbf{B}_{m \times n}.$$

In [22] there is an analog of Carathéodory Theorem for **m.f.** of class $\mathbf{B}$, different from Theorem 6.

From Theorem 7 it follows

**THEOREM 8.** *Let* $W(z) \in \mathbf{B}_J$, $W^*(z)JW(z) < J$ *for* $z \in E$, $\zeta_0 \in \partial E$; *suppose that the limit* $W_0 = \lim\limits_{z \to \zeta_0} W(z)$ *exists. Then the limits*

$$N(\zeta_0) = \lim\limits_{z \to \zeta_0} 2|z - \zeta_0| [J - W^*(z)JW(z)]^{-1},$$

$$N_{W_o^*}(\zeta_o) = \lim_{z \to \zeta_o} |z - \zeta_o| [J - W_o^* J W(z)]^{-1}.$$

exist, such that $N_{W_o^*}(\zeta_o) = N(\zeta_o)$ and

$$N(\zeta_o) = \lim_{z \to \zeta_o} 2^{-1} |z - \zeta_o| [J - W^*(z)JW_o]^{-1}[J - W^*(z)JW(z)][J - W_o^* J W(z)]^{-1}.$$

From this theorem it follows the analog of Carathéodory Theorem for **m.f.** W in the class $\mathbf{B_J}$ which appeared in [23].

### § 5. PROOF OF THEOREMS 5–8

**1.** Let $E = D$, $S \in \mathbf{B}_{n \times m}$, $S^*(z)S(z) < I$ $(z \in E)$. The limit considered in (14) exists, because for $E \in \mathbf{B}_{m \times n}$, we have $[I - ES]^{-1} \in C_m$. In particular, such a limit also exists for $E(z) \equiv \mathbf{const.} = S_o^*$, where $S_o$ is a matrix of order $n \times m$ with $S_o^* S_o \leq I$.

**2.** For any $z_n$ and $z$ in $D$, the Pick inequality

$$\begin{bmatrix} \dfrac{I - S^*(z_n)S(z_n)}{1 - \bar{z}_n z_n} & \dfrac{I - S^*(z_n)S(z)}{1 - \bar{z}_n z} \\[4mm] \dfrac{I - S^*(z)S(z_n)}{1 - \bar{z}z_n} & \dfrac{I - S^*(z)S(z)}{1 - \bar{z}z} \end{bmatrix} \geq 0$$

holds. We shall consider $z = |z|\zeta_o$, $z_n = |z_n|\zeta_o$, $\zeta_o \in \partial D$. Let us put

$$\Phi(z) = 2|z - \zeta_o| [I - S^*(z)S(z)]^{-1}, \qquad \Psi(z;E) = |z - \zeta_o| [I - E(z)S(z)]^{-1},$$

$$\chi(z;E) = \Psi^*(z;E)\Phi^{-1}(z)\Psi(z;E).$$

By Pick inequality, it follows that for the considered $z_n$ and $z$, we have

(18)     $((1 + |z_n|)/2)\, \Phi(z_n) \leq [2(1 - |z_n||z|)^2/(1 + |z|)(1 - |z|)^2]\chi(z;S^*(z_n))$.

Using that $S^*(z)S(z_n)S^*(z_n)S(z) \leq S^*(z)S(z)$, we obtain

(19)                      $\chi(z;S^*(z_n)) \leq -((1 - |z|)/2)\, I + \mathbf{Re}\, \Psi(z;S^*(z_n))$.

By Pick inequality it also follows that the family of matrices $\Phi(z)$ $(z \in D)$ is bounded $(\|\Phi(z)\| \leq 2(1 - \|S(0)\|)^{-2})$. Therefore, for some sequence $z_n'$ $(\in D)$ the limit

$M_1(\zeta_o) = \lim\limits_{z'_n \to \zeta_o} \phi(z'_n)$ exists. The existence of the limit in (13) will be proved as soon as it will be shown that for any other sequence $z''_n$, the existence of $M_2(\zeta_o) = \lim\limits_{z''_n \to \zeta_o} \phi(z''_n)$ implies $M_2(\zeta_o) = M_1(\zeta_o)$.

**3.** Let $z_n$ be a subsequence of the sequence $z'_n$ for which there exists the limit $S_1 = \lim\limits_{z_n \to \zeta_o} S(z_n)$. Taking the limit in the inequalities (18) and (19) for $z_n \to \zeta_o$, we obtain

(20) $\qquad M_1(\zeta_o) \le (2/(1 + |z|))\chi(z;S_1^*) \le -[(1 - |z|)/(1 + |z|)]I + \mathbf{Re}\,\psi(z;S_1^*).$

Since for $z \to \zeta_o$ the limit of the left hand side of this double inequality is equal to $M_{S_1^*}(\zeta_o)$, we get $M_1(\zeta_o) \le M_{S_1^*}(\zeta_o)$.

**4.** We show now that relation (16) holds, when we replace $M(\zeta_o)$ with $M_1(\zeta_o)$. For this, consider

$$C_E(z) = [I - E(z)S(z)]^{-1} \ (\in \mathbf{C}_m), \quad E \in \mathbf{B}_{m \times n}.$$

We have (omitting the argument z for a while)

$$ESC_E = C_E - I, \quad (C_E^* - I)(C_E - I) \le C_E^* S^* S C_E,$$

$$C_E^*(I - S^*S)C_E - C_E - C_E^* \le 0.$$

We get

$$V_1^* V_1 \le I, \quad \text{where } V_1 := (I - S^*S)^{\frac{1}{2}} C_E (I - S^*S)^{\frac{1}{2}} - I;$$

$$C_E(z) = [I - S^*(z)S(z)]^{-\frac{1}{2}}[I + V_1(z)][I - S^*(z)S(z)]^{-\frac{1}{2}}.$$

Let us multiply both sides of the last eqality by $|z - \zeta_o|$ and take the limit by the subsequence $z_n$ of the sequence $z'_n$; the limit $V_1(\zeta_o) = \lim\limits_{z_n \to \zeta_o} V_1(z_n)$ exists. We have

$$M_E(\zeta_o) = M_1^{\frac{1}{2}}(\zeta_o)2^{-1}[I + V_1(\zeta_o)]M_1^{\frac{1}{2}}(\zeta_o).$$

Because $V_E(\zeta_o) = \mathbf{Re}\,2^{-1}[I + V_1(\zeta_o)]$ and taking into account that $M_E(\zeta_o) \ge 0$, we get

$$M_E(\zeta_o) = M_1^{\frac{1}{2}}(\zeta_o)V_E(\zeta_o)M_1^{\frac{1}{2}}(\zeta_o), \quad 0 \le V_E(\zeta_o) \le I.$$

This means that $M_E(\zeta_o) \le M_1(\zeta_o)$ ($E \in \mathbf{B}_{m \times n}$). Putting here $E(z) \equiv S_1^*$, we obtain that $M_{S_1^*}(\zeta_o) \le M_1(\zeta_o)$; but previously we get $M_1(\zeta_o) \le M_{S_1^*}(\zeta_o)$. Consequently $M_{S_1^*}(\zeta_o) = M_1(\zeta_o)$. Analogously, we get that

$$M_E(\zeta_0) \leq M_2(\zeta_0) = M_{S_2^*}(\zeta_0), \qquad S_2 := \lim_{z''_{n_k} \to \zeta_0} S(z''_{n_k}).$$

Therefore $M_1(\zeta_0)$ $(= M_{S_1^*}(\zeta_0)) \leq M_2(\zeta_0)$. In the same way we obtain that $M_2(\zeta_0) \leq$ $\leq M_1(\zeta_0)$, so $M_2(\zeta_0) = M_1(\zeta_0)$. Therefore the limit in (13) exists; $M(\zeta_0) = M_1(\zeta_0)$ holds and (16) is true. Taking the limit in (20) for $z \to \zeta_0$, we obtain (15). Theorem 7 is proved.

**5.** Theorem 8 follows from Theorem 7 since if $\tilde{S}(z)$ is related to $W(z)$ $(\in \mathbf{B_J})$ by (1), then:

$$[\mathbf{J} - W^*(z)\mathbf{J}W(z)]^{-1} = [P_+ + P_-\tilde{S}(z)][I - \tilde{S}^*(z)\tilde{S}(z)]^{-1}[P_+ + P_-\tilde{S}(z)]^*,$$

$$[\mathbf{J} - W^*(\zeta_0)\mathbf{J}W(z)]^{-1} = [P_+ + P_-\tilde{S}(z)][I - \tilde{S}^*(\zeta_0)\tilde{S}(z)]^{-1}[P_+ + P_-\tilde{S}(\zeta_0)]^*,$$

**6.** We prove now Theorem 5. Let $A(z) = [a_{jk}(z)]_1^2$ satisfying the conditions of the theorem. Consider

$$B(z) = [b_{jk}(z)] := A(z)U, \qquad U^* = (1/\sqrt{2}) \begin{bmatrix} -I_n & I_n \\ I_n & I_n \end{bmatrix};$$

$$(21) \qquad f_E(z) = [b_{11}(z)E(z) + b_{12}(z)][b_{21}(z)E(z) + b_{22}(z)]^{-1}, \qquad E \in \mathbf{B_n}.$$

For obtaining the family $\{f_{p,q}\}$ from (11) we can restrict ourselves to pairs $\xi = [I - E, I + E]$ where $E \in B_n$; it turns out that the family $\{f_{p,q}\}$ coincides with the family of **m.f.** $\{f_E\}$ obtained in (21). The **m.f.** $B(z)$ is $(\mathbf{j}, \mathbf{J_r})$-inner and meromorphic on E:

$$B^*(z)\mathbf{J_r}B(z) \leq \mathbf{j}, \qquad z \in E$$

$$B^*(z)\mathbf{J_r}B(z) = \mathbf{j}, \qquad \text{for } z \in \partial E \text{ a.e.}$$

Condition (12) for $B(z)$ can be written as

$$[b_{21}(z), b_{22}(z)]\mathbf{j}[b_{21}(z), b_{22}(z)]^* < 0, \qquad z \in E.$$

Therefore

$$\theta(z) := -b_{22}^{-1}(z)b_{21}(z) \in \mathbf{B_n},$$

$$I - \theta^*(z)\theta(z) > 0, \qquad z \in E.$$

Applying Theorem 7 to $S(z) = \theta^*(\bar{z})$ we obtain the existence of the finite limit

$$M(\zeta_0) = \lim_{z \to \zeta_0} 2|z - \zeta_0|[I - \theta(z)\theta^*(z)]^{-1},$$

$$M_E(\zeta_o) = \lim_{z \to \zeta_o} |z - \zeta_o| [I - \Theta(z)E(z)]^{-1}, \qquad E \in \mathbf{B}_n;$$

$$M_E(\zeta_o) \leq M(\zeta_o) = M_{E_o}(\zeta_o), \qquad E_o(z) := \Theta^*(\zeta_o).$$

From the existence of $M(\zeta_o)$ and $\mathbf{B}(\zeta_o)$ it follows the existence of the finite limit

$$m(\zeta_o) = \lim_{z \to \zeta_o} 2|z - \zeta_o| [b_{22}(z)b_{22}^*(z) - b_{21}(z)b_{21}^*(z)]^{-1}$$

and that

$$m(\zeta_o) = [b_{22}^{-1}(\zeta_o)]^* M(\zeta_o) b_{22}^{-1}(\zeta_o).$$

For the **m.f.** $f_E(z)$, defined by (21), we have:

$$f_E(z) = b_{11}(z)b_{21}^{-1}(z) + \Delta(z)[I - \Theta(z)E(z)]^{-1} b_{22}^{-1}(z),$$

where $\Delta(z) := b_{12}(z) - b_{11}(z)b_{21}^{-1}(z)b_{22}(z)$. From the equality $B(\zeta_o)jB^*(\zeta_o) = \mathbf{J}_r$ it follows that

$$\lim_{z \to \zeta_o} \Delta(z) = [b_{22}^{-1}(\zeta_o)]^*.$$

We obtain

$$m_{f_E}(\zeta_o) = [b_{22}^{-1}(\zeta_o)]^* M_E(\zeta_o) b_{22}^{-1}(\zeta_o),$$

$$m_{f_E}(\zeta_o) \leq m(\zeta_o) \quad \forall E; \ m_{f_E}(\zeta_o) = M(\zeta_o) \text{ for } E(z) \equiv \Theta^*(\zeta_o).$$

The assertion of Theorem 5 follows now by returning from the **m.f.** B(z) to the **m.f.** A(z), by taking into account the relation between the pairs $\xi = [p,q] \in O_{\mathbf{J}_r}$ and $E \in \mathbf{B}_n$ such that $f_{p,q} = f_E$; $f_{p_o,q_o} = f_{E_o}$ for $E_o(z) \equiv \Theta^*(\zeta_o)$, $[p_o,q_o] \equiv [b_{22}^*(\zeta_o), -b_{21}^*(\zeta_o)]$.

**7.** Let us prove now Theorem 6. By virtue of the assumption made on the ordered set X we can consider that $X = \mathbf{N}$ and that the limit $m(\zeta_o) = \lim_{n \to \infty} m(\zeta_o;n)$ exists since $m(\zeta_o;n) \leq m(\zeta_o;k)$ for $n > k$. Because $M = \bigcap_{n=1}^{\infty} M_n$, then $m_f(\zeta_o) \leq m(\zeta_o;n)$ $(f \in M)$ and therefore $m_f(\zeta_o) \leq m(\zeta_o)$ $(f \in M)$. Let $m(\zeta_o;n) = m_{f_n}(\zeta_o)$, $f_n \in M_n$. The family $M_1$ is compact, and thus there exists a convergent subsequence $f_{n_k}$ of $f_n$. Let $f_\infty(z) = \lim_{k \to \infty} f_{n_k}(z)$. From the description of the intersection $M$ of the family $M_n$ (made in [24]), it follows that $f_\infty \in M$. By Helly's Theorem we can assume that $\sigma_{f_\infty}(\mu) = \lim_{k \to \infty} \sigma_{f_{n_k}}(\psi)$, and therefore from the inequality $\sigma_{f_n}(\mu + \epsilon) - \sigma_{f_n}(\mu) \geq m(\zeta_o)$ it follows that

$\sigma_{f_\infty}(\mu + \varepsilon) - \sigma_{f_\infty}(\mu) \geq m(\zeta_0)$. On the other hand, since $f_\infty \in M$, then $m_{f_\infty}(\zeta_0) \leq m(\zeta_0)$. Consequently $m_{f_\infty}(\zeta_0) = m(\zeta_0)$. We obtained:

$$m_f(\zeta_0) \leq m_{f_\infty}(\zeta_0) = m(\zeta_0) \qquad f \in M.$$

The theorem is proved.


### REFERENCES


1. **Arov, D.Z.** : On regular and singular **J**-inner matrix-functions and related extrapolation problems (Russian), *Functional Anal. i Prilozhen* **22**(1988), 57–59.

2. **Arov, D.Z.** : Regular **J**-inner matrix-functions and related continuation problems (Russian), deposited in *Ukr. NIINTI*, no. **406** Uk. 87 Dep., 25 pages, 1987.

3. **Arov, D.Z.** : $\gamma$-derived matrices, **J**-inner matrix-functions and related extrapolation matrix-functions problems (Russian), deposited in *Ukr. NIINTI*, no. **726**- Uk. 86 Dep., 1986.

4. **Arov, D.Z.** : Three problems about **j**-inner matrix-function, in *Lecture Notes in Mathematics*, **1043**(1984), pp. 164–168.

5. **Arov, D.Z.** : Some problems in the theory of linear stationary passive systems, *in Operators in indefinite metric spaces, scattering theory and other topics*, Birkhauser Verlag, 1987, pp. 17–37.

6. **Helton, J.W.** : Non-Euclidean analysis and electronics, *Bull. Amer. Math. Soc.* **7**(1982), 1–64.

7. **Simakova, L.A.** : On meromorphic plus-matrix functions (Russian), *Mat Issled.* **10**:1(1975), 287–292.

8. **Krein, M.G.** : Basic considerations on the representation of Hermitian operators with defect index (m,m) (Russian), *Ukrain. Mat. Zh.* **1**:2(1949), 3–66.

9. **Krein, M.G. ; Langer, H.** : On some continuation problems which are closely related to the theory of operators in spaces $\Pi_\kappa$. IV: continuous analogues of orthogonal polynomials on the unit circle with respect to an indefinite weight and related continuation problems for some classes of functions, *J. Operator Theory* **13**(1985), 299–417.

10. **Adamjan, V.M. ; Arov, D.Z. ; Krein, M.G.** : Infinite Hankel matrices and related continuation problems (Russian), *Izv. Akad. Nauk. Armyan. SSR, Ser. Mat.* **6**:2-3(1971), 87–112.

11. **Adamjan, V.M.** : Non-degenerate unitary intertwinings of semi-unitary operators (Russian), *Funktional Anal. i. Prilozhen* **7**(1973) 1–16.

12. **Katznelshon, V.E.** : The regularizations of the basic matrix inequality problems on the decompositions of positive definite kernels in elementary products, *Dokl. Akad. Nauk Ukrain. SSR.* **3**(1984).

13. **Potapov, V.P.** : The multiplicative structure of **j**-contractive matrix-functions (Russian), *Trudy Moskov. Mat. Obshch.* **4**(1955), 125–236.

14.  **Sz.-Nagy, B. ; Foiaş, C. :** *Harmonic analysis of operators on Hilbert space* (Russian transl.), Moscow, Mir, 1970.

15.  **Hoffman, K.:** *Banach spaces of analytic functions* (Russian transl.), Moscow, Inn. Lit., 1963.

16.  **Shmuljan, Yu.L. :** Operatorial sphere, in *Functions theory, functional analysis and their applications, Harkov* **6**(1968), 68-81.

17.  **Krein, M.G. ; Nudelman, A.A. :** The Markov moment problems and extremal problems (Russian), Moscow, Nauka, 1973.

18.  **Krein, M.G. :** The Chebyshev-Markov inequality in the spectral function theory of strings (Russian), *Mat. Issled.* **5:**1(1970), 77-101.

19.  **Sahnovich, A.L. :** The factorization problem and operatorial identities (Russian), *Uspekhi Mat. Nauk* **41:**1(1986), 3-55.

20.  **Sahnovich, A.L. :** On a class of extremal problems (Russian), *Izv. Akad. Nauk. SSSR Ser. Mat.* **51:**2(1987), 436-443.

21.  **Carathéodory, C. :** *Conformal representations* (Russian transl.), Moscow, 1934.

22.  **Kovalishina, I.V. :** Carathédory-Julia theorem for matrix-function (Russian), in *Functions theory, function analysis and its applications, Harkov* **43**(1985), 70-82.

23.  **Melamud, E.Ia. :** Carathéodory theorem and Nevanlinna boundary interpolation problem for **j**-contractive matrix-functions (Russian), *Dokl. Akad. Nauk Armyan SSR Sec. Mat.* **80:**1(1985), 12-16.

24.  **Orlov, S.A. :** The parametrization of the limit matrix circles analyticaly depending on the parameter (Russian), in *Functions theory, functional analysis and its applications, Harkov* **41**(1984), 96-107.

25.  **Dym, H. :** *J-contractive matrix functions, reproducing kernel Hilbert spaces and interpolation,* to appear.

26.  **Arov, D.Z. :** γ-derived matrices, **J**-inner matrix-functions and related extrapolation matrix-functions problems. I, II, III (Russian), in *Functions theory, functional analysis and their applications, Harkov,* **51**(1989), 61-67; **52**(1989); **53**(1989).

**D.Z. Arov**

Pr. Dimitrova 15, kv. 18
270104, Odessa
USSR.

Operator Theory:
Advances and Applications, Vol. 43
© 1990 Birkhäuser Verlag Basel



# SPECTRAL ANALYSIS FOR SIMPLY CHARACTERISTIC OPERATORS BY MOURRE'S METHOD. I

G. Arsu

## 1. INTRODUCTION

The purpose of this paper is to give a time dependent scattering theory for operators of the form $p_o(D) + V$, where $p_o(D)$ is a convolution operator with the symbol not satisfying the condition $\lim_{|\xi| \to \infty} p_o(\xi) = \infty$ and V is a short range perturbation.

The methods used here are essentially the same as those used in [1] and [5]. However, if in [1] and [5] the homogeneity property of the symbol of the free Hamiltonian was intensely used, in this more general case the constructions of the auxiliary operators must be made with care, such that the results we obtain should not be affected by the absence of the homogeneity. As we shall see, these constructions are natural and give operators with nice commutation properties with functions of the free Hamiltonian.

### HYPOTHESES

I. The free Hamiltonian $H_o$ is a self-adjoint operator on the Hilbert space $H = L^2(\mathbf{R}^n)$, with the domain $D(H_o) = \{u \in H; p_o\hat{u} \in H\}$, $H_o u = F^{-1} p_o \hat{u}$ is the Fourier transform of u and $p_o$ is a real valued function which satisfies:

(i) $p_o : \mathbf{R}^n \to \mathbf{R}$ is a continuous function.

(ii) If we denote by S the following set $\{\xi \in \mathbf{R}^n; p_o$ is not $C^\infty$ in any neighborhood of $\xi$, or $\nabla p_o(\xi) = 0\}$, then $\overline{p_o(S)}$ is a countable subset of $\mathbf{R}$.

(iii) For any compact interval $I \subset \mathbf{R} \setminus \overline{p_o(S)}$ we have

$$\inf \{|\nabla p_o(\xi)|; \xi \in p_o^{-1}(I)\} > 0 .$$

(iv) (local compactness). For any compact interval $I \subset \mathbf{R} \setminus \overline{p_o(S)}$ and for each $r > 0$, the operator

$$F(|x| < r)E_o(I)$$

is compact. Here F(M) denotes the indicator function of the set M and $E_o(I)$ denotes the spectral projection for $H_o$ onto the interval I.

**II.** Let $V : \hat{D} \to \hat{D}$ be a symmetric operator such that

(v)   For some $\varepsilon > 0$ the operator $V\phi(H_o)\langle x\rangle^{1+\varepsilon}$ has a bounded extension to the whole of $H$ for each $\phi$ in $C_o^\infty(\mathbf{R})$.

We used the notations: $\hat{D}$ for the image of $D$ (the space of test functions defined on $\mathbf{R}^n$) by the Fourier transform and $\langle x\rangle = (1 + |x|^2)^{\frac{1}{2}}$, $x \in \mathbf{R}^n$.

**III.** (vi) The operator $H_o + V$ with the domain $\hat{D}$ has a self-adjoint extension H.
(vii) For any $\phi \in C_o^\infty(\mathbf{R})$ the operator $\phi(H) - \phi(H_o)$ is compact.

The main result is the following

**THEOREM 1.1.** *Assume that the hypotheses (i)-(vii) are satisfied. Then*

(a) *The wave operators* $W_\pm = \text{s-}\lim\limits_{t \to \pm\infty} e^{iHt} e^{-iH_o t} E_{ac}(H_o)$ *exist;*

(b) **Range** $W_\pm = H_c(H)$, *the continuous subspace of* H;

(c) $\sigma_{sc}(H) = \emptyset$;

(d) *Any eigenvalue of* H *not in* $\overline{p_o(S)} \cup \{0\}$ *is of finite multiplicity. The eigenvalues of* H *can accumulate only at points of* $\overline{p_o(S)} \cup \{0\}$.

Before proving the main theorem we wish to make a few remarks about the hypotheses we made and about the connections between the present paper and others related with this subject.

**REMARK 1.2.** a) The free Hamiltonian $H_o$ is a convolution operator with a continuous real symbol which satisfies the conditions (i) – (iii). The growth conditions which are commonly imposed (see [3], [4], [7], [8], [9]) are replaced with the condition (iii). This condition can be read as follows:

(iii)' If the free energy lies in a compact interval disjoint from thresholds, then the velocity is bounded from below by a positive constant.

b) If we replace the condition (iii) by the stronger condition:

(iii)"   $\lim\limits_{|\xi| \to \infty, \, \xi \notin S} (|p_o(\xi)| + |\nabla p_o(\xi)|) = \infty$,

then the local compactness property of $H_o$ (i.e. condition (iv)) is fulfilled (see the appendix).

c) In the same way one can prove a similar theorem with the condition (v) replaced by the condition

(v)' For some $\varepsilon > 0$ the operator $\phi(H)V\langle x\rangle^{1+\varepsilon}$ has a bounded extension to the whole of $H$ for each $\phi$ in $C_o^\infty(\mathbf{R})$.

This condition is always true when $V$ is a symmetric $H_o$-compact operator and there is an $\varepsilon > 0$ such that the operator

$$(H_o + i)^{-1}V\langle x\rangle^{1+\varepsilon}$$

has a bounded extension.

d) By taking into account the above remarks one can compare this paper with [7] and [8].

PROOF OF THEOREM 1.1. (a) The proof of the existence of wave operators is standard (see [3], [4], [9]), so that we only sketch it.

By Cook's argument, it suffices to show that

$$\int_{-\infty}^{\infty} \| V\phi(H_o)e^{-iH_o t}u\| \, dt < \infty$$

for any $\phi \in C_o^\infty(\mathbf{R})$ and $u \in S(\mathbf{R}^n)$ with $\mathbf{supp}\ \hat{u}$ a compact set, disjoint from S. By using (v) this follows from

$$\int_{-\infty}^{\infty} \| \langle x\rangle^{-1-\varepsilon}e^{-iH_o t}u\| \, dt < \infty \quad ,$$

which can be proved for instance by writing

$$\langle x\rangle^{-1-\varepsilon}e^{-iH_o t}u = \langle x\rangle^{-1-\varepsilon}F(|x| \le \delta t)e^{-iH_o t}u + \langle x\rangle^{-1-\varepsilon}F(|x| \ge \delta t)e^{-iH_o t}u = B + C$$

and estimating $\| B\|$ by means of stationary phase method and $\| C\|$ by $c_{\delta,\varepsilon}\langle t\rangle^{-1-\varepsilon}$.

The other parts of Theorem 1.1 will be proved below by means of time dependent methods.

## 2. PRELIMINARIES

In this section we shall make the constructions already announced. As a consequence they will provide operators which will serve to prove some propagation estimates which are the main tools in the proof of Theorem 1.1.

We pass now to define the operators which we mentioned at the begining. Let $\gamma \in C_o^\infty(\mathbf{R} \setminus p_o(S))$. We define the smooth vector field $v$ in phase space by

$$(2.1) \qquad v(\xi) = p_o(\xi)\gamma(p_o(\xi))\,|\,\nabla p_o(\xi)\,|^{-2}\nabla p_o(\xi), \qquad \xi \in \mathbf{R}^n.$$

Then the condition (iii) implies that there exists $0 < c < \infty$ such that

$$(2.2) \qquad |\,v(\xi)\,| \le c, \qquad \forall \xi \in \mathbf{R}^n.$$

From this relation it follows that the Cauchy problem

$$(2.3) \qquad \begin{cases} (d/d\alpha)\Gamma(\alpha,\xi) = v(\Gamma(\alpha,\xi)) \\[2mm] \Gamma(0,\xi) = \xi \end{cases}$$

defines a group of $C^\infty$-diffeomorphisms $\Gamma(\alpha,\cdot):\mathbf{R}^n \to \mathbf{R}^n$. To this group of diffeomorphisms $\{\Gamma(\alpha,\cdot)\}_{\alpha \in \mathbf{R}}$ we associate a group of unitary operators $\{V(\alpha)\}_{\alpha \in \mathbf{R}}$ on $L^2(\mathbf{R}^n,d\xi)$ by

$$(2.4) \qquad (V(\alpha)\psi)(\xi) = |\,\mathbf{det}\,\partial\,\Gamma(\alpha,\xi)/\partial\,\xi\,|^{\frac{1}{2}}\psi(\Gamma(\alpha,\xi)), \qquad \psi \in L^2(\mathbf{R}^n,d\xi).$$

If we denote by $F$ the Fourier transform on $L^2(\mathbf{R}^n)$, then we obtain another group of unitary operators on $L^2(\mathbf{R}^n,dx)$ defined by

$$(2.5) \qquad U(\alpha) = F^{-1}V(\alpha)F \qquad \text{on } L^2(\mathbf{R}^n,dx).$$

Let now $A = A_{H_o,\gamma}$ be the self-adjoint operator on $H = L^2(\mathbf{R}^n,dx)$ such that

$$U(\alpha) = e^{-iA\alpha}$$

By taking into account the definition of $U(\alpha)$ one obtains in a straightforward manner the following

**LEMMA 2.1.** $\hat{D}$ *is a core of* A *and*

$$(2.6) \qquad A = \sum_{j=1}^{n} (v_j(D)x_j + x_j v_j(D))/2 \qquad \text{on } \hat{D}.$$

Next we shall establish some relations which give the commutators $i[f(H_o),A]$ and $i[f(H_o),(A + i)^{-m}]$.

**LEMMA 2.2.** *Let* $f \in C^1(\mathbf{R})$ *be a bounded function. Then the form* $i[f(H_o),A] = i(f(H_o)A - Af(H_o))$ *defined on* $D(A)$ *has a bounded extension and*

$$(2.7) \qquad i[f(H_o),A] = H_o f'(H_o)\gamma(H_o).$$

PROOF. The proof of this lemma is elementary and it is based on the relation

$$U(\alpha)f(H_o)U(-\alpha) = F^{-1}M_{f\circ p_o \circ \Gamma(\alpha,\,\cdot\,)}F.$$

**Q.E.D.**

**LEMMA 2.3.** *Let* $m \in \mathbf{N}$ *and let* $f \in C^\infty(\mathbf{R})$ *be a bounded function. Then*

(2.8)
$$(A + i)^{-m}f(H_o) = \Big\{ \sum_{k=0}^{m} \binom{m}{k}(A + i)^{-k}f_k(H_o)\Big\}(A + i)^{-m}$$

*with* $f_k \in C^\infty(\mathbf{R})$ *given by*

$$f_k(\lambda) = (-i\lambda\gamma(\lambda)(d/d\lambda))^k f(\lambda).$$

PROOF. The proof of this lemma is also elementary and is made by induction.

$$(A + i)^{-1}f(H_o) = f(H_o)(A + i)^{-1} + (A + i)^{-1}[f(H_o),A](A + i)^{-1} =$$

$$= \{f_o(H_o) + (A + i)^{-1}f_1(H_o)\}(A + i)^{-1} \qquad \text{(by (2.7))}$$

Assume that the statement is true for m. Then

$$(A + i)^{-m-1}f(H_o) = (A + i)^{-1}\Big\{ \sum_{k=0}^{m} \binom{m}{k}(A + i)^{-k}f_k(H_o)\Big\}(A + i)^{-m}.$$

But (2.7) of Lemma 2.2 implies that

$$(A + i)^{-1}f_k(H_o) = \{f_k(H_o) + (A + i)^{-1}f_{k+1}(H_o)\}(A + i)^{-1}.$$

So we obtain

$$(A + i)^{-m-1}f(H_o) = \Big\{ \sum_{k=0}^{m} \binom{m}{k}(A + i)^{-k}f_k(H_o) + (A + i)^{-1}f_{k+1}(H_o)\Big\}(A + i)^{-m-1} =$$

$$= \Big\{ \sum_{k=0}^{m} \binom{m}{k}(A + i)^{-k}f_k(H_o) + \sum_{k=1}^{m+1} \binom{m}{k-1}(A + i)^{-k}f_k(H_o)\Big\} \cdot (A + i)^{-m-1}.$$

Since $\binom{m}{k} + \binom{m}{k-1} = \binom{m+1}{k}$, it follows that

$$(A + i)^{-m-1}f(H_o) = \Big\{ \sum_{k=0}^{m+1} \binom{m+1}{k}(A + i)^{-k}f_k(H_o)\Big\}(A + i)^{-m-1}.$$

**Q.E.D.**

One can use the above results to prove another needed lemma.

**LEMMA 2.4.** *For* $0 \leq \alpha \leq 2$,

$$\langle A\rangle^{\alpha}(H_o + i)^{-1}\langle x\rangle^{-\alpha} \equiv J\alpha$$

*is a bounded operator on H. Here* $\langle A \rangle = (1 + A^2)^{\frac{1}{2}} = |A + i|$.

PROOF. We need only to prove the case $\alpha = 2$ and then use the complex interpolation. Thus we must to prove that

$$A^2(H_o + i)^{-1}\langle x \rangle^{-2}$$

is bounded. For a suitable function f, we obtain from Lemma 2.2 that

$$Af(H_o) = iH_o\gamma(H_o)f'(H_o) + f(H_o)A .$$

By iterating this formula we get

$$A^2 f(H_o) = -H_o\gamma(H_o)(\gamma(H_o)f'(H_o) + H_o\gamma'(H_o)f'(H_o) + H_o\gamma(H_o)f''(H_o)) +$$

$$+ 2iH_o\gamma(H_o)f'(H_o)A + f(H_o)A^2 .$$

By taking $f(\lambda) = (\lambda + i)^{-1}$ we obtain the conclusion of Lemma 2.4 by using the explicit formula for A (Lemma 2.1).                                                    **Q.E.D.**

We can now prove the basic estimate which we shall use in the proof of the asymptotic completeness. Since we shall work with functions belonging to the space $C_o^\infty((a,b))$, where (a,b) is an open interval such that $[a,b] \subset \mathbf{R}^+ \setminus \overline{p_o(S)} \cup \{0\}$ or $[a,b] \subset$ $\subset \mathbf{R}^- \setminus \overline{p_o(S)} \cup \{0\}$, we shall consider, as an auxiliary operator, the self-adjoint operator $A = A_{H_o,\gamma}$ associated to a function $\gamma \in C_o^\infty((\alpha,\beta))$, $0 \leq \gamma \leq 1$, $\gamma = 1$ in a neighborhood of [a,b]. Here $(\alpha,\beta)$ is another open interval such that $[a,b] \subset (\alpha,\beta)$ and $[\alpha,\beta] \subset \mathbf{R}^+ \setminus$ $\setminus \overline{p_o(S)} \cup \{0\}$ in the first case and $[\alpha,\beta] \subset \mathbf{R}^- \setminus \overline{p_o(S)} \cup \{0\}$ in the second case. Let $P^+$ and $P^-$ be the spectral projectors of A on the positive and negative parts of its spectrum, $\langle A \rangle$ the usual operator $(1 + A^2)^{\frac{1}{2}} = |A + i|$ and $\chi^\pm$ the indicator function of $\mathbf{R}^\pm \setminus \{0\}$. Then we have the following

**THEOREM 2.5.** *Let* $0 \leq \mu' < \mu$. *Assume that (a,b) is an open interval such that* $[a,b] \subset \mathbf{R}^+ \setminus p_o(S) \cup \{0\}$. *Let* $g \in C_o^\infty((a,b))$. *Then there is a constant* $c = c(g,\mu,\mu')$ *such that*

(2.9)                         $\| \chi^\pm(t)\langle A \rangle^{-\mu} e^{-iH_o t} g(H_o)P^\pm \| \leq c|t|^{-\mu'} .$

PROOF. 1° The proof of this theorem follows in almost the same way as the proof of Theorem 2.1 of [1] or the proof of Lemma 2.1 of [5]. However, the absence of homogeneity of $p_o$ requires some changes which we shall point out at the right time.

We shall give the proof in the case $t > 0$. As a result of the first three steps of the proof of Theorem 2.1 of [1] we obtain that

$$(2.10) \quad \langle A \rangle^{-m} e^{-iH_o t} g(H_o) P^+ = m'!(it)^{-m'} e^{\epsilon t}(2\pi i)^{-1} \cdot$$

$$\cdot \int_{-\infty}^{\infty} e^{-iEt} \langle A \rangle^{-m}(H_o - E - i\epsilon)^{-m'-1} g(H_o) P^+ dE ,$$

$$(2.11) \quad \left\| \int_{\mathbf{R}\backslash[a,b]} e^{-iEt} \langle A \rangle^{-m}(H_o - E - i\epsilon)^{-m'-1} g(H_o) P^+ dE \right\| \leq c_o(g,m,m'), \qquad \forall\, 0 \leq \epsilon \leq 1,$$

and we see that we are reduced to study the family of operators

$$\left\{ \langle A \rangle^{-m}(H_o - E - i\epsilon)^{-m'-1} P^+ \right\}_{E\in[a,b];\ \epsilon\in(0,1]} \cdot$$

(Here we use Lemma 2.3 to prove that $\langle A \rangle^{-m} g(H_o) \langle A \rangle^{m}$ is a bounded operator for $m \in \mathbf{R}$.)

$2°$ If $m,n \in \mathbf{N}$, $m > n$, $0 < \epsilon \leq 1$, $a \leq E \leq b$, $0 \leq \theta \leq \pi/2$, we define

$$F(\epsilon,E,\theta) = \langle A \rangle^{-m}(H_o e^{-i\theta} - E - i\epsilon)^{-n} e^{-\theta A} P^+$$

with

$$F(\epsilon,E,0) = \langle A \rangle^{-m}(H_o - E - i\epsilon)^{-n} P^+$$

and

$$\mathrm{s\text{-}lim}_{\theta\to 0+}\ F(\epsilon,E,\theta) = F(\epsilon,E,0) .$$

Next we shall prove that the following estimate holds

$$(2.12) \quad \left\| (\partial/\partial\theta) F(\epsilon,E,\theta) \right\| \leq c(g,m,n) \left\| \langle A \rangle^{-m+1}(H_o e^{-i\theta} - E - i\epsilon)^{-n} e^{-\theta A} P^+ \right\|$$

for $0 < \epsilon \leq 1$, $a \leq E \leq b$, $0 < \theta \leq \delta$ with $\delta$ sufficiently small.

By using Lemma 2.2 it follows easily that

$$(\partial/\partial\theta) F(\epsilon,E,\theta) = ine^{-i\theta} \langle A \rangle^{-m} H_o(1 - \gamma(H_o))(H_o e^{-i\theta} - E - i\epsilon)^{-n-1} e^{-\theta A} P^+ -$$

$$- \langle A \rangle^{-m} A(H_o e^{-i\theta} - E - i\epsilon)^{-n} e^{-\theta A} P^+.$$

To prove (2.12) we must estimate the norm of the operator

$$e^{-i\theta} \langle A \rangle^{-m} H_o(1 - \gamma(H_o))(H_o e^{-i\theta} - E - i\epsilon)^{-n-1} e^{-\theta A} P^+$$

which we shall write in the form

$$(A + i)^{m} \langle A \rangle^{-m}(A + i)^{-m} f(H_o)(H_o e^{-i\theta} - E - i\epsilon)^{-n} e^{-\theta A} P^+$$

with

$$f(H_o) = H_o(1 - \gamma(H_o))(H_o - Ee^{i\theta} - i\epsilon e^{i\theta})^{-1} .$$

Since we have $0 < \alpha < a_1 < a < b < b_1 < \beta$, $[\alpha,\beta] \cap \overline{p_o(S)} = \emptyset$, **supp** $g \subset (a,b)$, **supp** $\gamma \subset (\alpha,\beta)$, $\gamma = 1$ on $(a_1,b_1)$, it follows from Lemma 2.3 by choosing $\delta$ sufficiently small that

$$(A + i)^{-m} f(H_o) = \left\{ \sum_{k=0}^{m} \binom{m}{k}(A + i)^{-k} f_k(H_o) \right\}(A + i)^{-m}$$

with $\sum_{k=0}^{m} \binom{m}{k}(A + i)^{-k} f_k(H_o)$ uniformly bounded for $0 < \epsilon \leq 1$, $a \leq E \leq b$ and $0 \leq \theta \leq \delta$, so the proof of (2.12) is complete.

   $3°$ Following now the arguments of the steps $3°$, $4°$, $5°$ and $6°$ of the proof of Theorem 2.1 of [1], one proves that for every $(m,m') \in \mathbf{N} \times \mathbf{N}$, $m > m' + 1$ there exists $c = c(g,m,m')$ such that

(2.13)                    $$\| <A>^{-m} e^{-iH_o t} g(H_o)P^+ \| \leq ct^{-m'}.$$

From this relation we deduce that for every $(m,m') \in \mathbf{R} \times \mathbf{N}$, $m > m' + 2$ there is $c = c(g,m,m')$ such that (2.13) holds.

   Now the general case follows by interpolation.                       **Q.E.D.**

   Fo the case $[a,b] \subset \mathbf{R}^- \setminus \overline{p_o(S)} \cup \{0\}$ we have the following

   **COROLLARY 2.6.** *Let* $0 \leq \mu' < \mu$. *Assume that $(a,b)$ is an open interval such that* $[a,b] \subset \mathbf{R}^- \setminus \overline{p_o(S)} \cup \{0\}$. *Let* $g \in C_o^\infty((a,b))$. *Then there is a constant* $c = c(g,\mu,\mu')$ *such that*

(2.9)'                    $$\| \chi^{\mp}(t)<A>^{-\mu} e^{-iH_o t} g(H_o)P^\pm \| \leq c|t|^{-\mu'}.$$

   PROOF. Apply the above theorem to the operator $-H_o$ (i.e. $-p_o(D)$) and the functions $\check{g}$, $\check{\gamma}$, and observe that $A_{-H_o,\check{\gamma}} = A_{H_o,\check{\gamma}}$. Here $\check{f}$ means the function defined by $\check{f}(x) = f(-x)$.                       **Q.E.D.**

### 3. ASYMPTOTIC COMPLETENESS

   For the proof of the asymptotic completeness we need two compactness results. We start with the case of an open interval $(a,b)$ such that $[a,b] \subset \mathbf{R}^+ \setminus \overline{p_o(S)} \cup \{0\}$.

   **LEMMA 3.1.** *Assume that the hypotheses (i)-(vii) are fulfilled, and let $(a,b)$ be an open interval such that* $[a,b] \subset \mathbf{R}^+ \setminus \overline{p_o(S)} \cup \{0\}$. *Then for every* $g \in C_o^\infty((a,b))$ *the operators* $(W_\pm - 1)g(H_o)P^\pm$ *are compact on* $H$.

PROOF. We have

$$(W_+ - 1)g(H_o)P^+ = i\int_o^\infty e^{iHs}Ve^{-iH_o s}g(H_o)P^+ds$$

and for any $s > 0$, $Ve^{-iH_o s}g(H_o)P^+$ is a compact operator as it follows from the hypotheses (iv) and (v).

Furthermore the integral

$$\int_o^\infty \| Ve^{-iH_o s}g(H_o)P^\pm \| \, ds$$

is well defined since

$$\| Ve^{-iH_o s}g(H_o)P^+ \| = \| V\gamma(H_o)(H_o + i)^{-m}e^{-iH_o s}(H_o + i)^m g(H_o)P^+ \| \leq$$

$$\leq \| V\gamma(H_o)\langle x\rangle^{1+\epsilon} \| \, \| \langle x\rangle^{-1-\epsilon}(H_o + i)^{-m}\langle A\rangle^{1+\epsilon} \| \, \cdot$$

$$\cdot \| \langle A\rangle^{-1-\epsilon}e^{-iH_o s}(H_o + i)^m g(H_o)P^+ \|.$$

From Theorem 2.5 and the condition (v), it suffices to verify that $m > 0$ can be chosen such that $\langle x\rangle^{-1-\epsilon}(H_o + i)^{-m}\langle A\rangle^{1+\epsilon}$ is a bounded operator on $H$.

By Lemma 2.4 this is true for $m = 1$, because we always may suppose that $\epsilon \leq 1$ in (v).                                                                                               **O.E.D.**

**COROLLARY 3.2.** *Assume that the hypotheses* (i)-(vii) *are fulfilled, and let* (a,b) *be an open interval such that* $[a,b] \subset \mathbf{R}^- \setminus \overline{p_o(S)} \cup \{0\}$. *Then for every* $g \in C_o^\infty((a,b))$ *the operators* $(W_{\overset{-}{+}} - 1)g(H_o)P^\pm$ *are compact on H.*

Now it is clear that the conclusions of Theorem 1.1 can be obtained by using the Enss argument. Since there are many papers on this method ([1], [3], [9], [10]) we shall not repeat Enss' argument here.

### APPENDIX

We commented in Remark 1.2 b) that the local compactness property of $H_o$ is implied if we assume that the function $p_o$ satisfies conditions (i), (ii) and (iii)". In this appendix we shall prove this compactness result which we shall state as

**PROPOSITION A.1.** *Assume that* $p_o : \mathbf{R}^n \to \mathbf{R}$ *is a function which satisfies the conditions* (i), (ii) *and* (iii)". *Let* $I \subset \mathbf{R} \setminus \overline{p_o(S)}$ *be a compact interval and let* $r > 0$. *Then*

$$F(|x| \leq r)E_o(I)$$

*is a compact operator on* $H$.

We assume that $\mathbf{R}^n$ is divided into unit "cubes" $C_k$, $k \in \mathbf{N}$ so that

$$\mathbf{R}^n = \bigcup_k \overline{C}_k \quad \text{and} \quad C_k \cap C_\ell = \emptyset \quad \text{for } k \neq \ell .$$

Then, in order to prove Proposition A.1, it suffices to show (cf. Corollary 3 of [2]) that the following lemma is true.

**LEMMA A.2.** *Assume that* $p_o$ *satisfies* (i), (ii) *and* (iii)". *Let* $I \subset \mathbf{R} \setminus \overline{p_o(S)}$ *be a compact interval. Then*

$$\lim_{k \to \infty} | C_k \cap p_o^{-1}(I)| = 0 .$$

*Here* $|A|$ *denotes the Lebesgue measure of the measurable set* A.

PROOF. If we denote by

$$\beta_k = \inf \left\{ | \nabla p_o(\xi)| \, ; \, \xi \in p_o^{-1}(I) \cap C_k \right\}$$

then the compactness of I and the condition (iii)" imply that

$$\lim_{k \to \infty} \beta_k = \infty.$$

Therefore the proof of the lemma is completed by the following estimate:

$$(A.1) \qquad\qquad | p_o^{-1}(I) \cap C_k| \leq n\sqrt{n}|I|\beta_k^{-1}, \qquad k \in \mathbf{N} .$$

Let $B_j = \{\xi \in \mathbf{R}^n \setminus S; \, | \nabla p_o(\xi)| \leq \sqrt{n}| \partial_j p_o(\xi)|\}$ and $\Phi_j : \mathbf{R}^n \setminus S \to \mathbf{R}^n$ defined by

$$\Phi_j(\xi) = (\xi_1, \dots, \xi_{j-1}, p_o(\xi), \xi_{j+1}, \dots, \xi_n)$$

for $j = 1, \dots, n$. Then $\Phi_j$ is a local diffeomorfism at every point in $p_o^{-1}(I) \cap B_j$.

Since $p_o^{-1}(I) \cap C_k = \bigcup_{j=1}^{n} p_o^{-1}(I) \cap C_k \cap B_j$, then (A.1) follows from

$$(A.1)' \qquad | p_o^{-1}(I) \cap C_k \cap B_j| \leq \sqrt{n}|I|\beta_k^{-1}, \qquad k \in \mathbf{N}, \qquad j = 1, \dots, n .$$

This estimate can be obtained by making a change of variable. Let us write $p_o^{-1}(I) \cap C_k \cap B_j$ as a disjoint union $\bigcup_\ell \Phi_j^{-1}(C_{kI}) \cap C_k \cap B_j \cap M_\ell$, where $C_{kI} = \pi_1(C_k) \times \dots$ $\dots \times \pi_{j-1}(C_k) \times I \times \pi_{j+1}(C_k) \times \dots \times \pi_n(C_k)$ and $M_\ell$, $\ell \in \mathbf{N}$ are disjoint measurable sets which have neighborhoods on which $\Phi_j$ is a diffeomorphism.

Then

$$p_o^{-1}(I) \cap C_k \cap B_j = \bigcup_\ell \Phi_j^{-1}(A_{kj\ell})$$

where $A_{kj\ell}$, $\ell \in \mathbf{N}$ are disjoint measurable subsets of $C_{kI}$ such that $\Phi_j^{-1}(A_{kj\ell}) =$
$= \Phi_j^{-1}(C_{kI}) \cap C_k \cap B_j \cap M_\ell$ , $\Phi_j$ is a diffeomorfism in a neighborhood of $\Phi_j^{-1}(A_{kj\ell})$ and

$$|\det (\Phi_j^{-1})'(\eta)| \le \sqrt{n}\beta_k^{-1} \quad \text{for } \eta \in A_{kj\ell} .$$

Hence

$$|p_o^{-1}(I) \cap C_k \cap B_j| = \sum_\ell \int_{\Phi_j^{-1}(A_{kj\ell})} d\xi = \sum_\ell \int_{A_{kj\ell}} |\det (\Phi_j^{-1})'(\eta)| \, d\eta \le$$

$$\le \sqrt{n}\beta_k^{-1} \int_{C_{kI}} d\eta = \sqrt{n}|I|\beta_k^{-1}.$$

Q.E.D.

## REFERENCES


1. **Arsu, G.** : A time dependent scattering theory for strongly propagative systems with perturbations of short-range class, *Rev. Roumaine Math. Pures Appl.*, to appear.

2. **Davies, E.B. ; Muthuramalingam, Pl.** : Trace properties of some highly anisotropic operators, *J. London Math. Soc. (2)* **31**(1985), 137-149.

3. **Ginibre, J.** : La methode "dependante du temps" dans le problème de la complétude asymptotique, preprint, Univ. de Paris-Sud, 1980.

4. **Hörmander, L.** : *The analysis of linear partial differential operators. II*, Springer Verlag, Berlin, 1983.

5. **Mourre, E.** : Link between the geometrical and the spectral transformation approaches in scattering theory, *Comm. Math. Phys.* **68**(1979), 91-94.

6. **Mourre, E.** : Absence of singular continuous spectrum for certain self-adjoint operators, *Comm. Math. Phys.* **78**(1981), 391-408.

7. **Muthuramalingam, Pl.** : A note on time dependent scattering theory for $P_1^2 - P_2^2 + (1 + |Q|)^{-1-\varepsilon}$ and $P_1 P_2 + (1 + |Q|)^{-1-\varepsilon}$ on $L^2(\mathbf{R}^2)$, *Math. Z.* **188**(1985), 339-348.

8. **Muthuramalingam, Pl.** : A time dependent scattering theory for a class of simply characteristic operators with short range local potentials, *J. London Math. Soc. (2)* **32**(1985), 259-264.

9. **Simon, B.** : Phase space analysis of simple scattering systems: extensions of some work of Enss, *Duke Math. J.* **46**(1979), 119-168.

10. **Yafaev, D.R.** : On the proof of Enss of asymptotic completeness in potential scattering theory, preprint, Steklov Institute, Leningrad, 1979.



**Gruia Arsu**

Department of Mathematics, INCREST
Bd. Păcii 220, 79622 Bucharest
Romania.

Operator Theory:
Advances and Applications, Vol. 43
© 1990 Birkhäuser Verlag Basel



# NONCOMMUTATIVE UNIFORM ALGEBRAS

**Victor Arzumanian** and **Suren Grigorian**

## 0. INTRODUCTION

The algebras of continuous functions on a compact set which are called *uniform*, were introduced in connection with applications of methods of functional analysis in the classical theory of functions. Nowadays, the theory of uniform algebras is an independent branch of analysis which has its own objects and methods of investigation. Beginning from 70's some interest on noncommutative analogues of uniform algebras arose and a few generalizing results were obtained, [13], [14], [12]. These algebras are some closed subalgebras of a $C^*$-algebra of operator fields, introduced by Fell [7]. It seems natural to expect that the theory of noncommutative uniform algebras must include such standard notions as the maximal ideal space, Choquet boundary, maximality etc. It is not clear how legitimate are such formulations of questions for algebras with varying fibres. Our paper is an attempt to fulfil this program for a special case of uniform algebras of continuous $C^*$-valued functions on a compact set. A series of papers is devoted to these subjects, [2], [3], [4], [5].

Let T be a compact Hausdorff space (everywhere assumed metrizable), let A be a unital $C^*$-algebra (which is called fibre), and let C(T,A) be the $C^*$-algebra of all continuous A-valued functions on T (from algebraic point of view this algebra is the tensor product $C(T) \otimes A$). This algebra contains C(T) and A as subalgebras, and is generated by them; in particular, C(T,A) is an A-bimodule. By definition, the *uniform* (noncommutative) algebra $M$ is a closed subalgebra of C(T,A) which separates the points of T (i.e. for every $t_1, t_2 \in T$ and $a_1, a_2 \in A$ there exists $x \in M$ such that $x(t_1) = a_1$, $x(t_2) = a_2$) and contains all constants.

The most meaningful results are obtained when some restrictions on the fibre and on the uniform algebra are assumed. First of all, we frequently suppose that $M$ is an A-bimodule of C(T,A), i.e. $A \subset M$ (so A plays the role of the scalars of $M$). Such an algebra will be called uniform A-*algebra*. The simplest subclass of the class of A-algebras consists of algebras which we call *decomposable*: they are generated by A and by a uniform subalgebras of C(T). One can prove that every uniform A-algebra with matrix fibre is decomposable. The algebra of analytic matrix-functions in the unit circle is a typical example of decomposable algebra (the analogue of the classical disc-algebra). Further, the fibre algebra A will be assumed, at least, with trivial center. The most restrictive condition on A is simplicity of this algebra. An intermediate case between these two classes is that of the algebras which we shall describe now more precisely.

Let $\phi$ be a pure state on A, and u be a unitary element from A. Let $\phi_u$ denote the pure state which is defined as follows: for each $a \in$ A, $\phi_u(a) = \phi(u^{-1}au)$. A pure state $\phi$ on A will be called *total* if the set of all finite convex combinations of $\phi_u$, where u runs the set of all unitary elements of A, is dense in the state space of A. The algebra A will be called *weakly*$^*$-*transitive* if there exists a total state on A. It is easy to verify that every weakly$^*$-transitive C$^*$-algebra has a trivial center and that the algebra A is simple iff each pure state on A is total. As an example of weakly$^*$-transitive but not simple algebra we indicate B(H), the algebra of all bounded linear operators on an infinite dimensional Hilbert space. A characteristic property of weakly$^*$-transitive algebras A, which will be used in the sequel, is the following: the intersection of every two non-trivial two-sided ideals in A is also non-trivial.

For every uniform algebra $M \subset C(T,A)$ we introduce the notion of *relative spectrum* replacing the usual notion of maximal ideal space and reducing to it when the fibre A is trivial (i.e. A = **C**, the C$^*$-algebra of complex numbers). Some properties of this compact Hausdorff space are discussed in Section 1. Let us remark that the notion of relative spectrum essentially takes into account the structure of continuous fields, and differs from usual spectrum even for a commutative non-trivial fibre.

In the following Section 2 we consider in detail the geometrical structure of the relative spectrum. We introduce the notion of *Choquet boundary* for a noncommutative uniform algebra, and give its characterization in terms of representing measures and *weak peak* points. The last notion, which replaces the important notion of pure state for non-involutive algebras, does not permit a direct generalization to the noncommutative case.

The rich structure of uniform algebras on compact Abelian groups invariant with respect to all group shifts gives the possibility to revise some early obtained results. In the center of Section 3 are the questions of maximality of invariant uniform algebras. Here we widely use the notion of Hilbert C$^*$-module and a generalization of GNS-construction for conditional expectations.

All general topics and results relating to uniform and operator algebras can be found in the monographs [6] and [8].

## 1. RELATIVE SPECTRUM

Let us denote by P(T,A) the set of all conditional expectations from the algebra C(T,A) to the algebra A. It is clear that each probability measure on T determines a unique conditional expectation on C(T,A).

**DEFINITION 1.1.** Let $M$ be a uniform algebra on T. The space $\mathbf{Sp}_A\, M$ of all homomorphisms $p : M \to A$ which can be extended to conditional expectations from P(T,A), equipped with the topology of pointwise convergence, is called the *relative spectrum* of $M$.

It is obvious that each point $t \in T$ (atomic measure) determines a A-bilinear homomorphism $p_t$, $p_t(x) = x(t)$, where $x \in C(T,A)$ and $p_t \in P(T,A)$. Thus T is a subset of $\mathbf{Sp}_A\, M$ for every uniform algebra $M$.

**THEOREM 1.2.** *Let the center of the algebra A be trivial. Then the relative spectrum of any uniform algebra is compact.*

PROOF. Each conditional expectation $p \in P(T,A)$ is determined by its values on C(T), and the A-bilinearity of p implies that p(C(Y)) is contained in the center of A, thus by the assumption of the theorem, p determines a unique state on C(T). It is possible to show that the inverse statement is also true. Then P(T,A) is compact and therefore $\mathbf{Sp}_A\, M$ is compact too.                                    ☐

Overcoming certain technical difficulties we can describe the spectrum of C(T,A) in the following terms: the space $\mathbf{Sp}_A\, C(T,A)$ is homeomorphic to the space of all continuous mappings of maximal ideal space of the center of A to the compact set T, endowed with the topology of pointwise convergence. Thus, if the center of A is trivial, the relative spectrum of C(T,A) coincides with T.

If the uniform algebra $M$ is decomposable, $M = M \otimes A$, where M is a uniform algebra from C(T), and the center of A is trivial, the relative spectrum of $M$ is then homeomorphic to the maximal ideal space of the algebra M.

For a large class of algebras, the concept of relative spectrum can be described in a more convenient way:

**THEOREM 1.3.** *Let A be a weakly[*]-transitive $C^*$-algebra and let $M$ be a uniform A-algebra on T. Then the relative spectrum of $M$ coincides with the set of all non-trivial A-linear continuous homomorphisms from $M$ to A.*

PROOF. Let $p : M \rightarrow A$ be such a homomorphism, and let $\phi$ be a total (pure) state on A. Then, obviously, $\| p \| = 1$ and $\phi \circ p$ is a continuous normalized functional on $M$ which is therefore extendable to a state $\Phi$ on C(T,A). For every such state, there exists some probability measure $\mu$ on T such that for any $x \in C(T,A)$ we have:

$$\Phi(x) = \int \phi(x(t)) d\mu(t).$$

Let $p_\mu$ be the conditional expectation on C(T,A) which is uniquely determined by $\mu$; then $\Phi = \phi \circ p_\mu$. For each $x \in \mathbf{Ker}\, p$ and every unitary element u from the algebra A we have

$$\phi_u(p_\mu(x)) = \phi(u^{-1} p_\mu(x) u) = \Phi(u^{-1} x u) = \phi_u(p(x)) = 0.$$

Hence, by virtue of weakly$^*$-transitivity of A, we obtain that $\psi(p_\mu(x)) = 0$ for each state $\psi$ on the algebra A. Therefore $\mathbf{Ker}\, p \subset \mathbf{Ker}\, p_\mu$ and since $M = A + \mathbf{Ker}\, p$ (the consequence of A-linearity of p) the restriction of $p_\mu$ on $M$ coincides with p.            $\square$

## 2. CHOQUET BOUNDARY

Let $M$ be a uniform algebra on T, the fibre algebra A being assumed without center. Then (see Theorem 1.2) P(T,A) is a compact convex space. We denote by P($M$) the restriction of P(T,A) to the algebra $M$. The set of extreme points of the compact P($M$) is called the *Choquet boundary* of the uniform algebra $M$.

Now we describe the set $\mathbf{ext}\, P(M)$ in terms of representing measures. It is easy to see that for each $p \in P(M)$ there exists a probability measure $\mu$ on T, such that for $f \in C(T)$,

$$\hat{p}(f) = \int f d\mu,$$

where $\hat{p} \in P(T,A)$, $p = \hat{p} | M$.

Every such measure (not unique, of course) is called *representing* for p.

**PROPOSITION 2.1.** *Let A be a unital C$^*$-algebra with trivial center, M be a uniform algebra on* T. *Then* $\mathbf{ext}\, P(M)$ *is imbedded in* T *and coincides with those points of* T *for which* $p_t$ *has a unique (atomic) representing measure.*

PROOF. It is clear that every extreme point of P($M$) determines an atomic measure on T; thus $\mathbf{ext}\, P(M) \subset T$. Since the algebra $M$ separates the points of T we obtain the uniqueness of the representing measure.            $\square$

Now we give the functional description of the Choquet boundary.

**DEFINITION 2.2.** A point $t_o \in T$ is called an *weak peak* point for a uniform algebra $M \subset C(T,A)$ iff for every $d > 1$ and a neighborhood $U$ of $t_o$ there exists $x \in M$ such that $\| x \| < d$, $x(t_o) \geq 0$, $\| x(t_o) \| = 1$ and $\| x(t) \| < (2d)^{-1}$ when $x \notin U$.

**THEOREM 2.3.** *Every weak peak point of a uniform algebra $M$ on $T$ belongs to the Choquet boundary of $M$. If $M$ is an $A$-algebra with weakly*-transitive fibre, then the converse is also true.*

PROOF. If $t_o$ is a weak peak point for $M$, then it is easy to check that $p_{t_o}$ has a unique representing measure, thus by Proposition 2.1, $p_{t_o} \in \mathbf{ext}\ P(M)$.

Now, let $M$ be an $A$-algebra, $A$ being weakly*-transitive and let $p_{t_o} \in \mathbf{ext}\ P(M)$. We choose a number $d$ such that $1 < d < (e/2)^{\frac{1}{2}}$ and some neighbourhood $U$ of $t_o$. There exists a function $f \in C(T)$ such that $f(t_o) = 1$, $0 < f(t) < 1$, when $t \neq t_o$ and $f(t) < 1 - \ln 2d^2$, for $t \notin U$. It is possible to show that for each $\varepsilon > 0$ there exists a $u \in \mathbf{Re}\,M$, $0 \leq u \leq f$ for which $\phi(u(t_o)) > 1 - \varepsilon$, $\phi$ being a total state on the algebra $A$. If $y$ is chosen in $M$ such that $\mathbf{Re}\,y = u$, then it remains to verify that the element $x$ from $M$,

$$x = \mathbf{exp}(y - i\,\mathbf{Im}\,y(t_o)) \| u(t_o) \|^{-1},$$

realizes the weak peak at the point $t_o$.                                    □

This result establishes the existence of weak peak points.

One can define the *peak point* notion in the usual sense: there exists a function $x \in M$ such that $x(t_o) = 1$ and $\| x(t_o) \| < 1$ for $t \neq t_o$. Even the uniform $A$-algebra with simple fibre could have no peak points. A description of peak points in terms of operator-valued measures (without a discussion of their existence) was given in [13] (for the case of general uniform algebras of operator fields). Let us note that in [14] and [12] a general Hoffman-Wermer theorem was proved under the assumption of the existence of a peak point.

For certain classes of algebras (see, for example, Theorem 3.2) the notions of peak point and weak peak point coincide.

## 3. INVARIANT ALGEBRAS

Let $T$ be a compact Abelian group, and let $S = \hat{T}$ be its dual. A uniform algebra on $T$ is called $T$-invariant if it is invariant with respect to all group shifts.

For each conditional expectation $p \in P(T,A)$ one can produce, by analogy with usual GNS-construction, a Hilbert $A$-module (about the notion of Hilbert $C^*$-module, see [8] or [11]) which we denote by $L^2(T,A,p)$. Among all conditional expectations, there exists a unique T-invariant one, corresponding to the Haar measure on T, and which we will denote by $p_0$. Every x from $L^2(T,A,p_0)$ has a unique orthogonal (with respect to the A-product $\langle x,y \rangle = p_0(y^*x)$) expansion

$$x = \sum a_s \gamma^s,$$

where $s \in S$, $\gamma^s$ is the natural representation of S in C(T), $a_s \in A$ and the convergence of the series is understood in the norm of $L^2(T,A,p_0)$. It is clear that the "coeficients" $a_s$ are determined by the formula

$$a_s = p_0(x\gamma^{-s}).$$

It is not difficult to verify that if the center of the algebra A is trivial and the algebra $M$ is T-invariant, then the function $x \in C(T,A)$ belongs to $M$ iff all $a_s\gamma^s$ belong to $M$.

For each $s \in S$ we denote

$$A_s(M) = \{a \in A, \ a\gamma^s \in M\},$$

$$S_M = \{s \in S, \ A_s(M) \neq \{0\}\}.$$

It is easy to see that $A_s$ is a closed two-sided ideal of the algebra A. The set $\gamma^{S_M}$ is not, in general, a subset of $M$ (in contrast with the commutative case). If, however, $\gamma^{S_M} \subset M$ then $M$ is decomposable. When the algebra A is simple, every T-invariant A-algebra is decomposable. Further, the set $S_M$ is not in general a semigroup.

The following statement contains, in particular, an extension of one of the results of Arens and Singer [1].

**THEOREM 3.1.** *Let* A *be a weakly*[*]*-transitive unital algebra, and let* $M$ *be a* T-*invariant* A-*algebra on* T. *Then*

i) $S_M$ *is a subsemigroup of* S;

ii) *there exists a canonical homeomorphism of* $\mathrm{Sp}_A M$ *to the compact set (in the standard topology) of all homomorphisms of* $S_M$ *to the unit circle.*

PROOF. If $s, u \in S_M$ then $A_s A_u \subset A_{s+u}$. Since the algebra A is weakly[*]- -transitive we have $A_s \cap A_u \neq \{0\}$ (see Section 0), hence the first statement is true.

Let M be a uniform algebra from C(T) which is generated by $\gamma^s$, $s \in S$ and $\hat{M}$ be a uniform A-algebra from $C(T,A)$ generated by M and A (decomposable by the definition). Then $\mathbf{Sp}_A \hat{M}$ coincides with $\mathbf{Sp}\, M$ (see Section 1), therefore, by Arens-Singer theorem, it is homeomorphic to the compact set of all homomorphisms of the semigroup $S_M$ to the unit circle. It remains to check only that $\mathbf{Sp}_A \hat{M}$ coincides with $\mathbf{Sp}_A M$. $\quad\Box$

The following result establishes the existence of peak points for T-inavarint uniform algebras.

**THEOREM 3.2.** *Let* **M** *be a uniform* T*-invariant* A*-algebra on a compact Abelian group* T. *Then each point of* T *is a peak point for* **M**.

PROOF. Since **M** is T-invariant, it is sufficient to check the existence of a function $x \in M$ such that $x(1) = 1$, $\| x(t) \| < 1$, $t \neq 1$. We note that the set of all finite linear combinations of the form $a\gamma^s$, where $a \in A_s(M)$ and $s \in S_M$ is dense in **M**. Since $A_s(M)$ is a closed two-sided ideal in A we can suppose that $a \in A_+$.

For each $t_o \in T$, $t_o \neq 1$, and for each state $\phi$ on A, there exist elements $a \in A_+$ and $s \in S_M$ such that $\phi(a) \neq 0$ and $\gamma^s(t_o) \neq 1$.

Therefore, using the weak compactness of the state space of A we can establish the existence of $a_1, a_2, \ldots, a_n \in A_+$ and $s_1, s_2, \ldots, s_n$ from $S_M$ such that $a_k \gamma^{s_k} \in M$ and for each state $\psi$ on A, we have $\psi(a_i) \neq 0$, $\gamma^{s_i}(t_o) \neq 1$ for some couple $(a_i, s_i)$.

We denote by y the element from **M** defined as

$$ y = 1 + (2\sum \| a_k \|)^{-1} \sum a_k (\gamma^{s_k} - 1) $$

which has the following properties: $\| y \| = 1$, $y(1) = 1$ and $\| y(t_o) \| < 1$.

Let $U_k$ be a countable system of neighbourhoods of the identity. Then there exists a system of elements of **M** with the properties: $\| x_k \| = 1$, $x_k(1) = 1$ and $\| x_k(t) \| < 1$ for $t \in T \setminus U_k$ (we use the compactness of T and the properties of y). It remains only to verify that the element $x = \sum 2^{-k} x_k$ realizes the peak at identity. $\quad\Box$

A uniform algebra **M** is called *maximal* if $C(T,A)$ is the only uniform algebra which contains **M**.

For each closed two-sided ideal I of A we put

$$ S_M(I) = \{ s \in S_M, \; A_s(M) = I \}. $$

Even in the case when $S_M$ is a semigroup, $S_M(I)$ could not be a subsemigroup of $S_M$. However, $S_M(A)$ is always a semigroup.

The following statement presents a description of the maximal invariant uniform algebras and, in particular, it extends one of results of Hoffman and Singer [9].

**THEOREM 3.3.** *Let $M$ be a uniform T-invariant maximal A-algebra with weakly[*]-transitive fibre. Then $M$ is a Dirichlet algebra (i.e. $\overline{\mathrm{Re}\,M} = \mathrm{Re}\,C(T,A)$), and a semigroup $S_M(A)$ defines on $S$ an Archimedian order. In addition:*

*i) If $S_M \neq S$, then $S_M = S_M(A)$ (hence $M$ is decomposable). In this case the algebra $A$ is simple and*

$$\mathrm{Sp}_A\,M = T \times [0,1]/T \times \{0\}.$$

*ii) If $S_M = S$ then there exists a maximal two-sided ideal $I$ of $A$, such that $S_M(I) \cup \{0\} = -S_M(A)$. In this case $\mathrm{Sp}_A\,M = T$.*

PROOF. We note that if $J$ is a closed two-sided ideal of $A$ and $\hat{M}$ is a uniform T-inavariant A-algebra generated by $(A_s(M) + J)\gamma^s$ where $s \in S_M$, then $A_s(\hat{M}) = A_s(M) + J$.

i) If $S_M \neq S$, then, by the previous remark, $S_M = S_M(A)$ and, therefore, $M$ is generated by $A$ and some uniform algebra $M$ from $C(T)$, which is also maximal and T-invariant. Thus, the essential part of the statement i) is a consequence of the quoted theorem by Hoffman and Singer.

ii) If $S_M = S$ then $S_M(A) \neq S$. It is not difficult to verify that $S_M(A)$ is the maximal subsemigroup of $S$ and does not contain any subgroup of $S$. Hence $S_M(A)$ defines on $S$ an Archimedian order. The statement on relative spectrum is an evident consequence of Theorem 3.1 ii).

Since $M_1 + M_1^*$ is dense in $C(T,A)$, where $M_1 \subset M$ is a uniform algebra generated by $S_M(A)$ and $A$, then in both cases the algebra $M$ is Dirichlet. $\qquad\square$


## REFERENCES


1.  **Arens, R. ; Singer, I.** : Generalized analytic functions, *Trans. Amer. Math. Soc.* **81**(1956), 379-393.

2.  **Arzumanian, V.A. ; Grigorian, S.A.** : Uniform algebras of operator fields (Russian), *Zap. Nauchn. Sem. LOMI* **123**(1983), 185-189.

3.  **Arzumanian, V.A. ; Grigorian, S.A.** : The spectrum of uniform algebras of operator fields (Russian), *Izv. Acad. Sci. Arm. SSR, Matematica*, **21**:1(1986), 63-79; English trans.: *Soviet J. of Contemp. Math. Anal.* **21**:1(1986), 64-81.

4.  **Arzumanian, V. ; Grigorian, S.** : The boundaries of uniform algebras of operator fields, *Izv. Acad. Sci. Arm. SSR, Matematica* **23**:5(1988), 422-438.

5.  **Arzumanian, V. ; Grigorian, S.** : Invariant algebras of operator fields on compact Abelian groups, *Izv. Acad. Sci. Arm. SSR, Matematica*, to appear.

6.   **Dixmier, J.** : *Les C*-algèbres et leurs représentations*, Gauthier-Villars, Paris, 1969.

7.   **Fell, J.** : The structure of algebras of operator fields, *Acta Math.* **106**(1961), 233-280.

8.   **Gamelin, T.W.** : *Uniform algebras*, Prentice-Hall, New-York, 1969.

9.   **Hoffman, K. ; Singer, I.** : Maximal subalgebras of $C(\Gamma)$, *Amer. J. Math.* **79** (1957), 295-305.

10.  **Kasparov, G.G.** : Hilbert $C^*$- modules: theorems of Stinespring and Voiculescu, *J. Operator Theory* **4**(1980), 133-150.

11.  **Paschke, W.L.** : Inner product modules over $B^*$- algebras, *Trans. Amer. Math. Soc.* **182**(1973), 443-468.

12.  **Sallas, A.** : Une extension d'un théorème de K. Hoffman et J. Wermer aux algèbres de champe continus d'opérateurs, *C. R. Acad. Sci. Paris, Ser.A, Math.* **284**(1977), 1049-1051.

13.  **Taylor, D.C.** : Interpolation in algebras of operator fields, *J. Funct. Anal.* **10**(1972), 159-190.

14.  **Taylor, D.C.** : A general Hoffman-Wermer theorem for algebras of operator fields, *Proc. Amer. Math. Soc.* **52**(1975), 212-216.

**V. Arzumanian** and **S. Grigorian**

Institute of Mathematics, Armenian Academy of Sciences

Marshal Bagramian av. 24-b

Yerevan 375019, Armenian S.S.R.

U.S.S.R.

Operator Theory:
Advances and Applications, Vol. 43
© 1990 Birkhäuser Verlag Basel



# OPERATOR THEORY IN KREIN SPACES AND OPERATOR PENCILS

T.Ya. Azizov

The present paper is a detailed presentation of the results announced in [1].

1. Let $\mathfrak{H}$ be a Hilbert space, $A_i$ $(i = 0, 1, 2)$ be operators acting in $\mathfrak{H}$ and satisfying the following conditions:

1) $A_0 = A_0^*$ is an invertible operator, $A_0 = J|A_0|$ is its polar decomposition;

2) $A_2 = A_2^*$ is an invertible operator whose negative spectrum consists of a finite set $\kappa \geq 0$ of eigenvalues counting the multiplicity and $H := J|A_0|^{-\frac{1}{2}} A_2 |A_0|^{-\frac{1}{2}} \in$ $\in \mathfrak{S}_p$ $(p < \infty)$;

3) $T := J|A_0|^{-\frac{1}{2}} A_1 |A_0|^{-\frac{1}{2}} \in \mathfrak{S}_\infty$;

4) the linear pencil

$$(1) \qquad\qquad L_1(\lambda) := A_0 + A_1 + \lambda A_2$$

is invertible at least in one point $\lambda_0$.

It should be noted that the conditions 1) – 4) are independent. In particular, the condition 4) is not always realized in the case $\kappa = 0$.

Using standard methods the investigation of the completeness problem for root vectors of the pencil (1) and the study of the properties of its spectrum are reduced to the analogous problem for the pencil

$$(2) \qquad\qquad L_2(\lambda) := I + T + \lambda H,$$

and these problems in turn are reduced to the investigation of the same questions for the operator

$$(3) \qquad\qquad A = H(I + S),$$

where $S = (I + T + \lambda_0 H)^{-1} - I$.

Operator pencils of the type (2) and operators of the type (3) with an invertible selfadjoint operator $H \in \mathfrak{S}_p$ and arbitrary $T \in \mathfrak{S}_\infty$ were studied for the first time by M.V. Keldysh (see [3], Theorem V.8.1 and Theorem V.8.2). In our situation the operator $H$ is selfadjoint iff the operator $H$ commutes with the operator $J$. In the general case the operator H is J-selfadjoint in a Krein space[*] $\mathfrak{H}$ with respect to an indefinite J-metric $[x,y] = (Jx,y)$. Moreover as the negative spectrum of the operator $A_2$ consists of a finite set of eigenvalues, the operator H may be represented as a sum of J - positive and finite-dimensional J - selfadjoint operators. By virtue of Jonas-Langer theorem [4] the operator H is definitizable, i.e. there exists a polinomial p such that $[p(H)x,x] \geq 0$.

In particular, if $\kappa = 0$, then H is a J-positive operator: $[Hx,x] > 0$ for $x \neq 0$. It is known that the spectrum of a compact J-positive operator H is real (Ju.P. Ginzburg), the system of its eigenvectors is complete in the space $\mathfrak{H}$ (T.Ja. Azizov and V.A. Shtraus), and if its J-spectral function is regular then the operator H is similar to some selfadjoint operator. This shows that in some sense selfadjoint and J-positive operators are close one to another.

A.S. Marcus conjectured that the operator (3) with a J-positive H, just as in case $H = H^*$, has a complete system of root vectors and its characteristic numbers, excepting at most a finite set, are located for every $\varepsilon > 0$ in the union of angles $|k\pi - \arg\lambda| < \varepsilon$, $k = 0, 1$.

It will be showed below that this conjecture is not always correct and the sufficient conditions for its correctness will be given.

**2.** In this section we shall give three examples. The operator (3) of the first example is of the Keldysh type with the J-positive $H \in \mathfrak{S}_p$, the compact S, and has the spectrum located on an arbitrary given denumerable system of rays; in the second example the operator (3) is of the Volterra type, and in the third example, which is a combination of the two previous examples, the operator will have an incomplete system of root vectors and its characteristic numbers will not be concentrated in the neighborhood of a finite set of rays.

Let $\mathfrak{H} = \mathfrak{H}_1 \oplus \mathfrak{H}_1$, where $\mathfrak{H}_1$ is an infinite-dimensional separable Hilbert space. Put

$$J = \begin{bmatrix} 0 & I \\ I & 0 \end{bmatrix} \qquad H = \begin{bmatrix} 0 & F^\alpha \\ F & 0 \end{bmatrix} \qquad S = \begin{bmatrix} 0 & G \\ 0 & 0 \end{bmatrix}$$

where $F \in \mathfrak{S}_p$ is a positive operator, $\alpha > 1$ and $G \in \mathfrak{S}_\infty$. Hence $H \in \mathfrak{S}_p$ is a J-positive operator, $S \in \mathfrak{S}_\infty$ and the operator (3) has the matrix representation

---

[*] The theory of indefinite spaces and properties of operators acting in these spaces are set forth, for example, in [2].

$$A = H(I + S) = \begin{bmatrix} 0 & F^{\alpha} \\ F & FG \end{bmatrix}.$$

It is known, that the spectrum of the operator A coincides with the spectrum of the quadratic pencil

$$L_3(\lambda) := \lambda^2 I - \lambda FG - F^{1+\alpha}.$$

After the change of variables $\lambda \to i\lambda$ we obtain the pencil

(4)
$$L(\lambda) := \lambda^2 I + i\lambda FG + F^{1+\alpha}.$$

We shall show below that using F, G and $\alpha$ one can construct the desired examples.

**EXAMPLE 1.** Let $F_1 \in \mathfrak{S}_p$ be a positive operator, $\omega_j$ $(j = 1, 2, ..)$ be its eigenvalues, $\{f_j\}$ be a orthonormal basis of its eigenvectors: $F_1 f_j = \omega_j f_j$. Choose the sequence $\{\lambda_j\}$ in such a way that

(5)
$$\omega_j^2/\lambda_j \to 0, \quad \lambda_j^2/\omega_j \to 0 \quad (j \to \infty).$$

Assume in the pencil (4) that $\alpha = 2$, $F = F_1$, $G = G_1$, where $G_1 : G_1 f_j = i\,\lambda_j^{-1}\omega_j^{-1} \times$ $\times (\lambda_j^3 + \omega_j^3) f_j$. The condition (5) implies the compactness of the operator $G_1$. In this example the numbers $\lambda_j$ $(j = 1, 2, ...)$ are the eigenvalues of the pencil (4), i.e. the numbers $(-i\lambda_j)$ are eigenvalues of the operator

$$A_1 := \begin{bmatrix} 0 & F_1^2 \\ F_1 & F_1 G_1 \end{bmatrix}.$$

The property (5) depends only on the absolute values of the numbers $\lambda_j$. Therefore **arg** $\lambda_j$ may be chosen in an arbitrary way and the numbers $\lambda_j$ may be located on any preassigned denumerable set of rays going out of origin.

**EXAMPLE 2.** Let $\mathfrak{H}_1 = L_2(0,1)$, and the operator $Z := 2i\int_o^t \cdot\, ds$. Z is the nuclear Volterra operator $Z \in \mathfrak{S}_1$, $\sigma_p(Z) = \emptyset$.

If $\alpha_2 > 3$ then

$$(|Z|^{-2/(1+\alpha_2)} Zf)(t) = 2i(\pi/4)^{(1-\alpha_2)/(1+\alpha_2)} \sum_k (2k + 1)^{(1-\alpha_2)/(1+\alpha_2)} \times$$

$$\times \int_o^t f(\tau)(\sin \pi(2k+1)\pi/2 - (-1)^k)d\tau \cos \pi(2k+1)t/2.$$

From here it follows that the operator $|Z|^{-2/(1+\alpha_2)}Z$ is compact. The combination of this fact and the compactness of the operator $|Z|^{-2/(1+\alpha_2)}Z^*$ implies the relation $G_2 := -i|Z|^{-2/(1+\alpha_2)}(Z + Z^*) \in \mathfrak{S}_\infty$.

Assume $F = F_2 := |Z|^{-2/(1+\alpha_2)}$, $G = G_2$. Hence it follows that the pencil $L(\lambda)$ has the factorisation

$$L(\lambda) = \lambda^2 + \lambda(Z + Z^*) + Z^*Z = (\lambda I + Z^*)(\lambda I + Z)$$

and therefore $\sigma_p(L) = \emptyset$, i.e. $\sigma_p(A_2) = \emptyset$, where

$$A_2 = \begin{bmatrix} 0 & F_2^{\alpha_2} \\ F_2 & F_2 G_2 \end{bmatrix}.$$

**EXAMPLE 3.** Put $\tilde{\mathfrak{H}} = \mathfrak{H}^{(1)} \oplus \mathfrak{H}^{(2)}$, where $\mathfrak{H}^{(1)}$ is the Krein space from Example 1 and $\mathfrak{H}^{(2)}$ is from Example 2. Analogously we define the operators $\tilde{J} = J^{(1)} \oplus J^{(2)}$ and $\tilde{A} = A_1 + A_2$. According to its construction the operator $\tilde{A}$ is of type (3) with J-positive principal part, but has an incomplete system of root vectors and its characteristic numbers do not concentrate in the neighborhood of a finite set of rays.

**3.** The examples mentioned above generate the following question.

**QUESTION.** Under what conditions on the operators H and S have we completeness for the root vectors of the operator $A = H(I + S)$ with the definitizable $H \in \mathfrak{S}_p$?

Before we give a series of sufficient conditions for the positive solution of this question we shall formulate a general result (Theorem 1 below), where the operator H is not supposed to be definitizable. Its proof differs from the proof of Theorem V.8.1 in [3] only in details, for example it does not use an analog of the deep Lemma V.7.1 from [3] etc. Nevertheless, we consider that it is worth to be presented here.

**THEOREM 1.** *Let* $A = H(I + S)$ *be an invertible operator of the Keldysh type, let* $\lambda = 0$ *be a point of its continuous spectrum* $(0 \in \sigma_c(A))$, *let S be a compact operator, and the operator* $H \in \mathfrak{S}_p$ *satisfies the following conditions:*

*a) there exists a finite number of rays* $P_1, P_2, \ldots P_n$ *going out of the origin and forming with the positive semi-axis angles* $\phi_1, \phi_2, \ldots, \phi_n$ *respectively,* $0 \le \phi_1 <$

$< \phi_2 < \ldots < \phi_n < 2\pi$, *such that in the union of any closed angles* $\Lambda_1$, $\Lambda_2, \ldots, \Lambda_n$ *with vertices in the origin and located strictly between the rays* $P_1$ *and* $P_2$, $P_2$ *and* $P_3, \ldots, P_n$ *and* $P_1$ *respectively, there is situated at most a finite set of eigenvalues of the operator* H;

> b) $\|(I - \lambda H)^{-1} x\| \to 0$ *for all* $x \in \mathfrak{H}$, $\lambda \to \infty$ *and* $\lambda \in \bigcup\limits_{j=1}^{n} \Lambda_j$.

*Then the system of root vectors of the operator* A *is complete in* $\mathfrak{H}$, *and its spectrum* $\sigma(A)$ *satisfies requirement of* a) *with the same rays* $P_1, P_2, \ldots, P_n$.

PROOF. Let $\varepsilon > 0$ be an arbitrary given number, and the angles $\Lambda_j$ be such that the opening between their sides and nearest rays $P_K$ does not exceeds $\varepsilon$. The symbol T will denote the operator $T = I - (I + S)^{-1}$. As $S \in \mathfrak{S}_\infty$, then $T \in \mathfrak{S}_\infty$. The condition b) of Theorem 1 and the compactness of the operator T implies that the operators $(I - \lambda H)^{-1} T$ tend to 0 in the uniform operator topology as $\lambda \to \infty$, $\lambda \in \bigcup\limits_{1}^{n} \Lambda_j$. Therefore there exists a number $r > 0$ such that $\|(I - \lambda H)^{-1} T\| < q < 1$ for $|\lambda| \geq r$, $\lambda \in \bigcup\limits_{1}^{n} \Lambda_j$. Such set of numbers $\lambda$ we shall denote by the symbol $F_\varepsilon$.

Since $I - \lambda A = (I - \lambda H)(I - (I - \lambda H)^{-1} T)(I + S)$, the operator $I - \lambda A$ is invertible in the domain $F_\varepsilon$

$$(6) \qquad (I - \lambda A)^{-1} = (I + S)^{-1} \sum_{0}^{\infty} ((I - \lambda H)^{-1} T)^k (1 - \lambda H)^{-1}.$$

It implies, in particular, from this equality that the spectrum of the operator A satisfies the condition a) of Theorem 1 with the same rays $P_1, P_2, \ldots, P_n$.

As from the condition b) of Theorem 1 it follows the boundedness of the operator-function $(I - \lambda H)^{-1}$, in the domain $F_\varepsilon$, then (6) implies the boundedness of the operator-function $(I - \lambda A)^{-1}$ in $F_\varepsilon$.

Let $\mathfrak{C}(A)$ be the closed linear span of all root vectors of A. We shall prove that $\mathfrak{C}(A) = \mathfrak{H}$.

Let us assume the contrary: $\mathfrak{C}(A) \neq \mathfrak{H}$. Let P be the orthogonal projection on $\mathfrak{C}(A)$ and $Q = I - P$.

The operator $A_1 := QAQ$ is of the Volterra type and the operator-function $(I - \lambda A_1)^{-1}$ is an entire function. As $(I - \lambda A_1)^{-1} = Q(I - \lambda A)^{-1} Q + P$, then the operator-function $(I - \lambda A_1)^{-1}$ is bounded in $F_\varepsilon$. By virtue of V.I. Matsaev theorem (see [3], Theorem V.5.2)

$$(7) \qquad \ln \|(I - \lambda A_1)^{-1}\| = o(|\lambda|^p) \qquad (\lambda \to \infty)$$

as $p > p(H)$. Choose then $\alpha < \pi/p$. Since on the sides of the angles $\Lambda_j$ the operator-
-function $(I - \lambda A_1)^{-1}$ is bounded, by virtue of (7) and Phragmen-Lindelöff theorem, it is
also bounded inside angles complementing $\overset{n}{\underset{1}{\cup}} \Lambda_j$. Thus the entire function $(I - \lambda A_1)^{-1}$ is
bounded in all the complex plane. It implies that $A_1 = 0$ which contradicts the condition
$0 \in \sigma_c(A)$.

Therefore the system of root vectors of A is complete in $\mathfrak{H}$ : $\mathfrak{C}(A) = \mathfrak{H}$.


**4.** It should be noted that the conditions a) and b) of Theorem 1 are satisfied for
invertible operators in the following classes: for definitizable operators with regular J -
- spectral function in $\lambda = 0$, and for J - selfadjoint operators of the class $K(H)$. Also it
should be noted that the condition a) implies the condition b) for J-normal operators in
the class $K(H)$, in particular, it is true for $\pi$ - selfadjoint and $\pi$ - normal operators in
Pontryagin spaces. In the proof of the following Theorem 2 we shall need only the last
of mentioned results and we shall explain it in the right place.

Further, the symbol $\widetilde{\mathfrak{S}}_\infty$ will denote the set of continuous operators whose
spectrum excepting the point $\lambda = 0$ consists of normal eigenvalues. In particular,
$\mathfrak{S}_\infty \subset \widetilde{\mathfrak{S}}_\infty$.

A continuous operator B, acting in a nondegenerate G-space, will be called G-
-normal if its G-adjoint operator $B^c$ exists and is defined on the whole space and also
$BB^c = B^c B$.

We should recall (see [2]) that by the term $G^{(\kappa)}$ - space is denoted a G-space,
generated by the operator G, whose negative spectrum consists of $\kappa < \infty$ (counting
multiplicity) eigenvalues.

By the symbol $\lambda_j(X)$, we shall denote the eigenvalues of the operator X.

**THEOREM 2.** *Let* $\mathfrak{H}$ *be* $G^{(\kappa)}$ *- space,* $A = H(I + S)$, *where* $H \in \widetilde{\mathfrak{S}}_\infty$ *is a* $G^{(\kappa)}$ -
*- normal operator whose spectrum satisfies the condition a) of Theorem 1 and*
$\underset{j}{\sum} |\lambda_j(H)|^p < \infty$ *for some finite p, and S is a continuous operator such that its* $G^{(\kappa)}$ -
*- adjoint operator* $S^c$ *is also continuous and their product is compact. Then if* $0 \in \sigma_c(A)$
*the lineal* $G^{(\kappa)}\mathfrak{C}(A)$ *is dense in* $\mathfrak{H}$ *and* $\sigma(A)$ *satisfies the condition a) of Theorem 1 with
the same rays as* $\sigma(H)$. *If, besides,* $H = FG^{(\kappa)}$, *where F is a continuous operator, then*
$\mathfrak{C}(A) = \mathfrak{H}$.


PROOF. Let $\Pi_\kappa$ be the standard completion of the $G^{(\kappa)}$ - space $\mathfrak{H}$ with respect
to the norm $(|G^{(\kappa)}|x,x)^{\frac{1}{2}}$ (see [2], Proposition I.6.14). Since the $G^{(\kappa)}$ - adjoints of H and
S are continuous, according to the known Krein theorem on operators acting in spaces

with two norms, the operators H and S are continuous in $\Pi_\kappa$ too. Let $\tilde{H}$ and $\tilde{S}$ be their closures. From the complete continuity of the product of the operator S and its $G^{(\kappa)}$ - - adjoint it follows that $\tilde{S} \in \mathfrak{S}_\infty$.

Since $H \in \tilde{\mathfrak{S}}_\infty$, we have $\tilde{H} \in \tilde{\mathfrak{S}}_\infty$, and in addition the operator $\tilde{H}$ has finite order. In fact since H is a $G^{(\kappa)}$ - normal operator, $\tilde{H}$ is $\pi$ - normal. From the Naimark theorem on invariant subspaces of commutative sets of $\pi$ - selfadjoint operators (see [2], Corollary III.4.12) it follows that the operator $\tilde{H}$ and its $\pi$ - adjoint have some generic $\kappa$ - - dimensional nonpositive invariant subspace $L$. Let $L_o$ be the isotropic part of $L$ and $M_o$ be the orthogonal complement of the $\pi$ - orthogonal complement $L_o^{[\perp]}$ of the subspace $L_o$, i.e. $M_o = (L_o^{[\perp]})^\perp$, and $\mathfrak{H}_o = L_o^\perp \cap L_o^{[\perp]}$.

Therefore

(8)
$$\Pi_\kappa = L_o + \mathfrak{H}_o + M_o.$$

With respect to this decomposition the operator $\tilde{H}$ has the matrix form:

(9)
$$\tilde{H} = \begin{bmatrix} \tilde{H}_{11} & \tilde{H}_{12} & \tilde{H}_{13} \\ 0 & \tilde{H}_{22} & \tilde{H}_{23} \\ 0 & 0 & \tilde{H}_{33} \end{bmatrix}.$$

Since the $\pi$ - normal operator $\tilde{H}_{22}$ and its $\pi$ - adjoint have the generic maximal negative invariant subspace $L \cap \mathfrak{H}_o$, the operator $\tilde{H}_{22}$ is similar to some Hilbert normal operator. Granting this and taking into account that

$$\sum_j |\lambda_j(\tilde{H}_{22})|^p \leq \sum_j |\lambda_j(\tilde{H})|^p = \sum_j |\lambda_j(H)|^p < \infty,$$

we have that $\tilde{H}_{22} \in \mathfrak{S}_p$. Since in the decomposition (8) $\dim(L_o \oplus M_o) < \infty$, the operator $\tilde{H}$ has finite order, too.

Let $\tilde{A}$ be the closure of A in $\pi_\kappa$. Then

(10)
$$\tilde{A} = \tilde{H}\,(I + \tilde{S}),$$

where $\tilde{H} \in \mathfrak{S}_p$ is a $\pi$ - normal operator and its spectrum satisfies the condition a) of Theorem 1, and $\tilde{S} \in \mathfrak{S}_\infty$.

We will show that the operator $\tilde{H}$ satisfies the condition b) of Theorem 1 too. In fact from (9) it follows that the operator-valued-function $(I - \lambda\tilde{H})^{-1}$ has the matrix form:

$$(11) \qquad (I - \lambda \tilde{H})^{-1} = \begin{bmatrix} \tilde{H}_{11}(\lambda) & \tilde{H}_{12}(\lambda) & \tilde{H}_{13}(\lambda) \\ 0 & \tilde{H}_{22}(\lambda) & \tilde{H}_{23}(\lambda) \\ 0 & 0 & \tilde{H}_{33}(\lambda) \end{bmatrix}$$

where

$$(\tilde{H}_{ii}(\lambda) = (I - \lambda \tilde{H}_{ii})^{-1} \qquad (i = 1, 2, 3),$$

$$\tilde{H}_{12}(\lambda) = \lambda \, \tilde{H}_{11}(\lambda)\tilde{H}_{12}\tilde{H}_{22}(\lambda),$$

$$\tilde{H}_{13}(\lambda) = -\lambda \tilde{H}_{12}(\lambda)\tilde{H}_{23}\tilde{H}_{33}(\lambda) + \lambda \tilde{H}_{11}(\lambda)\tilde{H}_{13}\tilde{H}_{33}(\lambda),$$

$$\tilde{H}_{23}(\lambda) = \lambda \tilde{H}_{22}(\lambda)\tilde{H}_{23}\tilde{H}_{33}(\lambda).$$

Since the operators $\tilde{H}_{11}$ and $\tilde{H}_{33}$ act in finite dimensional spaces and $\tilde{H}_{22}$ is similar to some Hilbert normal operator, the operator-valued functions $(I - \lambda \tilde{H}_{ii})^{-1}$ $(i = 1, 2, 3)$ satisfy the condition b) of Theorem 1. Hence, from the finite dimensionality of the operators $\tilde{H}_{12}$, $\tilde{H}_{13}$ and $\tilde{H}_{22}$, and taking into account the properties $\|\lambda \tilde{H}_{11}(\lambda)\| \to \|H_{11}^{-1}\|$ and $\|\lambda \tilde{H}_{33}(\lambda)\| \to \|H_{33}^{-1}\|$ for $\lambda \to \infty$, it follows that the operator-valued function $(I - \lambda \tilde{H})^{-1}$ satisfies also the condition b) of Theorem 1.

Consequently, from Theorem 1 we have $\mathfrak{C}(\tilde{A}) = \Pi_\kappa$. This is equivalent to the fact that in $\Pi_\kappa$ there exists no non-trivial vectors which are $\pi$ - orthogonal to $\mathfrak{C}(\tilde{A})$. Since the root vectors of the operators $A$ and $\tilde{A}$ coincide, in $\mathfrak{H}$ there exists no non--trivial vectors which are $G^{(\kappa)}$ - orthogonal to $\mathfrak{C}(A)$. Therefore the lineal $G^{(\kappa)}$ is dense in $\mathfrak{H}$.

Let $H = FG^{(\kappa)}$, then $\mathfrak{C}(A) = \mathfrak{H}$. In fact, suppose that some vector $x_0$ is orthogonal to $\mathfrak{C}(A)$. Then it is orthogonal to the lineal $A\,\mathfrak{C}(A)$ too. Therefore the vector $(I + S^c)F^*x_0$ is $G^{(\kappa)}$ - orthogonal to $\mathfrak{C}(A)$, i.e. $(I + S^c)F^*x_0 = 0$. From the conditions $0 \in \sigma_c(A)$ and $S^cS \in \mathfrak{S}_\infty$ it follows that $x_0 = 0$.

**REMARK 1.** If $G^{(\kappa)} = I$, then from the conditions of Theorem 2 we have that $H$ is a normal operator of finite order, and $S \in \mathfrak{S}_\infty$. Therefore Theorem 2 is a generalization of the known result on Keldysh type operators with normal principal part.

**COROLLARY 1.** *Let $\mathfrak{H}$ be a Krein J-space, and $A = H(I + S) = (I + S_0)H$, where $H \in \tilde{\mathfrak{S}}_\infty$ is a definitizable operator, $\sum_j |\lambda_j(H)|^P < \infty$, and $S, S_0$ be continuous operators with the conditions $S_0^cS_0 \in \mathfrak{S}_\infty$, and $0 \in \sigma_c(A)$.*

*Then we have* $\mathfrak{C}(A) = \mathfrak{H}$, *and for every* $\varepsilon > 0$ *the spectrum* $\sigma(A)$ *of the operator* A, *excepting a finite set, is contained in the angles* $|k\pi - \arg \lambda| < \varepsilon$ $(k = 0,1)$.

*If* $A \in \mathfrak{S}_p$ *and* $\mathfrak{H}$ *is not a Pontryagin space, then all but a finite set of eigenvectors of the operator* JA, *are contained either in the angle* $|\arg \lambda| < \varepsilon$ *or in the angle* $|\pi - \arg \lambda| < \varepsilon$.

PROOF. At first let $\mathfrak{H}$ be a Pontryagin space. Then the operator H is $\pi$- - selfadjoint and therefore it satisfies the conditions of Theorem 1. Here we have n = 2, $P_1 = [0,\infty)$ and $P_2 = (-\infty, 0]$, i.e. $\phi_1 = 0$, $\phi_2 = \pi$. Hence the statements of Corollary 1 follow from Theorem 1. It should be noted that if $P_1 \cap \sigma(JH)$ and $P_2 \cap \sigma(JH)$ are denumerable sets, then in the statement about $\sigma(JA)$ there are present both angles $|k\pi - \arg \lambda| < \varepsilon$ $(k = 0, 1)$.

Now suppose $\mathfrak{H} \neq \Pi_\kappa$. Since H is an invertible definitizable operator and $H \in \tilde{\mathfrak{S}}_\infty$, there is $H = H_1 + H_2$, where $H_1$ is a J - positive (or J - negative) operator and $H_2$ is a finite dimensional J - selfadjoint operator with $H_1^{-1}H_2 \in \mathfrak{S}_\infty$. Without loss of generality we will assume that $H_1$ is a J - positive operator. Therefore the negative spectrum of $JH = JH_1 + JH_2$ consists of $\kappa < \infty$ eigenvalues counting the multiplicity, and the form $(JHx,y)$ generates a $G^{(\kappa)}$ - space with $G^{(\kappa)} = JH$. The operator H satisfies the conditions of Theorem 2 and $H = FG^{(\kappa)}$, where $F = J$. Since $H(I + S) = (I + S_o)H$, there is $HS = S_o H$, i.e. $S_o^* = J(JH)^{-1}S^*(JH)J$. Consequently $(JH)^{-1}S^*(JH) = JS_o^*J$ is a continuous operator and $(JH)^{-1}S^*(JH)S = JS_o^*JS \in \mathfrak{S}_\infty$. Therefore the operator S also satisfies the conditions of Theorem 2.

From Theorem 2 we have $\mathfrak{C}(A) = \mathfrak{H}$.

Since the spectrum of the definitizable operator $H \in \tilde{\mathfrak{S}}_\infty$ is real excepting a finite set of eigenvalues counting the multiplicity, the spectrum of A satisfies the conditions of Theorem 2.

Let $A \in \mathfrak{S}_p$. Then $H \in \mathfrak{S}_p$. Since $JH = JH_1(I + H_1^{-1}H_2)$ where $JH_1$ is a positive operator and $H_1^{-1}H_2 \in \mathfrak{S}_\infty$, there is $JA = JH_1(I + \tilde{S}_1)$ with $\tilde{S}_1 = S + H_1^{-1}H_2(I + S) \in \mathfrak{S}_\infty$ and $JH_1 \in \mathfrak{S}_p$. From Theorem 1 we have $\mathfrak{C}(JA) = \mathfrak{H}$ and all but a finite set of eigenvectors of JA are contained in the angle $|\arg \lambda| < \varepsilon$.

If $H_1$ is a J - negative operator, then all but a finite set of eigenvectors of JA are contained in the angle $|\pi - \arg \lambda| < \varepsilon$.

**5.** In this section we shall prove two theorems on the completeness of root vectors of Keldysh type operators with a definitizable principal part and with a J - selfadjoint A or a J - selfadjoint S.

**THEOREM 3.** *Let* $A = H(I + S)$ *be a* $J$ *- selfadjoint operator,* $0 \in \sigma_c(A)$, $H \in \tilde{\mathfrak{S}}_\infty$ *be a definitizable operator and* $S \in \tilde{\mathfrak{S}}_\infty$.

*Then* $\mathfrak{C}(A) = \mathfrak{H}$ *and the nonreal spectrum of the operator* $A$ *consists (excepting a finite set) of normal eigenvalues.*

PROOF. From [2], Theorem IV.2.15 it follows that for the proof of Theorem 3 it suffices to verify the definitizability of $A$ and that $A \in \tilde{\mathfrak{S}}_\infty$.

If $\mathfrak{H}$ is a Pontryagin space, then $H \in \tilde{\mathfrak{S}}_\infty$ implies $H \in \mathfrak{S}_\infty$. Therefore $A \in \mathfrak{S}_\infty$ and $A$ is a definitizable operator as $\pi$ - selfadjoint operator.

Let $\mathfrak{H} \neq \Pi_\kappa$. Since $H \in \tilde{\mathfrak{S}}_\infty$ is a definitizable operator, we have $H = H_1 + H_2$, where $H_2$ (or $- H_1$) is a $J$ - positive operator and $H_2$ is a finite dimensional $J$ - selfadjoint operator. Let, without loss of generality, $H_1$ be a $J$ - positive operator. From V.A. Shtraus theorem [5] we have that $H_1^2 \in \mathfrak{S}_\infty$. Therefore $H^2 = H_1^2 + H_1 H_2 + H_2 H_1 + H_2^2 \in \mathfrak{S}_\infty$ and consequently $A^2 = H(I + S)H(I + S) = (I + JS^*J)H^2(I + S) \in \mathfrak{S}_\infty$. Hence $A \in \tilde{\mathfrak{S}}_\infty$.

Now we verify the definitizability of $A$. For this we introduce in $\mathfrak{H}$ the $G^{(\kappa)}$ - metric $[x,y]_1 = [Hx,y]$. Since $S = H^{-1}A - I$ and $A$ is a $J$ - selfadjoint operator, the continuous operator $S$ is $G^{(\kappa)}$ - selfadjoint. The continuous operator $H$ is $G^{(\kappa)}$ - selfadjoint too. Let now, by the proof of Theorem 2, the Pontryagin space $\Pi_\kappa$ be the standard completion of the $G^{\kappa)}$ - space $\mathfrak{H}$, $\tilde{H}$ and $\tilde{S}$ be the extensions of $H$ and $S$ respectively. Therefore $\tilde{A} = \tilde{H}(I + \tilde{S})$ is the extension of $A$. Since $H, S \in \tilde{\mathfrak{S}}_\infty$, we have $\tilde{H}, \tilde{S} \in \tilde{\mathfrak{S}}_\infty$. Granting that all $\pi$ - selfadjoint operators of the class $\mathfrak{S}_\infty$ are compact, we have $\tilde{H}, \tilde{S} \in \mathfrak{S}_\infty$. Hence the form $[x,y]_2 = [(I + S)x,y]_1$ generates a Pontryagin space too. In this space $\tilde{A}$ is a $\pi$ - selfadjoint operator, therefore $\tilde{A}$ is a definitizable operator, i.e. there exists a polynomial $p$ such that $[p(\tilde{A})x,x]_2 > 0$. For $x \in \mathfrak{H}$ we have $[Ap(A)x,x] = [p(\tilde{A})x,x]_2$ and therefore $A$ is definitizable in the Krein space $\mathfrak{H}$ too.

Since $\tilde{A}$ is a $\pi$ - selfadjoint operator, its nonreal spectrum consists (excepting a finite set) of normal eigenvalues. Hence the nonreal spectrum of $A$ is the same set.

**THEOREM 4.** *Let* $\mathfrak{H}$ *be a* $J$ *- space,* $\mathfrak{H} \neq \Pi_\kappa$, *and the operator* $A = H(I + S)$, *where* $H \in \mathfrak{S}_\infty$ *is an invertible definitizable operator and* $I + S$ *is a bounded and boundedly invertible* $J$ *- selfadjoint operator.*

*Then* $\mathfrak{C}(A) = \mathfrak{H}$ *and the nonreal spectrum of* $A$ *consists (excepting a finite set) of normal eigenvalues.*

PROOF. Since $(I + S)^{-1}$ and $I + S$ are bounded $J$ - selfadjoint operators, the form $[x,y]_1 := [(I + S)x,y] = (J(I + S)x,y)$ generate a $J(I + S)$ - Krein space. The operator $A$ is a compact $J(I + S)$ - selfadjoint. For the proof of Theorem 4 is suffices to verify the definitizability of A.

Suppose that $\mathfrak{H} \neq \Pi_\kappa$. Therefore $H = H_1 + H_2$, where $H_1$ is a J - definite operator and $\dim H_2 < \infty$. Hence $A = A_1 + A_2$, where $A_1 = H_1(I + S)$ is a $J(I + S)$ -definite operator and $A_2 = H_2(I + S)$ is a finite dimensional $J(I + S)$ - selfadjoint operator. From Jonas -
- Langer theorem [4] it follows the definitizability of A.

**REMARK 2.** The condition $\mathfrak{H} \neq \Pi_\kappa$ in Theorem 4 is essential. In fact, let A be a Volterra J - selfadjoint operator and $J_\kappa = J_\kappa^* = J_\kappa^{-1}$ generates a Pontryagin space $\Pi_\kappa$. Then $A = (AJJ_\kappa)(J_\kappa J)$, where $AJJ_\kappa$ is a compact $\pi$ - selfadjoint operator and $J_\kappa J$ is a bounded and boundedly invertible $\pi$ - selfadjoint operator. It is known that every $\pi$ -
- selfadjoint operator is definitizable. Thus all the conditions of Theorem 4 except $\mathfrak{H} \neq \Pi_\kappa$, are fulfilled, but A is a Volterra operator.

**6.** At the end we would like to formulate a conjecture.

**CONJECTURE.** Let $A = H(I + S)$ be a Keldysh type operator with a definitizable principal part $H \in \mathfrak{S}_p$, $S \in \mathfrak{S}_\infty$ and $0 \in \sigma_c(A)$. If $\lambda = 0$ is not a regular point of the J -
- spectral function of H, then there exists an operator S such that $\mathfrak{C}(A) \neq \mathfrak{H}$.

### REFERENCES


1.      **Azizov, T.Ya.** : On the completion of the systems of root vectors of operators in spaces with an indefinite metric (Russian), *Funktsional. Anal. i Prilozhen.* 21:1(1987), 66-67.

2.      **Azizov, T.Ya.** ; **Iohvidov, I.S.** : *Theory of linear operators in spaces with an indefinite metric* (Russian), Nauka, Moscow, 1986.

3.      **Gohberg, I.C.** ; **Krein, M.G.** : *Introduction to the theory of nonselfadjoint operators in Hilbert spaces* (Russian), Nauka, Moscow, 1965.

4.      **Jonas, P.** ; **Langer, H.** : Compact perturbations of definitizable operators, *J. Operator Theory* 2(1979), 63-77.

5.      **Shtraus, V.A.** : On the theory of selfadjoint operators on Banach spaces with a Hermitian form (Russian), *Sibirsk. Mat. Zh.* 19:3(1978), 685-692.



**T.Ya. Azizov**
394068 Voronezh-68
Shishkova 65, kv.58
USSR.

Operator Theory:
Advances and Applications, Vol. 43
© 1990 Birkhäuser Verlag Basel



# EXTENSIONS OF GROUPS AND SIMPLE $C^*$-ALGEBRAS

Fl. Boca and V. Niţică

Since Powers ([9]) proved that the reduced $C^*$-algebra of the free group on two generators $C_r^*(\mathbf{F}_2)$ is simple with unique trace, several other classes consisting of discrete groups G such that $C_r^*(G)$ is simple with unique trace have been produced (see e.g. [1], [4]).

There are two types of combinatorial conditions which imply the simplicity of $C_r^*(G)$ and the uniqueness of its trace. Examples of such groups were given by Akemann and Lee – the groups which contain a free normal subgroup, with trivial centralizer and by P. de la Harpe – the groups which satisfy a certain combinatorial property, called Powers property. Moreover, in [3] we have introduced a weaker combinatorial condition, called weak Powers property and we showed that for a given exact sequence of groups $1 \to G_1 \to G \to G_2 \to 1$ with $G_1$ and $G_2$ weak Powers groups, it follows that $C_r^*(G)$ is simple with unique trace, although the group G may be not weak Powers group.

In this paper, we prove that the above assertion is still true in the four cases when $G_1$ and $G_2$ satisfy either the Akemann-Lee condition or the weak Powers property.

We are grateful to P. de la Harpe for his letter about this subject.

Let G be a discrete group with unit 1, let $\ell^2(G)$ be the Hilbert space of the square summable complex functions on G. Denote by $\mathbf{C}[G]$ the group algebra of G, viewed as an operator algebra on $\ell^2(G)$. The elements of $\mathbf{C}[G]$ are finite sums $\sum_{g \in G} \lambda_g u(g)$, with $\lambda_g \in \mathbf{C}$, $u(g) \in B(\ell^2(G))$ unitaries

$$(u(g)\xi)(s) = \xi(g^{-1}s), \qquad \text{for } \xi \in \ell^2(G), s \in G.$$

The uniform closure $C_r^*(G)$ of $\mathbf{C}[G]$ in $B(\ell^2(G))$ coincides with the $C^*$-algebra generated by the unitaries $u(g)$, $g \in G$. The norm on $B(\ell^2(G))$ is denoted by $\| \cdot \|$ and the canonical trace on $C_r^*(G)$ is the unique extension of the positive functional

$$\tau : \mathbf{C}[G] \to \mathbf{C}, \qquad \tau\Big(\sum_{g \in G} \lambda_g u(g)\Big) = \lambda_1.$$

When $Y = \sum_{g \in G} \lambda_g u(g) \in \mathbf{C}[G]$, we denote $\mathbf{supp}\ Y = \{g \mid \lambda_g \neq 0\}$.

For $g_1, \ldots, g_n \in G$ one defines the unital complete positive map

$$\Phi : C_r^*(G) \to C_r^*(G), \qquad \Phi(a) = (1/n) \sum_{i=1}^{n} u(g_i) a u(g_i)^*.$$

Such maps are called ([1]) averaging processes. Clearly, if $\Phi$ and $\Psi$ are averaging processes, then $\Phi \circ \Psi$ is still an averaging process. In particular, for $g \in G$, $n \in \mathbf{N}^*$, consider the following averaging process:

$$\theta_{gn} : C_r^*(G) \to C_r^*(G), \qquad \theta_{gn}(a) = (1/n) \sum_{i=1}^{n} u(g)^i a u(g)^{-i}.$$

A subset $X$ of a group $G$ is called free if $X$ is a basis for a free subgroup of $G$. Recall the following important result ([2]).

(1)   For any free subset $\{x_1, \ldots, x_n\}$ of $G$, $\left\| \sum_{i=1}^{n} u(x_i) \right\| = 2\sqrt{n-1}$ in $C_r^*(G)$.

Recall also the following basic lemma (see e.g. [5, Proposition 1.9]).

(2)   Let $X$ be a subset of a group $G$. Then $X$ is free if and only if $X \cap X^{-1} = \emptyset$ and no product $w = x_1 \cdots x_n$ is trivial, where $n \geq 1$, $x_1, \ldots, x_n \in X^{\pm 1}$ and $x_i x_{i+1} \neq 1$.

We shall use two simple consequences of (2).

(3)   Let $G$ be a group, $H$ a normal subgroup of $G$, $\pi : G \to G/H$ the quotient map and $Y = \{y_i\}_{i \in I}$ a free subset of $G/H$. If $x_i \in G$ and all $\pi(x_i) = y_i$, then $X = \{x_i\}_{i \in I}$ is a free subset of $G$.

(4)   If $\{a, b\}$ is a free subset in $G$, then $\{a^n b^n\}_{n \geq 1}$ is also free.

Note that (4) is Exercise 12 in [6, page 51].

Using (3) and ideas from [1], we get the following lemma (for $H = \{1\}$, this lemma is just Theorem 3 in [1]).

**LEMMA.** *Let $G$ be a discrete group and let $H$ be a normal subgroup of $G$ such that $G/H$ contains a free normal subgroup $F$, with trivial centralizer in $G/H$. Then there exists a sequence $\{\theta_n\}_{n \geq 1}$ of averaging processes on $G$ such that*

$$\lim_{n \to \infty} \| \theta_n(u(g)) \| = 0, \qquad g \in G \setminus H.$$

PROOF. Assume first that $\{y_1, \ldots, y_k\}$ is a basis for the free group $F$. Let

$x_i \in \pi^{-1}(y_i)$, $i = 1, \ldots, k$ and set

$$(5) \qquad \theta_n = \theta_{x_1 n} \circ \theta_{x_{k-1} n} \circ \cdots \circ \theta_{x_1 n}.$$

Clearly, $\theta_n$ are averaging processes on G.

Let $g \in G \setminus H$ and set $w = \pi(g) \in G/H \setminus \{\pi(1)\}$. In view of the assumption that the centralizer of F in G/H is trivial, there exists $i \in \{1, \ldots, k\}$ such that $w^{-1} y_i w \neq y_i$. Choose p be the first such index and denote either $q = p + 1$, for $p < k$ or $q = 1$ for $p = k$.

When $\{w^{-1} y_p w, y_p\}$ is free in G/H, (3) and (4) tells us that $\{(g^{-1} x_p g)^i x_p^{-i}\}_{i \geq i}$ is free in G, so (1) applies and

$$\| \theta_n(u(g)) \| \leq \| \theta_{x_p n}(u(g)) \| = \| u(g^{-1}) \theta_{x_p n}(u(g)) \| =$$

$$= (1/n) \| \sum_{i=1}^{n} u((g^{-1} x_p g)^i) x_p^{-i} \| = 2\sqrt{n-1}/n.$$

When $\{w^{-1} y_p w, y_p\}$ is not free in G/H, $w^{-1} y_p w$ is a generator of F which commutes with $y_p$, hence $w^{-1} y_p w = y_p^{\pm 1}$. It follows that $w^{-1} y_p w = y_p^{-1}$ and $w^{-1} y_p^i w y_p^i = \pi(1)$. This implies that $g^{-1} x_p^i g x_p^i = h_i \in H$, for any i, and so

$$(6) \qquad \theta_{x_q n} \circ \theta_{x_p n}(u(g)) = (1/n^2) \sum_{i=1}^{n} \sum_{j=1}^{n} u(x_q^j x_p^i g x_p^{-i} x_q^{-j}) =$$

$$= (1/n^2) \sum_{i=1}^{n} \sum_{j=1}^{n} u(x_q^j g h_i x_p^{-2i} x_q^{-j}).$$

For every $i \in \{1, \ldots, n\}$ one obtains

$$(7) \qquad \| \sum_{j=1}^{n} u(x_q^j g h_i x_p^{-2i} x_q^{-j}) \| = \| \sum_{j=1}^{n} u(x_p^{2i} h_i^{-1} g^{-1} x_q^j g h_i x_p^{-2i} x_q^{-j}) \| =$$

$$= \| \sum_{j=1}^{n} u((x_p^{2i} h_i^{-1} g^{-1} x_q g h_i x_p^{-2i})^j x_q^{-j}) \| .$$

The element $y_p^{-2i} w^{-1} y_q w y_p^{-2i} \in F$ commutes with $y_q$ exactly when $y_p^{2i} w^{-1} y_q w y_2^{-2i} = y_q^s$, for an integer s and this is the case for at most one $i \in \{1, \ldots, n\}$. Let $i_o$ be such an index. Using (3) and (4) it follows that $\{(x_p^{2i} h_i^{-1} g^{-1} x_q g h_i x_p^{-2i})^j x_q^{-j}\}_{j \geq 1}$ is free in G, for $i \neq i_o$. Now, (1) and (7) imply

$$\| \sum_{i=1}^{n} u(x_q^{-j} g h_i x_p^{-2i} x_q^{-j}) \| = 2\sqrt{n-1},$$

and so, by (5):

$$\| \theta_n(u(g)) \| \leq \| \theta_{x_q n} \circ \theta_{x_p n}(u(g)) | \leq$$

$$\leq (1/n^2) \| \sum_{i=1}^{n} u(x_q^j g h_i{}_o x_p^{-2i}{}_o x_q^{-j}) \| + [2(n-1)/n^2]\sqrt{n-1} \leq (1/n) + [2(n-1)/n^2]\sqrt{n-1}.$$

Finally, in the case when $\{y_i\}_{i \geq 1}$ is a basis for F, take $x_i \in \pi^{-1}(y_i)$, $\theta_n =$

$= \theta_{x_n n} \circ \theta_{x_{n-1} n-1} \circ \ldots \circ \theta_{x_1 1}$. The same proof is still valid for $n > q = p+1$.  $\square$

A *weak Powers group* is a group G having the following property: given any non-empty finite subset F of $G \setminus \{1\}$ which is included into a conjugacy class, and any integer $N \geq 1$, there exist a partition $G = A \amalg B$ and elements $g_1, \ldots, g_N$ in G such that

$$fA \cap A = \varnothing, \qquad \text{for all } f \in F \quad \text{and}$$

$$g_j B \cap g_k B = \varnothing, \qquad \text{for } j, k = 1, \ldots, N, \ j \neq k.$$

Using the previous lemma and techniques from [3], we prove the main result in this paper.

**THEOREM.** *Let* $1 \to G_1 \to G \to G_2 \to 1$ *be an exact sequence of discrete groups such that one of the following conditions is fulfilled:*

i) $G_1$ *and* $G_2$ *are weak Powers groups;*

ii) $G_1$ *is weak Powers group and* $G_2$ *contains a free normal subgroup with trivial centralizer;*

iii) $G_1$ *contains a free normal subgroup with trivial centralizer and* $G_2$ *is weak Powers;*

iv) *Any* $G_i$ *contains a free normal subgroup with trivial centralizer.*

*Then* $C_r^*(G)$ *is simple, with unique trace.*

PROOF. $G_1$ is identified with a normal subgroup of G, $G_2$ with the quotient group $G/G_1$ and $\pi : G_1 \to G/G_1$ is the quotient map.

From a standard trick (see e.g. [4, Proposition 3]), it is enough to show that for any $Y = Y^* \in \mathbf{C}[G]$, $\tau(Y) = 0$ and for any $\varepsilon > 0$, there exists $g_1, \ldots, g_N \in G$ such that

$$(8) \qquad \| (1/N) \sum_{k=1}^{N} u(g_k) Y u(g_k)^* \| \leq \varepsilon.$$

It is clear that $Y = \sum_{j=1}^{r} \lambda_j u(g_j) + \overline{\lambda}_j u(g_j)^*$, where $g_j \neq 1$, $j = 1, \ldots, r$. We may assume without loss of generality that $g_1, \ldots, g_p \in G_1$ and $g_{p+1}, \ldots, g_r \in G \setminus G_1$. Set

$$\tilde{Y} = Y_1 + \ldots + Y_p.$$

When $G_1$ is a weak Powers group, Lemma 2.2 in [3] applies in a particular case ($A = \mathbf{C}$, the 2-cocycle $c$ and the action $\alpha$ are trivial, hence $A \rtimes_{\alpha,c} G = C_r^*(G)$) and we find $h_1, \ldots, h_n \in G_1$ such that

(9)
$$\left\| (1/n) \sum_{k=1}^{n} u(h_k) Y u(h_k)^* \right\| \leq p\varepsilon/r.$$

If $G_1$ contains a free normal subgroup with trivial centralizer, the previous lemma in this paper applied for $H = \{1\}$ (Theorem 3 in [1]) implies the same fact.

For $G/G_1$ weak Powers group, the statement follows as in [3, Proposition 2.10]. Take $\tilde{Y}_{p+1} = (1/n) \sum_{k=1}^{n} u(h_k) Y_{p+1} u(h_k)^*$. The support of $\tilde{Y}_{p+1}$ is clearly included in $G \setminus G_1$. Since $G/G_1$ is a weak Powers group, it is not hard to observe (see the second case in the proof of Proposition 1.5 in [3]) that for any finite set $M \subset G$ and for any integer $n_0 \geq 1$, there exist $G = A \amalg B$ and $\gamma_1, \ldots, \gamma_{n_0} \in G$ such that

$$\gamma A \cap A = \emptyset, \quad \text{for any } \gamma \in \{ g g_{p+1} g^{-1} \mid g \in M \} \text{ and}$$

$$\gamma_j B \cap \gamma_k B = \emptyset, \quad \text{for } j, k = 1, \ldots, n_0, \, j \neq k.$$

As in the first part of the proof of Lemma 2.2 in [3] we find $g_{11}, \ldots, g_{1n_1} \in G$ with

(10)
$$\left\| (1/n_1) \sum_{k_1=1}^{n_1} u(g_{1k_1}) \tilde{Y}_{p+1} u(g_{1k_1})^* \right\| \leq \varepsilon/r.$$

Putting this together with (9) we see that

$$\left\| (1/nn_1) \sum_{i=1}^{n} \sum_{k_1=1}^{n_1} u(g_{1k_1}) u(h_i) (\tilde{Y} + Y_{p+1}) u(h_i)^* u(g_{1k_1})^* \right\| \leq$$

$$\leq \left\| (1/n) \sum_{i=1}^{n} u(h_i) \tilde{Y} u(h_i)^* \right\| + \left\| (1/n_1) \sum_{k_1=1}^{n_1} u(g_{1k_1}) \tilde{Y}_{p+1} u(g_{1k_1})^* \right\| \leq (p+1)\varepsilon/r.$$

By an easy induction argument, we find $g_1, \ldots, g_N \in G$ such that (8) is fulfilled.

Assume finally that $G/G_1$ contains a free normal subgroup with trivial centralizer. Note that $\mathbf{supp} \sum_{k=1}^{n} u(h_k)(Y - \tilde{Y}) u(h_k)^* \subset G \setminus G_1$. The lemma applies and there are averaging processes $(\theta_n)_{n \geq 1}$ on $G$ such that

$$\lim_{k \to \infty} \| \theta_k (1/n) \sum_{k=1}^{n} u(h_k)(Y - \tilde{Y})u(h_k)^*) \| = 0,$$

from which we find $\gamma_1, \ldots, \gamma_m \in G$ with

$$(11) \qquad \| (1/m) \sum_{i=1}^{m} u(\gamma_i)((1/n) \sum_{k=1}^{n} u(h_k)(Y - \tilde{Y})u(h_k)^*)u(\gamma_i)^* \| \leq (r - p)\epsilon/r.$$

From (9) and (11) we see that

$$\| (1/nm) \sum_{i=1}^{m} \sum_{k=1}^{n} u(\gamma_i h_k)Yu(\gamma_i h_k)^* \| \leq \| (1/n) \sum_{k=1}^{n} u(h_k)\tilde{Y}u(h_k)^* \| +$$

$$+ \| (1/m) \sum_{i=1}^{m} u(\gamma_i)((1/n) \sum_{k=1}^{n} u(h_k)(Y - \tilde{Y})u(h_k)^*)u(\gamma_i)^* \| \leq p\epsilon/r + (r - p)\epsilon/r = \epsilon.$$

Letting $N = nm$, one obtains (8). $\qquad\qquad\qquad\qquad\qquad\qquad\qquad\qquad$ $\Box$

**REMARK.** The previous theorem is still true if in ii), iii) or iv) we assume only that $G_i$ contains a set of non-abelian, free, normal subgroups $\{F_j\}_{j \in J}$ (J countable), with $\bigcap_{j \in J} H_j = \{1\}$, where $H_j$ is the centralizer of $F_j$ in $G_i$ (using same arguments and Corollary 5 from [1]).

The two types of combinatorial properties of groups considered above are very useful in studying groups $G$ with $C_r^*(G)$ simple with unique trace. In several concrete cases it is easy to show that the group satisfies one of this conditions, but it is more difficult to decide the validity of the other.

**EXAMPLES.** 1) Consider the non-trivial automorphisms $\theta$ of $\mathbf{Z}_3$ of order 2, the action $\alpha : \mathbf{F}_2 \to \mathbf{Aut}(\mathbf{Z}_3)$, $\alpha_s = \mathbf{id}$, $\alpha_t = \theta$ (s, t are the generators of $\mathbf{F}_2$) and let G be the semi-direct product of $\mathbf{Z}_3$ by $\mathbf{F}_2$ relative to $\alpha$.

It is obvious that $\mathbf{F}_2$ is a free subgroup of G with trivial centralizer, of finite index $([G : \mathbf{F}_2] = 3)$, but $C_r^*(G)$ is not simple and has at least two traces because $\mathbf{Z}_3$ is normal in G and amenable (see [8, Proposition 1.6]). This shows us that in Theorem 3 from [1], the normality condition is essential. Also, Questions (1) and (3) in [4, page 234, respectively page 239] remains open only for *normal* subgroups.

2) The direct product $\mathbf{F}_2 \times \mathbf{F}_2$ is clearly a weak Powers group ([3, Proposition 1.4]) which is not free ([5, Observation at page 177]). Actually, $\mathbf{F}_2 \times \mathbf{F}_2$ does not contain a free normal subgroup with trivial contralizer. More precisely, we prove the following result:

**PROPOSITION.** *Let $G_1$ and $G_2$ be non-trivial groups without torsion, and with trivial center. Then their direct product $G = G_1 \times G_2$ does not contain a free normal subgroup with trivial centralizer.*

PROOF. Assume that G contains a free normal subgroup H with trivial centralizer. For any element $x \in G$ we denote $\langle x \rangle = \{gxg^{-1} | g \in G\}$. Since H is normal, it is clear that $H = \coprod_{i \in I} \langle h_i \rangle \times \langle h'_i \rangle$, where $h_i \in G_1$, $h'_i \in G_2$, $\forall i \in I$.

Let h, h' $\in$ G such that $\langle h \rangle \times \langle h' \rangle \subset H$. Since $G_1$ and $G_2$ have trivial center, we can find, for $h \neq 1$, $h' \neq 1$, some $y \in \langle h \rangle$, $y' \in \langle h' \rangle$ such that $y \neq h$, $y' \neq h'$. Note that $(yh^{-1},1) = (y,h')(h,h')^{-1}$ and $(1,y'h'^{-1})$ are elements of the free group H which commute, hence it follows (see e.g. [5]) that there exists $(w_1,w_2) \in H$, $(w_1,w_2) \neq (1,1)$, m, n $\in$ **N** such that $(yh^{-1},1) = (w_1,w_2)^m$ and $(1,y'h'^{-1}) = (w_1,w_2)^n$. These relations implies that $w_1^n = 1$, $w_2^m = 1$, hence $w_1 = 1$, $w_2 = 1$, which is a contradiction. We conclude that $h = 1$ or $h' = 1$. A similar argument shows that H does not contain two conjugacy classes $\langle h \rangle \times \{1\}$ and $\{1\} \times \langle h' \rangle$, with $h \neq 1$, $h' \neq 1$, hence $H \subset G_1 \times \{1\}$ or $H \subset \{1\} \times G_2$. In both cases, H has non-trivial centralizer. $\square$

It is easy to see that the last statement is still true for the weak infinite product $G = G_1 \times G_2 \times \ldots$ . Taking $G_i$ weak Powers groups without torsion, we obtain the weak Powers group G ([3]) which is weak commutative (for every finite set $F \subset G \setminus \{1\}$, there exists a $g \in G \subset \{1\}$ such that gf = fg, $\forall f \in F$), hence it is inner amenable ([7, Lemma 6.1.1]).

3) P. de la Harpe has pointed that $PSL_2(\mathbf{Q})$ is a simple Powers group, hence this is a group which do not belong to the Akemann - Lee class.

Unfortunatelly, we have not yet an example of a group which satisfies the Akemann-Lee conditions and is not a Powers group.

## REFERENCES


1.  **Akemann, C.A. ; Lee, T.Y. :** Some simple $C^*$-algebra associated with free groups, *Indiana Univ. Math. J.* **29**(1980), 505-511.

2.  **Akemann, C.A. ; Ostrand, P. :** Computing norms in group $C^*$-algebras, *Amer. J. Math.* **98**(1976), 1015-1047.

3.  **Boca, F. ; Niţică, V. :** Combinatorial properties of groups and simple $C^*$--algebras with a unique trace, *J. Operator Theory* **20**(1988), 183-196.

4.  **de la Harpe, P. :** Reduced $C^*$-algebras of discrete groups which are simple with a unique trace, in *Operator algebras and their connections with topology and ergodic theory*, Lecture Notes in Math., vol.**1132**, Springer-Verlag, 1983, pp. 230-253.

5.      **Lyndon, R.; Schupp, P.** : Combinatorial group theory, Springer-Verlag, 1977.

6.      **Magnus, W. ; Karrass, A. ; Solitar, D.** : *Combinatorial group theory* (Russian), Nauka, 1974.

7.      **Murray, F. ; von Neumann, J.** : Rings of operators. IV, *Ann. of Math.* **44**(1943), 716–808.

8.      **Paschke, W. ; Salinas, N** : $C^*$-algebras associated with free products of groups, *Pacific J. Math.* **82**(1979), 211–221.

9.      **Powers, R.** : Simplicity of the $C^*$-algebras associated with the free group on two generators, *Duke Math.J.* **42**(1975), 151–156.

**Florin Boca** and **Viorel Niţică**

Department of Mathematics, INCREST
Bdul Păcii 220, 79622 Bucharest

Romania.

Operator Theory:
Advances and Applications, Vol. 43
© 1990 Birkhäuser Verlag Basel



# THE KREIN SPACE $(\phi H^2)^{\perp}$ AND COEFFICIENTS OF ANALYTIC FUNCTIONS

**Douglas N. Clark** [1]

For $H^2 = H^2(U)$ the classical Hardy space of the unit disk and for $B(z)$ an inner function, the Hilbert space $BH^{2\perp} = H^2 \ominus BH^2$ is familiar to operator theorists as the basic space in the Sz.-Nagy–Foiaş model theory of contractions, [6].

The spaces $\phi H^{2\perp}$ in the title refer to the case where $\phi$ is a quotient of inner functions

$$\phi(z) = B(z)/C(z).$$

These generalizations of $BH^{2\perp}$ arise in the study of analytic functions on certain varieties in $U^N$ to be discussed below. No analogue of the representation $H^2 \ominus BH^2$ is available, since $\phi H^2 \not\subset H^2$. However, by noting that the reproducing kernels for $BH^{2\perp}$ are the functions

$$h_\zeta(z) = [1 - \overline{B(\zeta)}B(z)](1 - \overline{\zeta}z)^{-1}, \quad |\zeta| < 1,$$

we can require that the space $\phi H^{2\perp}$ be spanned by functions of the form

$$k_\zeta(z) = [1 - \overline{\phi(\zeta)}\phi(z)](1 - \overline{\zeta}z)^{-1},$$

(where $\zeta \in U$ is not a pole of $\phi$), with inner product $(\,,\,)_\phi$ satisfying

$$(1) \qquad (k_\zeta, k_\xi)_\phi = k_\zeta(\xi) = [1 - \overline{\phi(\zeta)}\phi(\xi)](1 - \overline{\zeta}\xi)^{-1}.$$

If we take $\zeta \in U$ such that $|\phi(\zeta)| > 1$, (1) implies $(k_\zeta, k_\zeta)_\phi < 0$. Therefore a sesquilinear form satisfying (1) can at best be an *indefinite* inner product on the span of $\{k_\zeta\}$.

A Krein space spanned by functions $k_\zeta(z)$ satisfying (1) need not be unique. Indeed, taking the *closure* (of the span of the $k_\zeta$) in an indefinite norm does not make sense. Here is a simple construction of one Krein space spanned by $\{k_\zeta\}$ satisfying (1). Let

$$K = C(z)^{-1}[B(z)H^{2\perp} + C(z)H^{2\perp}]$$

---

[1] Partially supported by an N.S.F. grant.

with the (Hilbert space) inner product of

$$f_i(z) = C(z)^{-1}[g_i(z) + h_i(z)], \quad g_i \in BH^{2\perp}, \quad h_i \in CH^{2\perp}, \quad i = 1, 2,$$

given by

$$\langle f_1, f_2 \rangle = (g_1, g_2) + (h_1, h_2)$$

(where the inner products on the right refer to $H^2$). Then $\{k_\zeta\}$, for $\zeta \in U$ not a pole of $\phi$, span a dense subspace of $K$ and the indefinite inner product

$$(f_1, f_2)_\phi = (g_1, g_1) - (h_1, h_2)$$

satisfies (1) when $f_1 = k_\zeta$, $f_2 = k_\xi$.

To prove the statements in the last sentence, note

$$k_\zeta(z) = C(z)^{-1}\{\bar{C}(\zeta)^{-1}[1 - \bar{B}(\zeta)B(z)](1 - \bar{\zeta}z)^{-1} - \bar{C}(\zeta)^{-1}[1 - \bar{C}(\zeta)C(z)](1 - \bar{\zeta}z)^{-1}\} \in K$$

and hence, for $f = C^{-1}(g + h) \in K$

$$\langle f, k_\zeta \rangle = \langle C^{-1}(g + h), k_\zeta \rangle = C(\zeta)^{-1}[g(\zeta) - h(\zeta)].$$

Thus if $f \perp k_\zeta$, for all $\zeta \in U$ not a pole of $\phi$, then $g = h \in BH^{2\perp} \cap CH^{2\perp} = (BH^2 + CH^2)^\perp = \{0\}$, if $B$ and $C$ are relatively prime inner functions. This proves $\{k_\zeta\}$ spans $K$. To prove (1), we have

$$(k_\zeta, k_\xi)_\phi = \bar{C}(\zeta)^{-1}C(\xi)^{-1}\{([1 - \bar{B}(\zeta)B(z)](1 - \bar{\zeta}z)^{-1}, [1 - \bar{B}(\xi)B(z)](1 - \xi z)^{-1}) -$$

$$-([1 - \bar{C}(\zeta)C(z)](1 - \bar{\zeta}z)^{-1}, [1 - \bar{C}(\xi)C(z)](1 - \bar{\xi}z)^{-1})\} = \bar{C}(\zeta)^{-1}C(\xi)^{-1}\{1 - \bar{B}(\zeta)B(\xi) -$$

$$- 1 + \bar{C}(\zeta)C(\xi)\}(1 - \bar{\zeta}\xi)^{-1} = k_\zeta(\xi).$$

In § 1 we generalize results proved for the case $BH^{2\perp}$ in [3], where $B$ and $C$ are finite Blaschke products. In § 2, we explain how finite dimensional results from [3] were applied in a recent paper [4] to analytic functions on varieties in $U^N$ and in § § 3,4, we prove new results for more general varieties, including an expansion theorem with coefficients in $\phi H^{2\perp}$ and a version of Cauchy's theorem.

It should be remarked that the spaces $\phi H^{2\perp}$ have arisen before in various generalizations of model theory. See Alpay and Dym [1] and Ball and Helton [2].

## 1. THE KREIN SPACE $\phi H^{2\perp}$

In order to explain the results of [3], let $B(z)$ be a finite Blaschke product.

Clearly the reproducing kernels $h_\zeta$ are analytic in the entire plane except at the poles of B. Thus, if $\lambda$ is a complex number of modulus 1, and if $\eta_1, \ldots, \eta_n$ are the points on $|z| = 1$ satisfying $B(\eta_j) = \lambda$, $j = 1, \ldots, n$, we have

$$(h_{\eta_i}, h_{\eta_j}) = \overline{\lambda}\eta_i B'(\eta_i)\delta_{ij} = |B'(\eta_i)|\,\delta_{ij}$$

so that $\{h_{\eta_j}\}$ form an orthogonal set in $BH^{2\perp}$. In particular, the operator of restriction to $\{\eta_j\}$ is a unitary map from $BH^{2\perp}$ onto $L^2(d\mu_\lambda)$, where $d\mu_\lambda$ is the purely atomic measure with masses

$$d\mu_\lambda(\{\eta_j\}) = |B'(\eta_j)|^{-1}, \quad j = 1, 2, \ldots, n.$$

Thus we have, for $f \in BH^{2\perp}$ and $\zeta \in U$,

$$f(\zeta) = (f, h_\zeta) = \int f(z)[1 - B(\zeta)\overline{B}(z)](1 - \zeta\overline{z})^{-1}d\mu_\lambda(z) =$$

(2)

$$= \sum_j f(\eta_j)[1 - B(\zeta)\overline{\lambda}](1 - \zeta\overline{\eta}_j)^{-1}|B'(\eta_j)|^{-1},$$

or, for f, $g \in BH^{2\perp}$,

$$(f,g) = \sum_j f(\eta_j)\overline{g}(\eta_j)|B'(\eta_j)|^{-1}.$$

It is possible to write down the analogue of (2) for $\phi$ a quotient of finite Blaschke products. Indeed, if $f \in \phi H^{2\perp}$ (which is unambiguously the linear span of $\{k_\zeta\}$, if B and C are finite Blaschke products), $|\lambda| = 1$ and $\zeta \in U$ is not a pole of $\phi$, then $f(z)(\phi(z) - \lambda)^{-1}$ is a rational function with poles only at the points $\{\eta_j\}$ where $\phi(\eta_j) = \lambda$. Expanding this function in partial fractions, we have

$$f(z)(\phi(z) - \lambda)^{-1} = \sum_j f(\eta_j)\phi'(\eta_j)^{-1}(z - \eta_j)^{-1}$$

so that

$$f(z) = \sum_j f(\eta_j)[\phi(z) - \lambda](z - \eta_j)^{-1}\phi'(\eta_j)^{-1},$$

or, if f, $g \in \phi H^{2\perp}$,

(3)

$$(f,g)_\phi = \sum_j f(\eta_j)\overline{g}(\overline{\eta}_j^{-1})\lambda\eta_j^{-1}\phi'(\eta_j)^{-1},$$

The sum in (3) is, of course, over all $\eta_j$ *in the plane* with $\phi(\eta_j) = \lambda$. The relation (3) does not quite give the Krein space inner product on $\phi H^{2\perp}$ as inner product with respect to a complex measure. Indeed inner product with respect to a nonreal complex measure does not give an inner product satisfying

$$(f,g) = \overline{(g,f)} \ .$$

This explains the necessity for $\overline{g}(\eta_j^{-1})$ instead of $\overline{g}(\eta_j)$ in (3).

**PROBLEM.** In [3], I obtained a generalization of the unitary restriction map from $BH^{2\perp}$ onto $L^2(d\mu_\lambda)$ to the case where B is an arbitrary inner function. The measure $d\mu_\lambda$, for the generalization, is defined, for an appropriate constant c, by

$$(4) \qquad [\lambda + B(z)]/[\lambda - B(z)] = c + \int_0^{2\pi} (e^{i\theta} + z)/(e^{i\theta} - z)d\mu_\lambda(z).$$

Is there a generalization of (3) to an integral formula, in case $\phi$ is a quotient of arbitrary inner functions? The representation (4) for an inner function B is, of course, just a consequence of the fact that a positive harmonic function lies in $h^1$. But this slick result does little to *explain* why the real part of the left side of (4) has $L^1$ type growth and gives no insight about $[\lambda + \phi(z)]/[\lambda - \phi(z)]$.

## 2. COEFFICIENTS OF ANALYTIC FUNCTIONS

In a recent papaer [4] dealing with analytic functions on a variety in $U^N$, I had occasion to apply the representation (2) for $BH^{2\perp}$ functions.

Let $V_o = \{(z_1,\ldots,z_N) \in U^N : B_1(z_1) = \ldots = B_N(z_N)\}$ where $B_1,\ldots,B_N$ are finite Blaschke products. In [4, Corollary 2.1], it was proved that a function $f(z_1,\ldots,z_N)$, analytic in $V_o$, has a convergent expansion of the form

$$f(z_1,\ldots,z_N) = \sum_{\nu=0}^{\infty} f_\nu(z_1,\ldots,z_N)B_1(z_1)^\nu,$$

where $f_\nu \in B_1H^{2\perp} \otimes \ldots \otimes B_NH^{2\perp}$. Now the representation (2) gives a Cauchy formula [4, Theorem 3.1], for $f(z_1,\ldots,z_N)$ analytic in $V_o$ and satisfying suitable growth conditions. Indeed, if $d\mu_\lambda$ is the measure with point masses

$$d\mu_\lambda(\{\eta_{1k}(\lambda),\ldots,\eta_{Nk}(\lambda)\}) = |B_1'(\eta_{1k})\ldots B_N'(\eta_{Nk})|^{-1}$$

at the points where $B_1(\eta_{1k}(\lambda)) = \ldots = B_N(\eta_{Nk}(\lambda)) = \lambda$, then, since $B_1(z_1) = \lambda$ a.e. $(d\mu_\lambda)$,

$$\int f(z_1,\ldots,z_N)\prod_j [1 - B_j(w_j)\overline{B}_j(z_j)](1 - w_j\overline{z}_j)^{-1}d\mu_\lambda(z_1,\ldots,z_N) =$$

$$= \sum_{\nu=0}^{\infty} \lambda^\nu \int f_\nu(z_1,\ldots,z_N)\prod_j [1 - B_j(w_j)\overline{B}_j(z_j)](1 - w_j\overline{z}_j)^{-1}d\mu_\lambda = \sum_{\nu=0}^{\infty} f_\nu(w_1,\ldots,w_N)\lambda^\nu,$$

so that, by Cauchy's formula for the disk,

$$f(w_1,\ldots,w_N) = \int_0^{2\pi} \int f(z_1,\ldots,z_N)\prod_j[1 - B_j(w_j)\overline{B}_j(z_j)](1 - w_j\overline{z}_j)^{-1} \times$$

$$\times (1 - e^{-i\theta}B_1(z_1))^{-1}d\mu_{e^{i\theta}}(z)d\theta.$$

The growth conditions on f which justify the use of Cauchy's formula for U will be made explicit in the next sections.

## 3. ANALYTIC FUNCTIONS WITH COEFFICIENTS IN $\phi H^{2\perp}$

We return now to the case where $\phi_1,\ldots,\phi_N$ are quotients of finite Blaschke products

$$\phi_j(z_j) = B_j(z_j)/C_j(z_j) \quad j = 1,\ldots,N,$$

and we consider the variety V in $U^N$ defined by

$$V = \{(z_1,\ldots,z_N) \in U^N : \phi_1(z_1) = \ldots = \phi_N(z_N)\}.$$

We deal first with

$$V_+ = \{(z_1,\ldots,z_N) \in V : \phi_1(z_1) \in U\}.$$

The relation with the spaces $\phi_j H^{2\perp}$ is contained in

**THEOREM.** *If* $f(z_1,\ldots,z_N)$ *is analytic in* $V_+$, *then f has an expansion of the form*

(5)
$$f(z_1,\ldots,z_N) = \sum_{\nu=0}^{\infty} f_\nu(z_1,\ldots,z_N)\phi_1(z_1)^\nu,$$

*where* $f_\nu \in \phi_1 H^{2\perp} \otimes \ldots \otimes \phi_N H^{2\perp}$. *The right side of (5) converges uniformly on compact subsets of* $V_+$.

In proving the theorem we shall make use of certain subspaces of $\phi_1 H^{2\perp} \otimes \ldots \otimes \phi_N H^{2\perp}$. Let $r$ $(0 < r \le 1)$ be such that $\phi_j(z_j)$ and $\phi_j(z_j/r)$ have the same number of zeros and the same number of poles in $|z_j| < 1$, for $j = 1,\ldots,N$. Let $\phi_j^r(z_j)$ be the quotient of Blaschke products whose zeros in the plane are r times the zeros of $\phi_j$:

$$\phi_j(z_j) = \prod_k (z_j - b_{jk})/(1 - \overline{b}_{jk}z_j), \qquad \phi_j^r(z_j) = \prod_k(z_j - rb_{jk})/(1 - r\overline{b}_{jk}z_j).$$

By assumption, $\phi_j$ and $\phi_j^r$ have the same number of zeros and the same number of poles in U, and hence in $|z_j| > 1$. Let $M_j^r$ be the subspace of $\phi_j^r H^{2\perp}$ spanned by

(6) $$[\phi_j(rz_j) - \lambda]/[rz_j - \eta_{jk}(\lambda)]$$

where $\eta_{j1}, \ldots, \eta_{jm}$ are the $\lambda$-points of $\phi_j$ in $|z_j| < 1$. The inclusion $M_j^r \subset \phi_j^r H^{2\perp}$ follows from the observation that a linear combination of functions (6) has the form

$$p(z_j)/\Pi(1 - r\bar{b}_{jk}z_j)$$

where $b_{j1}, \ldots, b_{jN}$ are the zeros of $\phi_j$ (in the entire plane) and $p$ is a polynomial of degree at most $n - 1$. The corresponding characterization of $\phi H^{2\perp}$ follows from its representation as $K = C^{-1}(BH^{2\perp} + CH^{2\perp})$ given above.

Let $B_{jr}(z_j)$ denote the finite Blaschke product with zeros $r^{-1}\eta_{j1}, \ldots, r^{-1}\eta_{jm}$ (where we now assume that $r$ is so close to 1 that $|\eta_{jk}(\lambda)| < r$, $j = 1, \ldots, N$, $k = 1, \ldots$ $\ldots, m$). The space $B_{jr}H^{2\perp}$ is spanned by the set of functions of the form

(7) $$B_{jr}(z_j)/[rz_j - \eta_{jk}(\lambda)].$$

From the forms of these spanning sets, (6), (7), the space $B_{jr}H^{2\perp}$ is seen to be mapped onto $M_j^r$ by the operator of multiplication by

$$e_j(z_j) = [\phi_j(rz_j) - \lambda]/B_{jr}(z_j).$$

Incidentally, $C_j(rz_j)e_j$ being the outer part of the $H^1$ function $B_j(rz_j) - \lambda C_j(rz_j)$, we have

(8) $$e_j(z_j) = C_j(rz_j)^{-1}\exp \int \log |B_j(re^{it}) - \lambda C_j(re^{it})|(e^{it} + z_j)/(e^{it} - z_j)dt.$$

Next we introduce

$$h(z_1, \ldots, z_N; \lambda) = \sum_k f(\eta_k) \prod_{j=1}^{N} [(\phi_j(z_j) - \lambda)/(z_j - \eta_{jk}(\lambda))][1/\phi_j'(\eta_{jk}(\lambda))]$$

where the sum is over all $(\eta_{1k}(\lambda), \ldots, \eta_{Nk}(\lambda))$ such that $|\eta_{jk}(\lambda)| < 1$ and $\phi_j(\eta_{jk}(\lambda)) = \lambda$. Because $h$ is a symmetric function of $(\eta_{1k}(\lambda), \ldots, \eta_{Nk}(\lambda)) \in V_+$, it is single valued for $(z_1, \ldots, z_N; \lambda) \in V_+ \times U$, except possibly at those points for which $\phi_j'(\eta_{jk}(\lambda)) = 0$ for some $j, k$.

**LEMMA.** *As a function of* $\lambda$, $h(z_1, \ldots, z_N; \lambda)$ *extends to a holomorphic* $\phi_1 H^{2\perp} \otimes \ldots \otimes \phi_N H^{2\perp}$ *-valued function in* U. *For fixed* $\lambda \in U$, $h(z_1, \ldots, z_N; \lambda) \in M_1^1 \otimes \ldots$

$\ldots \otimes M_N^1$ *and for* $(z_1, \ldots, z_N) \in V_+$ *we have*

(9)
$$h(z_1, \ldots, z_N; \phi_1(z_1)) = f(z_1, \ldots, z_N).$$

PROOF. Clearly for fixed $\lambda$, $h(z_1, \ldots, z_N; \lambda) \in M_1^1 \otimes \ldots \otimes M_N^1 \subset \phi_1 H^{2\perp} \otimes \ldots \otimes \phi_N H^{2\perp}$. Let $U_1$ be the set of $\lambda \in U$ such that $\phi_j'(\eta_{jk}(\lambda)) \neq 0$ for all j, k. Then for each $\lambda \in U_1$, $h(z_1, \ldots, z_N; \lambda)$ is the $M_1^1 \otimes \ldots \otimes M_N^1$ function equal to $f(z_1, \ldots, z_N)$ at the points $\{\eta_k(\lambda)\}$.

Let f be analytic in $V_+$. We have to show that $h(z_1, \ldots, z_N; \lambda)$ has a removable singularity at each $\lambda \in U \setminus U_1$. By a theorem of Cartan [5, Theorem 7.1.2], f extends to an analytic function (again denoted $f(z_1, \ldots, z_N)$), analytic in $U^N$. For r < 1,

$$f_r(z_1, \ldots, z_N) = f(rz_1, \ldots, rz_N)$$

is therefore in $H^2(U^N)$. The function

$$h_r(z_1, \ldots, z_N; \lambda) = h(rz_1, \ldots, rz_N; \lambda)$$

satisfies

$$h_r(r^{-1}\eta_{1k}(\lambda), \ldots, r^{-1}\eta_{Nk}(\lambda); \lambda) = h(\eta_{1k}(\lambda), \ldots, \eta_{Nk}(\lambda); \lambda) =$$

$$= f(\eta_{1k}(\lambda), \ldots, \eta_{Nk}(\lambda)) = f_r(r^{-1}\eta_{1k}(\lambda), \ldots, r^{-1}\eta_{Nk}(\lambda))$$

and therefore $h_r$ is the $M_1^r \otimes \ldots \otimes M_N^r$ – valued function (of $\lambda$) agreeing with $f_r$ at the points $\{r^{-1}\eta_k(\lambda)\}$. If $P_r$ is the projection from $H^2(U^N)$ on $B_{1r}H^{2\perp} \otimes \ldots \otimes B_{Nr}H^{2\perp}$, then for $g \in H^2(U^N)$, $P_r g$ agrees with g at the points $\{(r^{-1}\eta_{1k}(\lambda), \ldots, r^{-1}\eta_{Nk}(\lambda))\}$ and therefore the function

$$e_1(z_1) \ldots e_N(z_N) P_r e_1(z_1)^{-1} \ldots e_N(z_N)^{-1} f_r(z_1, \ldots, z_N)$$

lies in $M_1^r \otimes \ldots \otimes M_N^r$ and agrees with f at $\{(r^{-1}\eta_{1k}(\lambda), \ldots, r^{-1}\eta_{Nk}(\lambda))\}$. That is

$$h_r(z_1, \ldots, z_N; \lambda) = e_1(z_1) \ldots e_N(z_N) P_r e_1(z_1)^{-1} \ldots e_N(z_N)^{-1} f_r(z_1, \ldots, z_N).$$

By the characterization (8) of $e_j(z_j)$ and by

$$P_r g = \int \ldots \int g(e^{it_1}, \ldots, e^{it_N}) \Pi(1 - B_{jr}(z_j)\bar{B}_{jr}(e^{it_j}))/(1 - e^{-it_j}z_j) \, dt_1 \ldots dt_N$$

we see that $h_r$ extends continuously to each $\lambda \in U \setminus U_1$. Since $h_r$ is therefore analytic in $\lambda \in U$, for any r < 1, it follows that h is analytic in $\lambda \in U$.

The relation (9) follows from the fact that for fixed $\lambda, h(z_1, \ldots, z_N; \lambda) =$

$$= f(z_1, \ldots, z_N) \text{ if } \phi_1(z_1) = \ldots = \phi_N(z_N) = \lambda.$$

PROOF OF THEOREM. By the lemma, $h(z_1, \ldots, z_N; \lambda)$ is an analytic $\phi_1 H^{2\perp} \otimes \ldots \otimes \phi_N H^{2\perp}$ - valued function of $\lambda$ in $|\lambda| < 1$. Thus

$$h(z_1, \ldots, z_N; \lambda) = \sum_{\nu=0}^{\infty} f_\nu(z_1, \ldots, z_N) \lambda^\nu$$

where $f_\nu \in \phi_1 H^{2\perp} \otimes \ldots \otimes \phi_N H^{2\perp}$. Setting $\lambda = \phi_1(z_1)$, we have by (9),

$$f(z_1, \ldots, z_N) = h(z_1, \ldots, z_N; \phi_1(z_1)) = \sum_{\nu=0}^{\infty} f_\nu(z_1, \ldots, z_N) \phi_1(z_1)^\nu.$$

## 4. CAUCHY FORMULA FOR V

To obtain a Cauchy formula for $f(z_1, \ldots, z_N)$ analytic in $V_+$ we note first that the $H^1$ condition

$$(10) \qquad \sum_k \int_0^{2\pi} |f(\eta_{1k}(re^{i\theta}), \ldots, \eta_{Nk}(re^{i\theta}))| \, \Pi \, |\phi_j'(\eta_{jk}(re^{i\theta}))|^{-1} d\theta \leq c, \quad 0 \leq r < 1$$

where the sum is over all $(\eta_{1k}, \ldots, \eta_{Nk}) \in V_+$, implies that the integrals

$$\int_0^{2\pi} \|h(z_1, \ldots, z_N; re^{i\theta})\|_K d\theta$$

are bounded for $0 \leq r < 1$; i.e., that the $\phi_1 H^{2\perp} \otimes \ldots \otimes \phi_N H^{2\perp}$ - valued function h is of class $H^1_K(U)$, $K = \phi_1 H^{2\perp} \times \ldots \otimes \phi_N H^{2\perp}$. Thus if $\lambda \in U$ and $(z_1, \ldots, z_N) \in U^N$, we have

$$h(z_1, \ldots, z_N; \lambda) = \int_0^{2\pi} h(z_1, \ldots, z_N; e^{i\theta})(1 - \lambda e^{-i\theta})^{-1} d\theta.$$

Since, by the lemma of §3, $h(z_1, \ldots, z_N; e^{i\theta})$ is $M_1^1 \otimes \ldots \otimes M_N^1$ – valued, (3) shows that

$$h(z_1, \ldots, z_N; \lambda) = \int_0^{2\pi} \sum_k f(\eta_k(e^{i\theta})) \Pi_j [(\phi_j(z_j) - e^{i\theta})/(z_j - \eta_{jk}(e^{i\theta}))][1/(\phi_j'(\eta_{jk}))](1 - \lambda e^{i\theta})^{-1} d\theta$$

where the sum is over $(\eta_{1k}, \ldots, \eta_{Nk}) \in \partial V_+$. Thus

$$(11) \qquad f(z_1, \ldots, z_N) = h(z_1, \ldots, z_N; \phi_1(z_1)) =$$

$$= \int_0^{2\pi} \sum_k f(\eta_k(e^{i\theta})) \Pi_j [(\phi_j(z_j) - e^{i\theta})/(z_j - \eta_{jk}(e^{i\theta}))][1/\phi_j'(\eta_{jk})](1 - \phi_1(z_1)e^{-i\theta})^{-1} d\theta$$

which is our Cauchy formula for $V_+$.

Replacing $\phi_j(z_j)$ by $\psi_j(z_j) = 1/\phi_j(z_j)$, and writing the analogue of (11) for f analytic in $V_-$, one gets

$$f(z_1,\ldots,z_N) =$$

$$= \int_0^{2\pi} \sum_k f(\zeta_k(e^{i\theta}))\prod_j [(\psi_j(z_j) - e^{i\theta})/(z_j - \zeta_{jk}(e^{i\theta}))][1/\psi_j'(\zeta_{jk})])(1 - \psi_1(z_1)e^{-i\theta})^{-1}d\theta$$

where the sum is over $(\zeta_{1k},\ldots,\zeta_{Nk})$ satisfying $\psi_j(\zeta_{jk}(\lambda)) = \lambda$, lying in $\partial V_-$. Since $\phi_j(\zeta_{jk}(\lambda)) = \lambda^{-1}$, we have $\zeta_{jk}(\lambda^{-1}) = \eta_{jk}(\lambda)$; and since $\phi_j'(z_j) = -\phi_j(z_j)^{-2}\phi_j'(z_j)$, we have $\psi_j'(\zeta_{jk}(\lambda)) = -\lambda^2\psi_j'(\zeta_{jk})$. Thus we have for $(z_1,\ldots,z_N) \in V_-$

$$f(z_1,\ldots,z_N) =$$

$$(12)\quad = \int_0^{2\pi} \sum_k f(\eta_k(e^{i\theta}))\prod_j [(\psi_j(z_j) - e^{i\theta})/(z_j - \eta_{jk}(e^{i\theta}))][(-\phi_j(\eta_{jk})^2/\phi_j'(\eta_{jk})])(1 - e^{i\theta}/\phi_1(z_1))^{-1}d\theta =$$

$$= -\int_0^{2\pi} \sum_k f(\eta_k(e^{i\theta}))\prod_j [(\phi_j(z_j) - e^{i\theta})/(z_j - \eta_{jk}(e^{i\theta}))][1/\phi_j'(\eta_{jk})](1 - e^{-i\theta}\phi_1(z_1))^{-1}d\theta.$$

If $f(z_1,\ldots,z_N)$ is analytic in V and satisfies

$$\sum_k \int_0^{2\pi} [\,|f(\eta_k(re^{i\theta}))|\,\prod_j|\phi_j'(\eta_{jk}(re^{i\theta}))|^{-1} + |f(\eta_k(r^{-1}e^{-i\theta}))\prod_j|\,|\phi_j'(\eta_{jk}(r^{-1}e^{-i\theta}))|^{-1}]d\theta \leq c$$

for $0 < r < 1$, where the sum is over $\{\eta_k\} \subset V_+$, we have that the right side of (11) is 0 for $(z_1,\ldots,z_N) \in V_-$ and that the right side of (12) is 0 for $(z_1,\ldots,z_N) \in V_+$. We thus obtain the Cauchy formula for f analytic in V by adding (11) and (12). When we do this, the terms in (11) and (12) on $\partial V_+ \cap \partial V_-$ cancel and we have

$$f(z_1,\ldots,z_N) =$$

$$= \int_0^{2\pi} \sum_k f(\eta_k(e^{i\theta}))\prod_j [(\phi_j(z_j) - e^{i\theta})/(z_j - \eta_{jk}(e^{i\theta}))][1/(\phi_j'(\eta_{jk})](1 - \phi_1(z_1)e^{-i\theta})^{-1}d\theta$$

where the sum is over all $(\eta_{1k},\ldots,\eta_{Nk}) \in \partial V$.

## REFERENCES

1.      **Alpay, D. ; Dym, H.** : On applications of reproducing kernel spaces to the Schur algorithm and rational J-unitary factorization, in *I. Schur methods in operator theory and signal processing*, Birkhäuser-Verlag, 1986, pp. 89-159.

2.      **Ball, J.A. ; Helton, J.W.** : Shift   invariant   subspaces,   passivity,   reproducing

kernels and $H^{\infty}$-optimization, to appear.

3.  **Clark, D.N.** : One dimensional perturbations of restricted shifts, *J. Analyse Math.* **25**(1972), 169-191.

4.  **Clark, D.N.** : Restrictions of $H^p$ functions in the polydisk, *Amer. J. Math.,* **110** (1988), 1119-1152.

5.  **Rudin, W.** : *Function theory in polydisks*, W.A. Benjamin, 1969.

6.  **Sz.-Nagy, B. ; Foiaş, C.** : *Harmonic analysis of operators on Hilbert space*, Akademiai Kiado, 1970.

**Douglas N. Clark**

Department of Mathematics
University of Georgia
Athens, Georgia 30602
U.S.A.

Operator Theory:
Advances and Applications, Vol. 43
© 1990 Birkhäuser Verlag Basel



# COMPLETIONS AND EXTENSIONS

Tiberiu Constantinescu

The aim of this note is to explain the interplay between completions and extensions at its most elementary — but also generic — level. That is, as completion problem we take into account the following:

Question A
$\left\{\begin{array}{l}\text{"Given operators A, B, C it is required to find conditions for the}\\ \text{existence of an operator D such that the block-matrix } \begin{bmatrix} A & B \\ C & D \end{bmatrix} \text{ becomes a}\\ \text{contraction, and to determine all the solutions D".}\end{array}\right.$

– for applications and solutions, see [13], [18], [2], [17], [7], [10], [12], [11], [8]. As extension problem, we consider the following:

Question B
$\left\{\begin{array}{l}\text{"Given Hilbert spaces } G_0 \text{ and } G_1, E_1 \text{ a closed subspace of } G_1, F_0 \text{ a closed}\\ \text{subspace of } G_0 \text{ and } w_0 \text{ an isometry of } E_1 \text{ onto } F_0, \text{ it is required to decide}\\ \text{whether there exist unitary operators } W_0 \in L(K_1, K_0), K_1 \supset G_1, K_0 \supset G_0,\\ \text{extending } w_0 \text{ and satisfying a certain minimality condition, and to}\\ \text{determine the class of all these extensions".}\end{array}\right.$

– for solutions and applications, see [16], [14], [1], [15], [4], [20].

The existence part of Question B is almost obvious and it turns out that Question A is a case of Question B. In this work we render explicit all the connections and the final picture is a concentration of methods developed in [1], [2], [5], [6], [4], [8].

Some of the results are scattered in [9], in a more general formulation, but we think an explicit and detailed treatment of the generic case may be useful.

**I.** In this section we show that Question A is equivalent with the lifting of a certain commutation property.

We take into consideration a contraction $T_0 \in L(H_1, H_0)$. $H_0$, $H_1$ being Hilbert spaces and $L(H_1, H_0)$ is the set of bounded linear operators acting between $H_1$ and $H_0$. The fact that $T_0$ is a contraction means that $\| T_0 \| \leq 1$, or equivalently that $I - T_0^* T_0$ is a positive operators in the sense that $((I - T_0^* T_0)h_1, h_1) \geq 0$ for all $h_1 \in H_1$, $(\cdot, \cdot)$ being the inner product in $H_1$. We define the defect operator $D_{T_o} = (I - T_o^* T_o)^{\frac{1}{2}}$ and the defect

space $D_{T_o} = \bar{R}(D_{T_o})$, where $R$ is the range of the underlying operators and $\bar{R}$ is its

closure. We attach to $T_o$ the following elements: two Hilbert spaces

$$(1.1) \qquad\qquad K_o = H_o \oplus D_{T_o}$$

$$(1.2) \qquad\qquad K_1 = D_{T_o^*} \oplus H_1$$

and the unitary operator

$$(1.3) \qquad
\begin{cases}
W_o : K_1 \;\rightarrow\; K_o \\[2mm]
W_o = \begin{bmatrix} D_{T_o^*} & T_o \\[2mm] -T_o^* & D_{T_o} \end{bmatrix}.
\end{cases}$$

This operator is the core of dilation theory. We will refer to it as the Kolmogorov decomposition of $T_o$, while

$$J(T_o) = \begin{bmatrix} T_o & D_{T_o^*} \\[2mm] D_{T_o} & -T_o^* \end{bmatrix} : \;
\begin{matrix} H_1 \\ \oplus \\ D_{T_o^*} \end{matrix} \rightarrow
\begin{matrix} H_o \\ \oplus \\ D_{T_o} \end{matrix}$$

is usually called the elementary rotation of $T_o$. Moreover, we define

$$(1.4) \qquad\qquad K_o^+ = H_o \oplus D_{T_o}$$

$$(1.5) \qquad\qquad K_1^+ = H_1$$

and the isometric operator

$$(1.6) \qquad
\begin{cases}
W_o^+ : K_1^+ \;\rightarrow\; K_o^+ \\[2mm]
W_o^+ = W_o \,|\, K_1^+ .
\end{cases}$$

Let us consider now two contractions $T_o \in L(H_1, H_o)$ and $T_o' \in L(H_1', H_o')$; $W_o \in L(K_1, K_o)$ and $W_o' \in L(K_1', K_o')$ are their Kolmogorov decompositions and for two contractions $A_o \in L(H_o, H_o')$ and $A_1 \in L(H_1, H_1')$ satisfying the commutativity relation $T_o' A_1 = A_o T_o$, we define the set:

$$\mathbf{CID}(\{A_o, A_1\}) = \{\{B_o, B_1\} \mid B_o,\ B_1 \text{ are contractions in } L(K_o^+, K_o'^+) \text{ and respecti-}$$
$$\text{vely } L(K_1^+, K_1'^+), \text{ such that } W_o'^+ B_1 = B_o W_o^+ \text{ and}$$
$$P_{H_o'} B_o = A_o P_{H_o},\ P_{H_1'} B_1 = A_1 P_{H1} \}.$$

Return now to Question A. Fix the operators $A \in L(M_1, N_1)$, $B \in L(M_2, N_1)$, $C \in L(M_1, N_2)$. In order to exist an operator $D \in L(M_2, N_2)$ such that $\begin{bmatrix} A & B \\ C & D \end{bmatrix}$ be a contraction, it is obviously necessary that $[A \ B]$ and $[A \ C]^t$ be contractions. It turns out that this conditions are also sufficient — see [17]. Now, fix $T_o$, $T'_o$, $A_o$, $A_1$ and suppose $\{B_o, B_1\} \in \mathbf{CID}(\{A_o, A_1\})$; then the conditions $P_{H'_o} B_o = A_o P_{H_o}$ and $P_{H'_1} B_1 = A_1 P_{H_o}$ show that

$$B_o = \begin{bmatrix} A_o & 0 \\ X_1 & X_2 \end{bmatrix} \quad \text{and} \quad B_1 = A_1$$

where $X_1 \in L(H_o, D_{T_o^*})$ and $X_2 \in L(D_{T_o}, D_{T'_o})$. The relation $W'^+_o B_1 = B_o W^+_o$ implies that

$$D_{T'_o} A_1 = X_1 T_o + X_2 D_{T_o}.$$

Multiplying $B_o$ on the right with $J(T_o)$, we get that

$$(1.7) \qquad \hat{B}_o = \begin{bmatrix} A_o T_o & A_o D_{T_o^*} \\ D_{T'_o} A_1 & S \end{bmatrix}$$

must be a contraction, where $S = X_1 D_{T_o^*} - X_2 T_o^*$. But

$$I - [A_o T_o \ \ A_o D_{T_o^*}] \begin{bmatrix} T_o^* A_o^* \\ D_{T_o^*} A_o^* \end{bmatrix} = I - A_o A_o^* \geq 0$$

and

$$I - [A_1^* T'^*_o, \ A_1^* D_{T'_o}] \begin{bmatrix} T'_o A_1 \\ D_{T'_o} A_1 \end{bmatrix} = I - A_1^* A_1 \geq 0,$$

so that $\hat{B}_o$ satisfies the conditions for the solvability of Question A, and an operator $S$ with the property that $B_o$ is a contraction, exists indeed.

As a conclusion, knowing that Question A is solvable in the conditions that $(A \ B)$ and $(A \ C)^t$ are contractions, it results that $\mathbf{CID}(\{A_o, A_1\}) \neq \emptyset$. Conversely suppose $(A \ B)$ and $(A \ C)^t$ be contractions and that $\mathbf{CID}(\{A_o, A_1\}) \neq \emptyset$. We define

$$A_o = (A \ B), \quad A_1 = (A \ C)^t, \quad T'_o = (I \ 0), \quad T_o = (I \ 0)^t,$$

then $T'_o A_1 = A_o T_o$. Moreover, $W'^+_o = \begin{bmatrix} I & 0 \\ 0 & I \end{bmatrix}$, $W^+_o = (I \ 0)^t$ and as $A_o$, $A_1$ are supposed contractions, $\mathbf{CID}(\{A_o, A_1\}) \neq \emptyset$. Let $\{B_o, B_1\} \in \mathbf{CID}(\{A_o, A_1\})$, then $B_1 = A_1$ and $B_o = \begin{bmatrix} A & B \\ X & Y \end{bmatrix}$ is a contraction. But as $W_o^{'+} B_1 = B_o W_o^+$, we obtain $X = C$ and $Y$ is a solution of Question A with data A, B, C. Precisely in the sense described, we have the following result. $\qquad\square$

**1.1. THEOREM.** *The solvability of Question A under the conditions that* (A  B) *and* (A  C)$^t$ *are contractions is equivalent with* $\mathbf{CID}(\{A_o, A_1\}) \neq \emptyset$. $\qquad\square$

**1.2. REMARK.** The fact that Question A is a problem of lifting of commutants appears for the first time in [8]. In [9], we describe an equivalent form of this theorem, as a nonstationary variant of the lifting theorem of Sarason – Sz-Nagy – Foiaş. $\qquad\square$

We pass on to the description of all solutions of Question A (and, in view of Theorem 1.1, this means to describe the set $\mathbf{CID}(\{A_o, A_1\})$). The formulas for the solutions of Question A are the following ([7]), [10], [12]): there exist uniquely determined contractions $G_1 \in L(M_2, D_{A^*})$, $G_2 \in L(D_A, N_2)$ such that $B = D_{A^*} G_1$, $C = G_2 D_A$ and

$$(1.8) \qquad D = -G_2 A^* G_1 + D_{G_2^*} G D_{G_1}$$

where G runs the set of contractions in $L(D_{G_1}, D_{G_2^*})$. For parametrizing $\mathbf{CID}(\{A_o, A_1\})$, we have to determine the corresponding $G_1$ and $G_2$ for the operator $\hat{B}_o$ given by (1.7).

Denote $C_o = A_o T_o = T_o' A_1$, then there exist unitary operators:

$$(1.9) \quad \begin{cases} \omega_* : D_{C_o^*} \to F_* = \{D_{T_o^*} A_o^* h_o' \oplus D_{A_o^*} h_o' \mid h_o' \in H_o'\}^- \subset D_{T_o^*} \oplus D_{A_o^*} \\[2ex] \omega_* D_{C_o^*} = \begin{bmatrix} D_{T_o^*} A_o^* \\ D_{A_o^*} \end{bmatrix} \end{cases}$$

and

$$(1.10) \quad \begin{cases} \omega : D_{C_o} \to F = \{D_{T_o'} A_1 h_1 \oplus D_{A_1} h_1 \mid h_1 \in H_1\}^- \subset D_{T_o'} \oplus D_{A_1} \\[2ex] \omega D_{C_o} = \begin{bmatrix} D_{T_o'} A_1 \\ D_{A_1} \end{bmatrix} \end{cases}$$

and it is readily checked that

$$A_o D_{T_o^*} = D_{C_o^*} (\omega_*)^* P_{F^*}^{D_{T_o^*} \oplus D_{A_o^*}} \begin{bmatrix} I \\ 0 \end{bmatrix}$$

and

$$D_{T'_o} A_1 = (I\ \ 0) P_F^{\,D_{T'_o} \oplus D_{A_1}}\, \omega D_{C_o}$$

so that, we have to define $\hat{G}_2 = (I\ \ 0) P_F^{\,D_{T_o^*} \oplus D_{A_1}}\, \omega : D_{C_o} \to D_{T_o}$ and $\hat{G}_1 =$
$= (\omega_*)^* P_{F_*}^{\,D_{T_o^*} \oplus D_{A_o^*}} (I\ \ 0)^t : D_{T_o^*} \to D_{C_o^*}$, in order to get on the basis of (1.8) that

$$(1.11) \qquad\qquad S = -\hat{G}_2 C_o^* \hat{G}_1 + D_{\hat{G}_2^*}\, \hat{G} D_{\hat{G}_1}$$

where $\hat{G}$ is a contraction in $L(D_{\hat{G}_1}, D_{\hat{G}_2^*})$. We can adapt some results in [5] in order to express $D_{\hat{G}_1}$ and $D_{\hat{G}_2^*}$ in terms of the data. We have:

$$I_{D_{T'_o}} - \hat{G}_2 \hat{G}_2^* = (I\ \ 0)\,(I - P_F^{\,D_{T'_o} \oplus D_{A_1}})\begin{bmatrix} I \\ 0 \end{bmatrix}$$

and we define the unitary operator

$$(1.12) \quad \left\{ \begin{array}{l} \gamma_{\hat{G}_2^*} : D_{\hat{G}_2^*} \to (I - P_F^{\,D_{T'_o} \oplus D_{A_1}}) \begin{bmatrix} D_{T'_o} \\ 0 \end{bmatrix}^- \\[2ex] \gamma_{\hat{G}_2^*} D_{\hat{G}_2^*} d'_o = (I - P_F^{\,D_{T'_o} \oplus D_{A_1}}) \begin{matrix} d'_o \\ 0 \end{matrix}\ , \quad d'_o \in D_{T'_o}. \end{array} \right.$$

But, in view of a result in [5],

$$(I - P_F^{\,D_{T'_o} \oplus D_{A_1}}) \begin{bmatrix} D_{T'_o} \\ 0 \end{bmatrix}^- = R = (D_{T'_o} \oplus D_{A_1}) \ominus F.$$

In a similar way, we obtain a unitary operator

$$(1.13) \quad \left\{ \begin{array}{l} \gamma_{\hat{G}_1^*} : D_{\hat{G}_1^*} \to (I - P_{F_*}^{\,D_{T_o^*} \oplus D_{A_o^*}}) \begin{bmatrix} D_{T_o^*} \\ 0 \end{bmatrix}^- \\[2ex] \gamma_{\hat{G}_1} D_{\hat{G}_1} d_{*o} = (I - P_{F_*}^{\,D_{T_o^*} \oplus D_{A_o^*}}) \begin{matrix} d_{*o} \\ 0 \end{matrix}\ , \quad d_{*o} \in D_{T_o^*} \end{array} \right.$$

and

$$(I - P_{F_*}^{\,D_{T_o^*} \oplus D_{A_o^*}}) \begin{matrix} D_{T_o^*} \\ 0 \end{matrix}^- = R_* = (D_{T_o^*} \oplus D_{A_o^*}) \ominus F_*.$$

We obtain the following result.

**1.3. THEOREM.** *There exists a one-to-one correspondence between* $\mathbf{CID}(\{A_o, A_1\})$ *and the set of contractions in* $L(R_*, R)$. $\qquad\qquad\qquad\qquad\qquad$ □

Recalling the definition of regularity of a factorization in [21], we obtain the following criterion which resembles the one in [3].

**1.4. COROLLARY.** $CID(\{A_o, A_1\})$ *is a singleton if and only if one of the factorizations* $T'_o A_1$ *or* $A_o T_o$ *is regular.*                    ☐

Of course, we can round the table in order to deduce (1.8) from (1.11) together with the identifications (1.12) and (1.13), but this is now obvious. Instead, we can use this for adding to [7], [10], [12] the following criterion for the uniqueness of the solutions of Question A in terms of data.

**1.5. COROLLARY.** *Question A has a unique solution if and only if one of the factorizations* $(I\ 0)(A\ C)^t$ *or* $(A\ B)(I\ 0)^t$ *is regular.*                    ☐

**II.** In this section we take into account Question B. First of all, we have to explicitate some of the elements involved in its statement. An extension $W_o : K_1 \to \to K_o$, $K_o \supset G_o$, $K_1 \supset G_1$ is minimal if

$$(2.1) \qquad\qquad K_o = G_o \vee W_o G_1.$$

Then, two minimal extensions $W_o \in L(K_1, K_o)$, $W'_o \in L(K'_1, K'_o)$ are equivalent if there exist unitary operators $\phi_o \in L(K_o, K'_o)$ and $\phi_1 \in L(K_1, K'_1)$ such that $\phi_o|G_o = I_{G_o}$, $\phi_1|G_1 = I_{G_1}$ and $W'_o \phi_1 = \phi_o W_o$. We denote by $E(w_o) = E(E_1, F_o, G_o, G_1; w_o)$ the set of equivalent minimal unitary extensions of $w_o$. Thus, Question B means the study of $E(w_o)$. Define $P_1 = G_1 \ominus E_1$, $R_o = G_o \ominus F_o$. The solution of Question B is quite obvious.

**2.1. PROPOSITION.** *There exists a one-to-one correspondence between* $E(w_o)$ *and the set of contractions in* $L(P_1, R_o)$.

PROOF. Take G a contraction in $L(P_1, R_o)$ and define

$$(2.2) \qquad\qquad K_o = E_1 \oplus P_1 \oplus D_{G^*}$$

$$(2.3) \qquad\qquad K_1 = F_o \oplus R_o \oplus D_G$$

and

$$(2.4) \qquad\qquad \left\{ \begin{array}{l} W_o : K_1 \to K_o \\[2ex] W_o = w_o \oplus J(G). \end{array} \right.$$

The application $\Phi$ between the contractions in $L(P_1, R_0)$ and $E(w_0)$ given by $\Phi(G) = $ class of $W_0$ given by (2.4), established the asserted correspondence.    $\square$

Now, we show that $CID(\{A_0, A_1\})$ is an $E(w_0)$ for a certain $w_0$. The idea is taken after [1] and [4]. Consider the positive block-matrices

$$G_1 = \begin{bmatrix} I & T'_0 & T'_0 A_1 \\ T'^*_0 & I & A_1 \\ A_1^* T'^*_0 & A_1^* & I \end{bmatrix} \quad \text{on } H'_0 \oplus H'_1 \oplus H_1$$

and

$$G_1 = \begin{bmatrix} I & A_0 & A_0 T_0 \\ A_0^* & I & T_0 \\ T_0^* A_0^* & T_0^* & I \end{bmatrix} \quad \text{on } H'_0 \oplus H_0 \oplus H_1$$

$$\left( \text{for instance, } G_0 = F_0^* F_0 \text{ with } F_0 = \begin{bmatrix} I & A_0 & A_0 T_0 \\ 0 & D_{A_0} & D_{A_0} T_0 \\ 0 & 0 & D_{T_0} \end{bmatrix} \right).$$

Let $G_1$ be the space obtained after renorming $H'_0 \oplus H'_1 \oplus H_1$ with $G_1$ (i.e., we introduce a new inner product in $H'_0 \oplus H'_1 \oplus H_1$ setting $(k_0, k'_0)_{G_1} = (G_1 k_0, k'_0)$, $k_0$, $k'_0 \in H'_0 \oplus H'_1 \oplus H_1$; factoring out the subspace of elements with zero $G_1$-norm if necessary and then completing this space, we get $G_1$). In a similar way, we get the space $G_0$ by renorming $H'_0 \oplus H_0 \oplus H_1$ with $G_0$. Define the spaces

(2.5)      $$F_0 = \{(h'_0, 0_{H_0}, h_1) \mid h'_0 \in H'_0, h_1 \in H_1\} \subset G_0$$

(2.6)      $$E_1 = \{(h'_0, 0_{H'_1}, h_1) \mid h'_0 \in H'_0, h_1 \in H_1\} \subset G_1$$

and the unitary operator

(2.7)      $$\begin{cases} w_0 : E_1 \to F_0 \\ w_0(h'_0, 0_{H'_1}, h_1) = (h'_0, 0_{H_0}, h)_1). \end{cases}$$

**2.2. THEOREM.** $CID(\{A_0, A_1\}) = E(w_0)$, *where* $w_0$ *is given by* (2.7).

PROOF. First of all, we embed $H_1$ and $H'_1$ into $G_1$ and $H_0$ and $H'_0$ into $G_0$ in an obvious way and denote these embeddings by $\hat{H}_0$, $\hat{H}'_0$, $\hat{H}_1$, $\hat{H}'_1$. Taking $W_0 \in E(w_0)$, we

148                                                          Constantinescu

compute that

$$(W_o \hat{h}_1, \hat{h}_o)_{G_o} = (T_o h_1, h_o)$$

where $\hat{h}_1 = (0 \ \ 0 \ \ h_1)$, $h_1 \in H_1$, $\hat{h}_o = (0 \ \ h_o \ \ 0)$, $h_o \in H_o$, and

$$(W_o \hat{h}'_1, \hat{h}'_o)_{G_o} = (\hat{h}'_1, W^*_o \hat{h}'_o)_{G_1} = (T'_o h'_1, h'_o)$$

where $\hat{h}'_o = (0 \ \ h'_o \ \ 0)$, $h'_1 \in H'_1$, $\hat{h}'_o = (h'_o \ \ 0 \ \ 0)$, $h'_o \in H'_o$.

Then, we define

$$\hat{K}_1^+ = \hat{H}_1, \ \ \hat{K}'^+_1 = \hat{H}'_1$$

$$\hat{K}_o^+ = \hat{H}_o \vee W_o \hat{H}_1, \ \ \hat{K}'^+_o = \hat{H}'_o \vee W_o \hat{H}'_1$$

and all these relations show that $W_o | \hat{K}_1^+ : \hat{K}_1^+ \to \hat{K}_o^+$ is an identification of the operator $W_o^+$ given by (1.6) and $W_o | \hat{K}'^+_o : \hat{K}'^+_o \to \hat{K}'^+_o$ is the identification of $W_o'^+$, the operator defined by (1.6) for $T'_o$. We define

$$B_o = P^{\hat{K}_o}_{\hat{K}'^+_o} | \hat{K}'^+_o, \ \ \ B_1 = P^{\hat{K}_1}_{\hat{K}'^+_1} | \hat{K}'^+_1$$

and we show that $\{B_o, B_1\} \in \mathbf{CID}(\{A_o, A_1\})$. $B_o$, $B_1$ are obviously contractions. Then,

$$(W_o'^+ B_1, \hat{h}'_1, \hat{h}'_o) = (\hat{h}'_1, W^*_o \hat{h}'_o)_{G_1} = (T'_o A_1 h'_1, h'_o)$$

and

$$(B_o W_o^+ \hat{h}_1, \hat{h}_o) = (W_o \hat{h}_1, \hat{h}_o)_{G_o} = (A_o T_o h_1, h'_o)$$

so that, in view of the relation $T'_o A_1 = A_o T_o$ we also have

$$W_o'^+ B_1 = B_o W_o^+.$$

Finally, we simply verify that $P_{\hat{H}'_o} B_o = A_o P_{\hat{H}_o}$ and $P_{\hat{H}'_1} B_1 = A_1 P_{\hat{H}_1}$, that is $\{B_o, B_1\} \in \mathbf{CID}(\{A_o, A_1\})$.

Now, it is quite useful to have a realization of $w_o$ in terms of the data. For this purpose, we take into account the unitary operators:

$$(2.8) \quad \left\{ \begin{array}{l} \Omega_o : G_o \to H'_o \oplus D_{A_o} \oplus D_{T_o} \\[2mm] \Omega_o (h'_o \ \ h_o \ \ h_1)\hat{} = F_o (h'_o \ \ h_o \ \ h_1) \end{array} \right.$$

(here $(h'_o \ h_o \ h_1)^{\curvearrowright}$ means the class of the element $(h'_o \ h_o \ h_1) \in H'_o + H_o + H_1$ in $G_o$), and

$$(2.9) \qquad \begin{cases} \Omega_1 : G_1 \rightarrow H'_o \oplus D_{T'_o} \oplus D_{A_1} \\[2mm] \Omega_1(h'_o \ h'_1 \ h_1)^{\curvearrowright} = \begin{bmatrix} I & T'_o & T'_o A_1 \\ 0 & D_{T'_o} & D_{T'_o} A_1 \\ 0 & 0 & D_{A_1} \end{bmatrix} \begin{bmatrix} h'_o \\ h'_1 \\ h_1 \end{bmatrix}. \end{cases}$$

It is clear that $\Omega_o F_o = H'_o \oplus F^*$, where

$$F^* = \{D_{A_o} T_o h_1 \oplus D_{T_o} h_1 \mid h_1 \in H_1\} \subset D_{A_o} \oplus D_{T_o}.$$

and $\Omega_1 E_1 = H'_o \oplus F$ ($F$ being given by (1.10)). Now, the point is that

$$(2.10) \qquad \Omega_o w_o = (I_{H_o}) \oplus \sigma)\Omega_1$$

where

$$(2.11) \qquad \begin{cases} \sigma : F \rightarrow F^* \\[2mm] \sigma(D_{T'_o} A_1 h_1 \oplus D_{A_1} h_1) = (D_{A_o} T_o h_1 \oplus D_{T_o} h_1), \quad h_1 \in H_1. \end{cases}$$

Now, in view of some results in [5], there exists a unitary operator from $R_*$ onto $R^* = (D_{A_o} \oplus D_{T_o}) \ominus F^*$, consequently, there exists a one-to-one correspondence between $\mathbf{CID}(\{A_o, A_1\})$ and the set of contractions in $L(R^*, R)$. By the above identifications, (2.8), (2.9) and (2.10), we obtain that

$$\mathbf{CID}(\{A_o, A_1\}) = E(\sigma) = E(w_o)$$

where $\sigma$ is given by (2.11) and $w_o$ by (2.7). $\qquad\qquad\qquad\qquad\qquad\qquad$ $\square$

### REFERENCES


1. **Adamjan, V.M. ; Arov, D.Z. ; Krein, M.G.** : Infinite Hankel matrices and generalized Carathéodory – Fejér and Schur problems (Russian), *Funcktional Anal. i Prilozhen.* 2: 4(1968), 1-17.

2. **Adamjan, V.M. ; Arov, D.Z. ; Krein, M.G.** : Infinite Hankel block-matrices and related continuation problem (Russian), *Izv. Akad. Nauk. Armyan SSR Ser. Mat.* 6(1971), 87-112.

3. **Ando, T. ; Ceauşescu, Z. ; Foiaş, C.** : On intertwining dilations. II, *Acta Sci. Math. (Szeged)* **39**(1977), 3-14.

4. **Arocena, R.** : Generalized Toeplitz kernels and dilations of intertwining operators, *Integral Equations Operator Theory* **6**(1983), 759-778.

5.  **Arsene, Gr. ; Ceaușescu, Z. ; Foiaș, C.** : On intertwining dilations VII, in *Proc. Coll. Complex Analysis, Joensuu,* Lecture Notes in Math., **747**(1979), 24-45.

6.  **Arsene, Gr. ; Ceaușescu, Z. ; Foiaș, C.** : On intertwining dilations VIII, *J. Operator Theory* **4**(1980), 55-91.

7.  **Arsene, Gr. ; Gheondea, A.** : Completing matrix contractions, *J. Operator Theory* 7(1982), 179-189.

8.  **Ball, J.A. ; Gohberg, I.** : A commutant lifting theorem for triangular matrices with diverse applications, *Integral Equations Operator Theory* **8**(1985), 205-267.

9.  **Constantinescu, T.** : Some aspects of nonstationarity. I, INCREST Preprint Nr. 9(1988); II, INCREST Preprint Nr. **43**(1988).

10. **Davis, C. ; Kahan, W.M. ; Weinberger, H.F.** : Norm-preserving dilations and their applications to optimal error bounds, *SIAM J. Numer. Anal.* **19**(1982), 445-469.

11. **Foiaș, C. ; Frazho, A.E.** : Redheffer products and lifting contractions on Hilbert space, *J. Operator Theory* 11(1984), 193-196.

12. **Ivanovskaia, R.N. ; Smulian, Iu.L.** : On matrices whose entries are contractions, *Izv. Vyssh. Uchebn. Zaved. Mat.* 7(230)(1981), 72-75.

13. **Krein, M.G.** : The theory of self-adjoint extensions of semi-bounded Hermitian transformations and its applications (Russian), *Math. Sb.* **20**(1947), 431-495; **21**(1947), 365-404.

14. **Krein, M.G.** : On Hermitian operators whose deficiency indices are 1 (Russian), *Dokl. Akad. Nauk. SSSR.* **43**:8(1944), 323-326.

15. **Krein, M.G. ; Nudelman, A.A.** : *The Markov problem of moments and extremal problems* (Russian), Izd. Nauka, Moscow, 1973.

16. **Naimark, M.A.** : On self-adjoint extensions of the second kind of a symmetric operator, *Izv. Akad. Nauk. SSR* **4**(1940), 54-104.

17. **Parrott, S.** : On a quotient norm and the Sz.-Nagy – Foiaș lifting theorem, *J. Funct. Anal.* **30**(1978), 331-328.

18. **Redheffer, R.M.** : Inequalities for a matrix Ricatti equation, *J. Math. Mech.* **8**(1959), 349-377.

19. **Sarason, D.** : Generalized interpolation in $H_\infty$, *Trans. Amer. Math. Soc.* **127**(1967), 179-203.

20. **Sarason, D.** : Moment problems and operators in Hilbert space, in *Proceedings of Symposia in Applied Mathematics,* Vol. **37**(1987), pp. 54-70.

21. **Sz.-Nagy, B. ; Foiaș, C.** : *Harmonic analysis of operators on Hilbert spaces,* North-Holland, Amsterdam, 1970.



**Tiberiu Constantinescu**

Department of Mathematics, INCREST
Bdul Păcii 220, 79622 Bucharest,
Romania.

Operator Theory:
Advances and Applications, Vol. 43
© 1990 Birkhäuser Verlag Basel



# OPERATOR FACTORIZATIONS AND QUASI-SIMILARITY ORBITS [1]

**Raúl E. Curto** and **Lawrence A. Fialkow**

This article presents recent results in the theory of quasi-similarity orbits, with particular emphasis on a new factorization technique introduced by the authors in [9]. Briefly stated, the factorization technique studies pairs of quasi-affinities $(X,Y)$ whose product lies in the commutant of a given operator S, the *model* operator. In a manner described below, it is possible to use $(X,Y)$ to generate, in a systematic way, families of *test* operators which are good candidates to be in the quasi-similarity orbit of S. The technique also comes with a criterion for membership in the similarity orbit of S. Moreover, it is possible to characterize quasi-similarity and similarity in terms of invariant operator ranges, and to determine spectral inclusions. In the particular case of the unilateral shift $U_+$, the factorization method can be used to construct a new one-parameter family of non-contractive, non-hyponormal, shift-like operators in the similarity orbit of $U_+$. Before we can describe our results in detail, we need some notation and preliminaries.

Let S and T be (bounded) operators on Hilbert spaces $H$ and $H_1$, respectively. S is said to be *similar* (respectively, *quasi-similar*) to T, in symbols $S \sim T$ (resp., $S \underset{qs}{\sim} T$), if there exists an invertible operator $V : H \to H_1$ such that $TV = VS$ (resp., if there exist quasi-affinities X and Y such that $XT = SX$ and $TY = YS$). It is well known that $\sim$ preserves the spectral picture, but $\underset{qs}{\sim}$ does not. However, $\underset{qs}{\sim}$ does preserve cyclicity and transitivity, although not necessarily the hyperlattice or reflexivity. It is also well known that

$$T \underset{qs}{\sim} S \Longrightarrow \sigma(T) \cap \sigma(S) \neq \emptyset \ .$$

The similarity (resp., quasi-similarity) orbit of an operator S, in symbols $S(S)$ (resp., $QS(S)$) is the set of operators similar (resp., quasi-similar) to S. A fundamental result on the similarity orbit of $U_+$, the unilateral shift acting on $\ell^2(\mathbf{Z}_+)$, is given as follows:

---

[1] Research partially supported by grants from the National Science Foundation.

$T \sim U_+ \Longleftrightarrow$ 1) $T$ is cyclic; 2) $T^{*n} \to 0$ strongly; 3) $T$ is power bounded; 4) $T$ is power bounded below

([25]). It is natural to ask what necessary and sufficient conditions are required for the quasi-similarity orbit of $U_+$.

**PROBLEM 1.** Characterize the quasi-similarity orbit of $U_+$.

As a first approximation to a suitable answer to Problem 1, one is led to formulate the following questions:

If $T \underset{qs}{\sim} U_+$, does it follow that

(i) $\sigma(T) = \bar{\mathbf{D}}$ ?

(ii) $\sigma_e(T) = \mathbf{T}$?

(iii) $T$ is reflexive?

Partial answers to these questions have been found. For instance, D. Herrero [18] showed that $\sigma_e(T)$ is connected and $\sigma_e(T) \supseteq \mathbf{T}$. H. Bercovici and T. Takahashi showed in [3] that $T$ must be reflexive. Recently, Takahashi [27] has shown that if $T$ is a contraction, then $\sigma_e(T) = \mathbf{T}$.

Although the above results are important in helping to determine the structure of the quasi-similarity orbit of the shift, not many specific examples of operators in that orbit have been found. The known examples of quasi-similarity usually concern operators in restrictive classes. Here's a sample:

* (S. Clary [6]) Quasi-similar hyponormal operators have equal spectra.

* (S. Clary [5]) Let $T$ be subnormal. Then $T \underset{qs}{\sim} U_+ \Longleftrightarrow$ there exists a measure $\mu$ on $\bar{\mathbf{D}}$ such that

1) $\nu := \mu \,|\, \mathbf{T} \ll \lambda$   ($\lambda$ is Lebesgue measure on the circle)

2) $\log \frac{d\nu}{d\lambda} \in L^1(\lambda)$

3) $T \simeq M_z$ on $H^2(\mu)$.

In addition, $T \sim U_+ \Longleftrightarrow \nu = \lambda$ and $\eta := \mu - \lambda$ is a Carleson measure.

* (M. Raphael [24]) If $T$ and $S$ are cyclic subnormal operators, and $T \underset{qs}{\sim} S$, then $\sigma_e(T) = \sigma_e(S)$.

* (L. Fialkow [15]) If $T$ and $S$ are cyclic subnormal, and $T \underset{qs}{\sim} S$, then $S(T)^- = S(S)^-$ (and therefore $T$ and $S$ have identical spectral pictures).

* (J. Conway [7]) If $T$ and $S$ are subnormal operators, with minimal normal extensions $N$ and $M$, and scalar spectral measures $\nu$ and $\mu$, respectively, and if $T \underset{qs}{\sim} S$, then the map $p \mapsto p$ extends to a $w^*$-homeomorphic isomorphism of $P^\infty(\mu)$ onto $P^\infty(\nu)$.

* (L. Fialkow [15]) If $T \underset{qs}{\sim} U_+$, then $T \in S(U_+)^-$ .

* (B. Sz.-Nagy and C. Foiaş [26, II.3, Proposition 3.5]) Every $C_{11}$ contraction is quasi-similar to a unitary operator.

* (Sz.-Nagy and C. Foiaş [26, VI.4.2]) There exists a $C_{11}$ contraction $T$ with $\sigma(T) = \bar{\mathbf{D}}$ .

* (T. Hoover [22]) Let $q_k$ be the Jordan cell of dimension k. Clearly $q_k \sim$ $\sim (1/k)q_k$, so

$$S := \underset{k}{\oplus} q_k \underset{qs}{\sim} \underset{k}{\oplus} \left( (1/k)q_k \right) =: T .$$

However, $\sigma(S) = \bar{\mathbf{D}}$, while $\sigma(T) = \{0\}$.

* A sequence $(M_n)$ of subspaces is *basic* if $M_j$ and $\underset{k \neq j}{\vee} M_k$ are complementary (all j), and $\underset{j=1}{\overset{\infty}{\cap}} \underset{k \neq j}{\vee} M_k = \{0\}$. In [1], [2], [13], [14] and [19] examples are given of quasi--similar operators $S$ and $T$ with different spectra, where $S$ has a basic sequence of invariant subspaces.

* In [29], a characterization of the contractions with finite defect indices that are quasi-similar to $U_+$ is given.

Some general results on spectral relations of quasi-similar operators are the following:

* (D. Herrero [17]) If $T \underset{qs}{\sim} S$, then every component of $\sigma(T)$ meets $\sigma(S)$.
* (L. Fialkow [12]; L. Williams [28]) If $T \underset{qs}{\sim} S$, then $\sigma_e(T) \cap \sigma_e(S) \neq \emptyset$ .
* (D. Herrero [20]) If $T \underset{qs}{\sim} S$, then every component of $\sigma_e(T)$ meets $\sigma_e(S)$.

Typically, the literature deals with the problem of deciding when a *test* operator, $T$, is quasi-similar to a *given* model, $S$. The factorization approach that we investigate here is aimed at *generating* test operators.

We begin with the following easy observation. If $T \underset{qs}{\sim} S$, then $S(XY) = XTY =$ $= (XY)S$, so $XY \in (S)'$, the commutant of $S$. (Notation: For an operator $A$, $N(A)$ and $R(A)$ denote the kernel and range of $A$, respectively.)

## DEFINITIONS 1.

$$Z_o(S) := \left\{ (X,Y) \in L(H) \times L(H) : X \text{ is } 1\text{-}1, XY \text{ is a q.a., and } XY \in (S)' \right\}.$$

For $Z = (X,Y) \in Z_o(S)$ let

$$H_1 := \overline{R(Y)}, \quad X_1 := X | H_1, \quad \text{and} \quad Y_1 : H \rightarrow H_1 \ (x \rightarrow Yx).$$

$$Z(S) := \{Z = (X,Y) \in Z_o(S) : Yx \xrightarrow{\tilde{T}_Z} YSx, \ x \in H, \text{ extends } boundedly \text{ to } T_Z \in L(H_1)\}.$$

Note that if $T \underset{qs}{\sim} S$ via quasi-affinities $X$ and $Y$, then $T = T_Z$, where $Z = (X,Y)$.

In what follows, we shall be very sketchy about proofs and refer the reader to [9] for full details.

**PROPOSITION 1.** *If* $Z \in Z(S)$, *then* $T_Z \underset{qs}{\sim} S$ *via* $X_1$ *and* $Y_1$.

**PROPOSITION 2.** *Let* $Z = (X,Y) \in Z(S)$, *and let* $S_1 = S \,|\, H_1$. *The following state-ments are equivalent*

    (i)  $T_Z = S_1$

    (ii)  $Y \in (S)'$

    (iii)  $XS_1 = SX_1$.

The content of Proposition 2 is that in order to obtain non-trivial elements in $QS(S)$ we must restrict attention to factorizations $XY$ for which $Y$ does *not* commute with S. The next result establishes a sufficient condition for similarity.

**PROPOSITION 3.** *Let* $Z = (X,Y) \in Z_o(S)$. $\tilde{T}_Z$ *extends to* $T_Z \sim S$ *if and only if there exists* $J : H \twoheadrightarrow \overline{R(Y)}$ *invertible such that* $XJ \in (S)'$.

PROOF. $\Leftarrow$ ) $X_1(\tilde{T}_Z J) = SX_1 J$ (by Proposition 1) $= SXJ = XJS = X_1(JS)$, so $\tilde{T}_Z$ extends boundedly to $JSJ^{-1}$, and $T_Z$ is similar to S. $\qquad\qquad\square$

We would like to look at some examples before we proceed.

**EXAMPLE 1.** Let $S := U_+$, and let us look for pairs $(X,Y)$ such that $XY \in (U_+)'$, with $Y \notin (U_+)'$. As a non-trivial quasi-affinity in the commutant of $U_+$, let us take $(I + U_+)^2$. Consider

$$X := I + U_+ + U_+\Delta$$

$$Y := I + U_+ + U_+\Sigma$$

where $\Delta := \mathbf{diag}\,(\delta_n)$ and $\Sigma := \mathbf{diag}\,(\sigma_n)$ are diagonal operators on $\ell^2(\mathbf{Z}_+)$. Since $XY = (I + U_+)^2$, we must have

$$U_+\Sigma + U_+\Delta + U_+^2\Sigma + U_+\Delta U_+ + U_+\Delta U_+\Sigma = 0,$$

so that

$$\begin{cases} \Sigma + \Delta = 0 \\ U_+\Sigma + \Delta U_+ + \Delta U_+\Sigma = 0. \end{cases}$$

We therefore get

$$\begin{cases} \sigma_n = -\delta_n \\ \sigma_n + \delta_{n+1} + \delta_{n+1}\sigma_n = 0, \end{cases}$$

or

$$\delta_{n+1} = \frac{\delta_n}{1-\delta_n}.$$

Letting $\delta := \delta_o$, we obtain

$$\delta_n = \frac{\delta}{1-n\delta} \qquad (\delta \neq \tfrac{1}{n}).$$

As a special case, let us take $\delta = -1$; then $\delta_n = -\dfrac{1}{n+1}$ and

$$\tilde{T}_{-1}\left(e_n + \frac{n+2}{n+1}\,e_{n+1}\right) := e_{n+1} + \frac{n+3}{n+2}\,e_{n+2}.$$

Two basic questions are:

* Is $\tilde{T}_{-1}$ bounded ?
* Is $\tilde{T}_{-1} \sim U_+$ ?

We shall come back to this example later.

**EXAMPLE 2.** The following result appears in [18]: If $(S)''$ has finite strict multiplicity, then $QS(S) = S(S)$. We shall give here an alternate proof based on Proposition 3. Let $Z = (X,Y) \in Z(S)$. Then $X_1 Y \in (S)'$, so $R(X_1 Y)$ is dense and invariant under $(S)''$. By the assumption on S and an argument from [18], $X_1 Y$ is onto. Therefore $X_1 : \overline{R(Y)} \to H$ is invertible and $J := X_1^{-1}$ satisfies $XJ \in (S)'$. It follows that any $T_Z \in QS(S)$ is actually *similar* to S.

**EXAMPLE 3.** (Nilpotent operators of order 2). Let $S = \begin{bmatrix} 0 & A \\ 0 & 0 \end{bmatrix}$ be a nilpotent of order 2 (A injective). Let

$$X = \begin{bmatrix} X_{11} & 0 \\ 0 & X_{22} \end{bmatrix}, \qquad Y = \begin{bmatrix} Y_{11} & 0 \\ 0 & Y_{22} \end{bmatrix}$$

be quasi-affinities, with $XY \in (S)'$, i.e. $AX_{22}Y_{22} = X_{11}Y_{11}A$. Using Propositions 2 and 3 we can show that the following are equivalent:

1.  $\tilde{T}_Z$ is bounded

2.  There exists a bounded operator B such that $BY_{22} = Y_{11}A$

3.  $A^*R(Y_{11}^*) \subseteq R(Y_{22}^*)$

4.  $AR(X_{22}) \subseteq R(X_{11})$.

In that case, $\tilde{T}_Z$ extends to $\begin{bmatrix} 0 & B \\ 0 & 0 \end{bmatrix}$ . This describes an important piece of the quasi-similarity orbit of S. Concerning similarity, $T_Z \sim S$ if and only if there exist invertible operators $V_{11}$ and $V_{22}$ such that $BV_{22} = V_{11}A$, i.e., if and only if A and B are *equivalent*.

**PROPOSITION 4.** (Range Inclusion). *Let* $Z = (X,Y) \in Z_0(S)$. *The following statements are equivalent.*

(i)  $\tilde{T}_Z$ *is bounded*

(ii)  $SR(X_1) \subseteq R(X_1)$

(iii)  $S^*R(Y^*) \subseteq R(Y^*)$.

*Furthermore, if* $SR(X) \subseteq R(X)$, *then* $\tilde{T}_Z$ *is bounded.*

At this point, we would like to digress for a moment and consider the operator equation A = BC, where A and B are given, and C is the unknown. R.G. Douglas showed in [11] that such an equation has a *bounded* solution C if and only if $R(A) \subseteq R(B)$. Later P. Fillmore and J. Williams established that an *invertible* solution exists if and only if R(A) = R(B) and N(A) = N(B) ([16]). The following result gives a criterion for the existence of a *Fredholm* solution.

**THEOREM 1.** *The operator equation* A = BC *has a Fredholm solution* C *if and only if the following conditions are satisfied:*

(i)  $R(A) \subseteq R(B)$;

(ii)  $R(B)/R(A)$ *is finite dimensional; and*

(iii)  $N(A) < \infty$ *iff* $N(B) < \infty$.

An important application of Theorem 1 is to the equation $SX_1 = X_1\tilde{T}_Z$ on R(Y) (where we take $A = SX_1$, $B = X_1$ and $C = \tilde{T}_Z$). Although part of the proof of Theorem 2 below can be derived from Theorem 1, we shall present the main argument needed so the reader can visualize clearly what is the obstruction to the inclusion "$\sigma_e(T_Z) \subseteq \sigma_e(S)$"; at the same time, a similar argument is used in the proof of Theorem 1.

**THEOREM 2.** *Let* $Z = (X,Y) \in Z_0(S)$, *and let* $\tilde{T}_Z$ *be given by* $\tilde{T}_Z Yx = YSx$, $x \in H$. *Then*

(i) $\tilde{T}_Z$ *is bounded* $\Longleftrightarrow SR(X_1) \subseteq R(X_1)$.

(ii) $T_Z$ *is invertible* $\Longleftrightarrow SR(X_1) = R(X_1)$ *and* $N(SX_1) = 0$.

(iii) $T_Z$ *is Fredholm* $\Longleftrightarrow R(X_1)/SR(X_1)$ *and* $N(SX_1)$ *are finite dimensional.*

(iv) $\tilde{T}_Z$ *is an isomorphism* $\Longleftrightarrow S$ *is invertible.*

(v) $\tilde{T}_Z$ *is algebraically Fredholm* $\Longleftrightarrow S$ *is Fredholm.*

(A linear transformation $T : V \to W$ between two vector spaces is *algebraically Fredholm* if $N(T)$ and $W/R(T)$ are both finite dimensional.)

SKETCH OF PROOF. Consider the following diagram of spaces and maps.

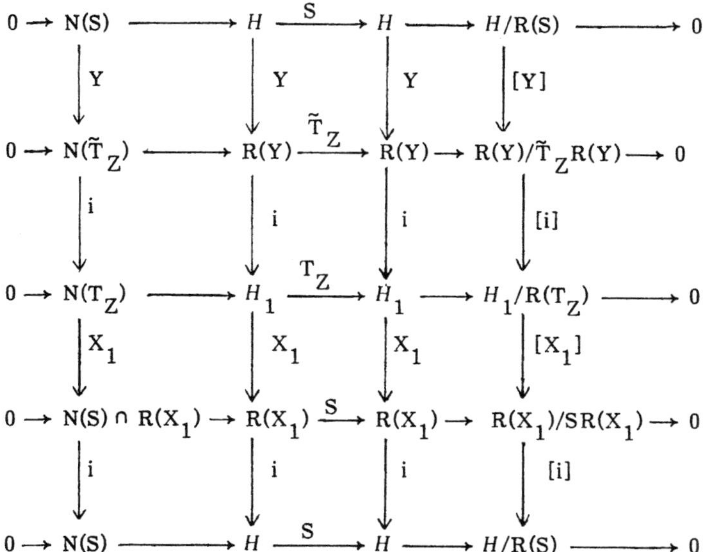

(Here i denotes the inclusion map and [ • ] stands for the maps induced in the quotient spaces.)

A close look at the diagram reveals that

$$N(S) = N(\tilde{T}_Z) \le N(T_Z) = N(S \mid R(X_1)) \le N(S),$$

so that in particular $N(T_Z) = N(\tilde{T}_Z)$. Also,

$$[Y] : H/R(S) \to R(Y)/\tilde{T}_Z R(Y)$$

is an isomorphism (so that S is Fredholm if and only if $\tilde{T}_Z$ is algebraically Fredholm, and S is invertible if and only if $\tilde{T}_Z$ is an isomorphism), and

$$[X_1] : H_1/R(T_Z) \rightarrow R(X_1)/SR(X_1)$$

is an isomorphism (so that $T_Z$ is Fredholm if and only if $R(X_1)/SR(X_1)$ and $N(SX_1)$ are both finite dimensional, and $T_Z$ is invertible if and only if $R(X_1) = SR(X_1)$ and $N(SX_1) = = 0$).                                                                    $\square$

We therefore see that the obstruction to the implication "S Fredholm $\Rightarrow T_Z$ Fredholm" is given by

$$\text{"}\tilde{T}_Z \text{ algebraically Fredholm } \overset{?}{\Rightarrow} T_Z \text{ Fredholm"},$$

or equivalently

$$\mathbf{dim}(R(Y)/\tilde{T}_Z R(Y)) < \infty \quad \Rightarrow \quad \mathbf{dim}(\overline{R(Y)}/T_Z \overline{R(Y)}) < \infty .$$

As we see, the containment $\sigma_e(T_Z) \subseteq \sigma_e(S)$ is equivalent to the condition

$$\mathbf{dim}(R(Y)/(T_Z - \lambda)R(Y)) < \infty \quad \Longrightarrow \quad \mathbf{dim}(\overline{R(Y)}/(T_Z - \lambda)\overline{R(Y)}) < \infty,$$

for every $\lambda \in \mathbf{C}$ (Fredholmness preserved under closure of the domain), which clearly shows why spectral inclusions of the above type are so hard to prove in particular examples.

Similarly,

$$\text{"S invertible} \Longrightarrow T_Z \text{ invertible"} \Longleftrightarrow (\tilde{T}_Z \text{ bijective} \Longrightarrow T_Z \text{ onto)},$$

$$\text{"}T_Z \text{ invertible} \Longrightarrow \text{S invertible"} \Longleftrightarrow (T_Z \text{ invertible} \Longrightarrow \tilde{T}_Z \text{ onto)},$$

and

$$\text{"}T_Z \text{ Fredholm} \Longrightarrow \text{S Fredholm"} \Longleftrightarrow (T_Z \text{ Fredholm} \Longrightarrow \tilde{T}_Z \text{ algebraically Fredholm)}.$$

The following theorem is a natural consequence of the previous discussion.

**THEOREM 3.** *Assume that* S *satisfies*

(∗)         $\mathbf{dim}(R(X_1)/(S - \lambda)R(X_1)) < \infty$ *for all* $(X,Y) \in Z(S)$ *and for all* $\lambda \notin \sigma_e(S)$.

*Then* $\sigma_e(T) \subseteq \sigma_e(S)$ *for all* $T \underset{qs}{\sim} S$.

**COROLLARY.** *If* S *satisfies*

(∗∗)    $\mathbf{dim}(M/(S - \lambda)M)$ *for every invariant operator range* M *and for every* $\lambda \notin \sigma_e(S)$,

*then* $\sigma_e(T) \subseteq \sigma_e(S)$ *for all* $T \underset{qs}{\sim} S$.

Since our main goal is a detailed description of the quasi-similarity orbit of the shift $U_+$, we would like to know whether $U_+$ satisfies condition $(*)$ or $(**)$. H. Salas has shown that $U_+$ does not satisfy $(**)$. Let $X := \mathbf{diag}(1, \frac{1}{2}, 1, \frac{1}{2}, \frac{1}{4}, 1, \frac{1}{2}, \frac{1}{4}, \frac{1}{8}, 1, \ldots)$.

Then $R(U_+X) \subseteq R(X)$ and $\mathbf{dim}(R(X)/R(U_+X)) = \infty$; also, $X$ is a quasi-affinitiy, but $XY \in (U_+)' \Longleftrightarrow Y = 0$, so for *no* $Z \in Z(U_+)$ is $Z = (X,Y)$.

**PROBLEM 2.** Does $U_+$ satisfy $(*)$ ?

**REMARK.** If $S = \alpha + F$, $\alpha \in \mathbf{C}$, $F$ of finite rank, then $\mathbf{dim}(M/(S - \lambda)M) < \infty$ for every $S$-invariant linear manifold $M$. Moreover, $S(S) = QS(S)$. More generally, D. Herrero has recently shown that property $(**)$ holds precisely when $S$ is an algebraic operator ([21, Theorem]). This yields a new proof of the result that quasi-similar algebraic operators have equal essential spectra ([9, Example 4.8]).

We are now ready to present our main application, that is, the construction of a one-parameter family of non-hyponormal, non-contractive, shift-like operators in the similarity orbit of $U_+$. As in Example 1, we would like to factor $(\alpha + U_+)(\beta + U_+)$ as the product $XY$ of two quasi-affinities, with

$$X := \alpha + U_+ + U_+\Delta ,$$

and

$$Y := \beta + U_+ + U_+\Sigma ,$$

where $|\alpha| = |\beta| = 1$ and $\Delta, \Sigma$ are diagonal. It follows that

$$XY = (\alpha + U_+)(\beta + U_+)$$

if and only if

$$\begin{cases} \alpha\Sigma + \beta\Delta = 0 \\ U_+^2\Sigma + U_+\Delta U_+ + U_+\Delta U_+\Sigma = 0 . \end{cases}$$

Let $\lambda := \frac{\alpha}{\beta}$. Two cases arise.

**CASE 1:** $\lambda = 1$ (i.e., $\alpha = \beta = 1$). Here $\Sigma = -\Delta$, so

$$\delta_n = \frac{\delta}{1-n\delta}, \quad \text{with} \quad \delta := \delta_0 \neq \frac{1}{n}, \quad (n = 1,2,\ldots) .$$

Assume for simplicity that $\delta := -1$. Then

$$\tilde{T}_{-1}(e_n + \frac{n+2}{n+1} \cdot e_{n+1}) = e_{n+1} + \frac{n+3}{n+2} \cdot e_{n+2}, \quad (n \geq 0) .$$

Since $\tilde{T}_{-1}Y = YU_+$ , we must have

$$\tilde{T}_{-1}(I + U_+(I - \Delta)) = (I + U_+(I - \Delta))U_+ = U_+(I + (I - \Delta)U_+) \; .$$

Let $\Delta' := I - \Delta$. To see that $\tilde{T}_{-1}$ is bounded, we must check that the map

$$(I + U_+\Delta')x \longmapsto (I + \Delta'U_+)x$$

is bounded on $\ell^2(\mathbf{Z}_+)$. We can answer this using Proposition 3 as follows: Let

$$J := \begin{bmatrix} 1 & 0 & 0 & 0 & \\ -\delta_0 & 1 & 0 & 0 & \\ \delta_1 & -\delta_1 & 1 & 0 & \\ -\delta_2 & \delta_2 & -\delta_2 & 1 & \\ & & & & \ddots \end{bmatrix} .$$

J is a Cesàro-like operator, and it is right Fredholm; moreover,

$$N(J^*) = N(Y^*) = \langle(1, -\tfrac{1}{2}, \tfrac{1}{3}, -\tfrac{1}{4}, \ldots)\rangle \; ,$$

and

$$XJ = I + U_+$$

Therefore $J : \ell^2 \to \overline{R(Y)}$ is invertible and $XJ \in (U_+)'$, so $T_{-1} \sim U_+$. Other properties of the family $\{T_\delta\}$ are listed below.

(i) $T_\delta \simeq U_+ + K$: This uses the Brown-Douglas-Fillmore theory ([4]) and the fact that the spectral picture of $T_\delta$ coincides with that of $U_+$. (To prove that $T_\delta$ is essentially normal one finds a finite dimensional extension of $T_\delta$ that differs from $U_+$ by a Hilbert-Schmidt operator.)

(ii) $T_\delta$ is *not* a weighted shift. For, since $T_\delta \in B_1(D)$, the Cowen-Douglas class [8], $T_\delta$ has a generalized Bergman kernel $k(z,w)$ (in the sense of [10]), and one checks that $\partial k(0,0) \neq 0$, thus violating the condition:

$$T \in B_1(D) \text{ is a weighted shift} \Longleftrightarrow \partial^\alpha \partial^\beta(0,0) = 0 \text{ for all } \alpha \neq \beta)$$

([10, Theorem 5.4]).

(iii) $\|T_{-1}\| > 1$, and therefore $T_{-1}$ is *not* hyponormal. Thus, $T_{-1}$ does not fall within the scope of Clary's work.

**CASE 2.** $\lambda \neq 1$

$\tilde{T}_\delta$ does *not* extend to a bounded operator on $\overline{R(Y)}$. For,

$$\delta_n = \frac{\delta}{\lambda^n - \delta \cdot \frac{\lambda^n - 1}{\lambda - 1}} \qquad (n \geq 0)$$

in this case, and if we take a sequence of integers $n_1 < n_2 < \ldots$ with $\lambda^{n_k} \to 1$ $(k \to \infty)$, then $\delta_{n_k} \to \delta$, which can be used to break the boundedness of $T_\delta$ (some values of $\delta$ must be omitted, of course, depending upon whether $\lambda$ is a root of unity or not).

**FINAL REMARKS.** (i) We have seen that our factorizations of quadratic functions of the shift lead only to $T_Z$'s that are either similar to the shift or unbounded. Similar techniques indicate that an analogous situation is encountered when polynomials of higher degree in $U_+$ are considered. Moreover, factorizations of linear functions in the form

$$X = \sqrt{I + U_+} + \text{perturbation}$$
$$Y = \sqrt{I + U_+} + \text{perturbation}$$

appear to behave analogously. For more general elements of the commutant we do not know the answer.

**PROBLEM 3.** Can non-similar quasi-similar operators be gotten by looking at non-commutative factorizations $XY = p(U_+)$, where $p$ is a polynomial for which $p(U_+)$ is a quasi-affinity ?

(ii) Recall that

$$X_1 = X \mid \overline{R(Y)}, \text{ and } Y_1 = Y : H \to \overline{R(Y)} .$$

Observe that $X_1 Y_1 \in (S)'$, but $Y_1 X_1 \in (T)'$. When $(S)'$ is *abelian*, one can show that

$$T_Z \sim S \Rightarrow X_1 Y_1 \sim Y_1 X_1 .$$

Thus, to break similarity it suffices to break similarity for $X_1 Y_1$ and $Y_1 X_1$. (Note: $X_1 Y_1 \underset{qs}{\sim} Y_1 X_1$ always.)

(iii) The results here, combined with those in [23], provide the following program for computing the quasi-similarity orbit of an operator S.

1) Choose $Z \in Z_0(S)$

2) Compute $H_1, X_1, Y_1, \tilde{T}_Z$

3) Polar decompose $X_1 : H_1 \to H_1$ as $X_1 = U|X_1|$, where $U$ is unitary and $|X_1|$ is a quasi-affinity.

4) Assume $|X_1|$ is not invertible (otherwise $T_Z \sim S$)

5) Test the inclusion

$$SR(X_1) \subseteq R(X_1),$$

or

$$(U^*SU)R(|X_1|) \subseteq R(|X_1|),$$

using the structure of

$$A(|X_1|) := \{L \in L(H_1) : LR(|X_1|) \subseteq R(|X_1|)\},$$

which is expressed in terms of the spectral measure of $|X_1|$.

We conclude this article by briefly indicating why the operator $T_{-1}$ constructed above is not a contraction. Recall that

$$T_{-1}(e_n + \frac{n+2}{n+1} \cdot e_{n+1}) = e_{n+1} + \frac{n+3}{n+2} \cdot e_{n+2}, \qquad (n \geq 0).$$

For $x = (x_o, \ldots, x_N, 0, 0, \ldots)$, we find the norms of $T_{-1}Yx$ and $Yx$.

$$\|T_{-1}Yx\|^2 = |x_o|^2 + \sum_{n=0}^{N-1} |x_{n+1} + a_{n+1}x_n|^2 + |a_{N+1}x_N|^2$$

and

$$\|Yx\|^2 = |x_o|^2 + \sum_{n=0}^{N-1} |x_{n+1} + a_n x_n|^2 + |a_N x_N|^2.$$

Now consider $A_N, B_N : \mathbf{C}^{N+1} \to \mathbf{C}^{N+2}$ given by

$$A_N = \begin{bmatrix} 1 & & & & \\ a_o & 1 & & & \\ & a_1 & 1 & & \\ & & & \ddots & \\ & & & & 1 \\ & & & & a_N \end{bmatrix}, \quad B_N = \begin{bmatrix} 1 & & & & \\ a_1 & 1 & & & \\ & a_2 & 1 & & \\ & & & \ddots & \\ & & & & 1 \\ & & & & a_{N+1} \end{bmatrix}.$$

Then, $\|T_{-1}\| \leq 1 \Longleftrightarrow A_N^* A_N \geq B_N^* B_N$ for all $N \geq 0$. Let

$$P_N := A_N^* A_N - B_N^* B_N = \begin{bmatrix} a_0^2 - a_1^2 & a_0 - a_1 & & \\ a_0 - a_1 & a_1^2 - a_2^2 & a_1 - a_2 & \\ & a_1 - a_2 & a_2^2 - a_3^2 & \\ & & & \ddots \end{bmatrix}$$

where $a_n := \dfrac{n+2}{n+1}$ .

In turns out that $P_N \geq 0$ for $1 \leq n \leq 2000$, $n \neq 35$, $n \neq 1410$, but $P_{35}$ has a negative eigenvalue (and so does $P_{1410}$). Therefore, $\| T_{-1} \| > 1$.

## REFERENCES


1. **Apostol, C.** : Operators quasisimilar to normal operators, *Proc. Amer. Math. Soc.* **53**(1975), 104–106.

2. **Bercovici, H. ; Kérchy, L.** : On the spectra of $C_{11}$ contractions, *Proc. Amer. Math. Soc.* **95**(1985), 412–418.

3. **Bercovici, H. ; Takahashi, T.** : On the reflexivity of contractions on Hilbert space, *J. London Math. Soc. (2)* **32**(1985), 149–156.

4. **Brown, L. ; Douglas, R.G. ; Fillmore, P.** : Extensions of $C^*$-algebras and K-homology, *Ann. of Math.* **105**(1977), 265–324.

5. **Clary, W.S.** : *Quasisimilarity and subnormal operators*, Ph. D. Thesis, University of Michigan, 1973.

6. **Clary, W.S.** : Equality of spectra of quasisimilar hyponormal operators, *Proc. Amer. Math. Soc.* **53**(1975), 88–90.

7. **Conway, J.B.** : *Subnormal Operators*, Res. Notes in Math., vol. **51**, Pitman Books Ltd., London–Boston–Melbourne, 1981.

8. **Cowen, M. ; Douglas, R.G.** : Complex geometry and operator theory, *Acta Math.* **141**(1978), 187–261.

9. **Curto, R. ; Fialkow, L.** : Similarity, quasisimilarity, and operator factorizations, *Trans. Amer. Math. Soc.*, to appear.

10. **Curto, R. ; Salinas, N.** : Generalized Bergman kernels and the Cowen-Douglas theory, *Amer. J. Math.* **106**(1984), 447–488.

11. **Douglas, R.G.** : On majorization, factorization and range inclusion of operators on Hilbert spaces, *Proc. Amer. Math. Soc.* **17**(1966), 413–415.

12. **Fialkow, L.** : A note on quasisimilarity of operators, *Acta Sci. Math. (Szeged)* **39** (1977), 67–85.

13. **Fialkow, L.** : A note on quasisimilarity of operators. II, *Pacific Math. J.* **70** (1977), 151–162.

14. **Fialkow, L.** : Weighted shifts quasisimilar to quasinilpotent operators, *Acta Sci. Math. (Szeged)* **42**(1980), 71–79.

15. **Fialkow, L.** : Quasisimilarity and closures of similarity orbitrs of operators, *J. Operator Theory* **14**(1985), 215–238.

16.    **Fillmore, P. ; Williams, J.** : On operator ranges, *Adv. in Math.* **7**(1971), 254-281.

17.    **Herrero, D.A.** : On the spectra of the restrictions of an operator, *Trans. Amer. Math. Soc.* **233**(1977), 45-58.

18.    **Herrero, D.A.** : Operator algebras of finite strict multiplicity. II, *Indiana Univ. Math. J.* **27**(1978), 9-18.

19.    **Herrero, D.A.** : Quasisimilar operators with different spectra, *Acta Sci. Math. (Szeged)* **41**(1979), 101-118.

20.    **Herrero, D.A.** : On the essential spectra of quasisimilar operators, preprint 1988.

21.    **Herrero, D.A.** : Algebraic operators and invariant linear manifolds, preprint 1988.

22.    **Hoover, T.B.** : Quasisimilarity of operators, *Illinois J. Math.* **16**(1972), 678-686.

23.    **Nordgren, E. ; Radjabalipour, M. ; Radjavi, H. ; Rosenthal, P.** : On invariant operator ranges, *Trans. Amer. Math. Soc.* **251**(1979), 389-398.

24.    **Raphael, M.** : Quasisimilarity and essential spectra for subnormal operators, *Indiana Univ. Math. J.* **31**(1982), 243-246.

25.    **Sz.-Nagy, B.** : On uniformly bounded linear transformations in Hilbert spaces, *Acta Sci. Math. (Szeged)* **11**(1947), 152-157.

26.    **Sz.-Nagy, B. ; Foias, C.** : *Analyse harmonique des opérateurs de l'espace de Hilbert*, Akademiai Kiadó-Masson et Cie., Budapest-Paris, 1967.

27.    **Takahashi, K.** : On quasiaffine transforms of unilateral shifts, *Proc. Amer. Math. Soc.* **100**(1987), 683-687.

28.    **Williams, L.** : *On quasisimilarity of operators on Hilbert spaces*, Ph.D. Thesis, Univ. of Michigan, 1976.

29.    **Wu, P.Y.** : When is a contraction quasisimilar to an isometry, *Acta Sci. Math. (Szeged)* **44**(1982), 151-155.

**Raúl E. Curto**
Department of Mathematics
The University of Iowa
Iowa City, Iowa 52242
U.S.A.

**Lawrence A. Fialkow**
Department of Mathematics and
Computer Science
S.U.N.Y. at New Paltz
New Paltz, New York 12561
U.S.A.

Operator Theory:
Advances and Applications, Vol. 43
© 1990 Birkhäuser Verlag Basel



# MULTIPLICATION OPERATORS ON BERGMAN SPACES ARE REFLEXIVE

Jörg Eschmeier

Since the fundamental paper [8] of S. Brown appeared in 1978, the Scott Brown technique has become a powerful instrument to prove invariant subspace results for many different classes of operators. For quite a while it seemed that the machinery developed by S. Brown to construct invariant subspaces needed Hilbert space techniques in an essential way. One of the first, who applied the ideas of S. Brown in a Banach space context, was C. Apostol [2]. The essential idea of C. Apostol in [2] and [3] was to demand the conditions necessary to make the S. Brown technique work not for the space, but for the operator. In [1] E. Albrecht and B. Chevreau followed the opposite direction. They relaxed the conditions on the operator, but to obtain their invariant subspace results, they worked on a rather special class of Banach spaces. In [10] and [11] the author was able to prove the results of E. Albrecht and B. Chevreau without any restriction on the underlying Banach space. The main new idea was to apply a finite dimensional factorization principle due to Ch. Zenger (cf. [7], p. 20).

In the present note we shall use similar methods to prove invariant subspace results for a concrete class of examples, namely for algebras of multiplication operators on Bergman spaces. The corresponding Hilbert space results are due to H. Bercovici [5].

## 0. PRELIMINARIES

Let $\Omega$ be a bounded open set in $\mathbf{C}^d$ for a fixed integer $d \geq 1$ and let $p$ be a real number with $1 \leq p < \infty$. We denote by $L^p(\Omega)$ the $L^p$-space on $\Omega$ with respect to the restricted 2d-dimensional Lebesgue measure. As usual $H^\infty(\Omega)$ is the Banach space of all bounded analytic functions on $\Omega$ equipped with the norm $\| f \| = \sup_{z \in \Omega} | f(z) |$. Each function $f \in H^\infty(\Omega)$ induces a continuous linear operator

$$M_f : L_a^p(\Omega) \to L_a^p(\Omega), \quad g \to fg,$$

where $L_a^p(\Omega)$ is the closed subspace of $L^p(\Omega)$ consisting of all analytic p-integrable functions. We shall prove that the invariant subspace lattice **Lat**$(B)$ of the algebra

$$B = \{M_f ; f \in H^\infty(\Omega)\}$$

is extremely rich and that the algebra $B$ is super-reflexive in the sense of [12].

For simplicity we use the abbreviations $X = L_a^p(\Omega)$, $Y = L^q(\Omega)/L_a^p(\Omega)^\perp$, where q is the conjugate number of p, i.e. $(1/p) + (1/q) = 1$ and $L^q(\Omega)$ is regarded as the dual space of $L^p(\Omega)$ via

$$\langle f,g \rangle = \int_\Omega f(z)g(z)dz \quad (f \in L^p(\Omega),\ g \in L^q(\Omega)).$$

The space Y is identified with the dual space of X. It is well known that $H^\infty(\Omega)$ is a $w^*$-closed subspace of $L^\infty(\Omega)$ relative to the duality $\langle L^1(\Omega), L^\infty(\Omega) \rangle$ and that a sequence in $H^\infty(\Omega)$ converges to zero in the $w^*$-topology if and only if it is bounded and converges to zero uniformly on all compact subsets of $\Omega$. Therefore the space $H^\infty(\Omega)$ can be regarded as the dual of the separable Banach space $Q(\Omega) = L^1(\Omega)/^\perp H^\infty(\Omega)$. The spaces X and Y are $H^\infty(\Omega)$-modules in a canonical way

$$H^\infty(\Omega) \times X \to X, \quad (f,x) \to fx,$$

$$H^\infty(\Omega) \times Y \to Y, \quad (f,[g]) \to f[g] = [fg].$$

For $x \in X$, $y \in Y$ and $\mu \in \Omega$ the functionals

$$x \otimes y : H^\infty(\Omega) \to \mathbf{C}, \quad g \to \langle gx,y \rangle,$$

$$E_\mu : H^\infty(\Omega) \to \mathbf{C}, \quad g \to g(\mu)$$

are $w^*$-continuous relative to the duality with $Q(\Omega)$. For $\alpha \in \mathbf{N}^d$ and $\mu \in \Omega$ a continuous linear functional is defined by

$$E_{\mu,\alpha} : X \to \mathbf{C}, \quad f \to (d^\alpha f/dz^\alpha)(\mu) \quad ([14]),\ \text{Corollary I.1.7}).$$

For pairwise distinct pairs $(\mu_1,\alpha_1),\dots,(\mu_r,\alpha_r)$ the functionals $E_{\mu_1,\alpha_1},\dots,E_{\mu_r,\alpha_r}$ are linearly independent. One way to prove this elementary fact is to construct a sequence of polynomials $p_1,\dots,p_r$ (in d variables) with

$$\langle p_i, E_{\mu_j,\alpha_j} \rangle = \delta_{ij} \quad (1 \le i,j \le r).$$

If $\Lambda$ is a finite subset of $\Omega$ and $k \ge 0$ is an integer, then

$$X_k(\Lambda) = \{f \in X ;\ (\partial^i f/\partial z_1^i)(\lambda) = 0 \ \text{ for all } \lambda \in \Lambda \text{ and } i = 0,\dots,k\}$$

is a closed finite codimensional subspace of X belonging to $\mathbf{Lat}(B)$. If r is the number of elements in $\Lambda$, then $N_k(\Lambda) = X_k(\Lambda)^{\perp}$ is a finite dimensional space of dimension $r(k+1)$ invariant for all operators in $B^* = \{(M_f)^* ; f \in H^{\infty}(\Omega)\}$.

A central role in our invariant subspace constructions will be played by the family $F = \{N_k(\Lambda) ; \Lambda \subseteq G \text{ finite and } k = 0, 1, 2, \ldots\}$.

## 1. INVARIANT SUBSPACES

Our first aim is to prove that an arbitrary linear combination of point evaluations $\sum_{i=1}^{r} c_i E_{\lambda_i}$ $(\lambda_1, \ldots, \lambda_r \in \Omega)$ can be written in the form

$$x \otimes y = \sum_{i=1}^{r} c_i E_{\lambda_i}$$

with suitable elements $x \in X$, $y \in Y$.

**LEMMA 1.1.** *If $(y_n)$ is a sequence in Y, which converges to zero weakly relative to the duality $\langle X,Y\rangle$, then*

$$\lim_{n \to \infty} x \otimes y_n = 0$$

*holds for all $x \in X$.*

PROOF. It suffices to notice that by a normal family argument and the $L^p$-dominated convergence theorem the set

$$\{fx ; f \in H^{\infty}(\Omega) \text{ with } \| f \| \leq 1\}$$

is a compact subset of X for each element $x \in X$.

The following is the first of a series of factorization results derived in this paper.

**PROPOSITION 1.2.** *Let $r,s \geq 1$ be integers. Consider non-negative real numbers $c_1, \ldots, c_r$ with $\sum_{i=1}^{r} c_i = 1$ and pairwise distinct points $\lambda_1, \ldots, \lambda_r \in \Omega$. If $a_1, \ldots, a_s \in X$, $F \in F$ and $\epsilon > 0$ are given, then there is a space $G \in F$ and there are vectors $x \in X$, $y \in G$ with $\mathbf{max}(\| x \|, \| y \|) \leq 2$ and*

$$x \otimes y = \sum_{i=1}^{r} c_i E_{\lambda_i},$$

$$x \otimes z = 0 \quad (z \in F), \qquad \| a_i \otimes y \| < \epsilon \quad (i = 1, \ldots, s).$$

PROOF. Assume that F is of the form

$$F = N_{k_o}(\Lambda) \qquad (k_o \in \mathbf{N}, \ \Lambda \subset \Omega \ \text{finite}).$$

With no loss of generality we may assume that $\Lambda = \{\lambda_1, \ldots, \lambda_r\}$. For simplicity we use the abbreviation $N_k = N_k(\Lambda)$ for $k \geq 0$.

If $k \geq 0$ is an integer, then for each $i = 1, \ldots, r$ we write

$$N_{ki} = \{f \in X_k(\Lambda) \, ; \, (\partial^{k+1} f / \partial z_1^{k+1})(\lambda_i) = 0\}^{\perp}$$

and choose a vector $y_{ki} \in N_{ki}$ with $\| y_{ki} \| = \| y_{ki} + N_k \| = 1$. Since $N_{k+1} = N_{k1} + \ldots + N_{kr}$ and since

$$\dim(N_{k+1}/N_k) = r, \qquad \dim(N_{ki}/N_k) = 1 \qquad (i = 1, \ldots, r),$$

the elements $(y_{ki} + N_k)_{i=1}^{r}$ form a basis of $N_{k+1}/N_k$. To reduce the number of indices let us fix for the moment the integer $k \geq 0$. By a result of Ch. Zenger (cf. [7], p. 20) there is a continuous linear form $\ell$ on $Y/N_k$ and there are complex numbers $\mu_1, \ldots, \mu_r \in \mathbf{C}$ with $\| \ell \| = 1$, $\| \sum_{i=1}^{r} \mu_i y_{ki} + N_k \| = 1$ and

$$\ell(\mu_i y_{ki} + N_k) = c_i \qquad (i = 1, \ldots, r).$$

We choose an element $w \in N_k$ such that $y_k = \sum_{i=1}^{r} \mu_i y_{ki} + w$ has norm 1. Since $N_{k+1}/N_k$ can be identified isometrically with the dual space of $^{\perp}N_k/^{\perp}N_{k+1}$, there is a vector $x_k \in {}^{\perp}N_k$ with $\| x_k \| < 2$ and

$$\langle x_k, y \rangle = \ell(y + N_k)$$

for $y \in N_{k+1}$. Because $^{\perp}N_k \in \mathbf{Lat}(B)$, it follows that $x_k \otimes z = 0$ for all $z \in N_k$. Notice that $(f - f(\lambda_i))N_{ki} \subset N_k$ for each $i = 1, \ldots, r$ and each $f \in H^{\infty}(\Omega)$. The observation

$$x_k \otimes y_k(f) = \sum_{i=1}^{r} \mu_i \langle x_k, (f - f(\lambda_i))y_{ki} \rangle + \sum_{i=1}^{r} \mu_i \langle x_k, y_{ki} \rangle f(\lambda_i) = \sum_{i=1}^{r} c_i f(\lambda_i)$$

valid for all $f \in H^{\infty}(\Omega)$ shows that our task is almost achieved. To finish the proof, choose a $w^*$-convergent subsequence $(y_{n_k})_{k \geq 0}$ of $(y_k)_{k \geq 0}$ and notice that

$$x_{n_k} \otimes (y_{n_k} - y_{n_{k-1}}) = \sum_{i=1}^{r} c_i E_{\lambda_i}$$

holds for all $k \geq 1$. If k is chosen large enough, then we have

$$x_{n_k} \otimes z = 0 \quad (z \in F),$$

$$\max_{1 \leq i \leq s} \| a_i \otimes (y_{n_k} - y_{n_{k-1}}) \| < \varepsilon.$$

A decomposition into real and imaginary part, positive and negative part allows us to prove the complex version of this result.

**COROLLARY 1.3.** *Let* $r, s \geq 1$ *be integers. Consider complex numbers* $c_1, \ldots, c_r$ *with* $\sum_{i=1}^{r} |c_i| = 1$ *and let* $\lambda_1, \ldots, \lambda_r \in \Omega$ *be pairwise distinct. If* $a_1, \ldots, a_s \in X$, $F \in \mathcal{F}$ *and* $\varepsilon > 0$ *are given, then there is a space* $G \in \mathcal{F}$ *and there are vectors* $x \in X$, $y \in G$ *with*

$$\max(\| x \|, \| y \|) \leq 8 \text{ and}$$

$$\| x \otimes y - \sum_{i=1}^{r} c_i E_{\lambda_i} \| < \varepsilon,$$

$$x \otimes z = 0 \quad (z \in F), \qquad \| a_i \otimes y \| < \varepsilon \quad (i = 1, \ldots, s).$$

PROOF. Write for $i = 1, \ldots, r$

$$c_i = \sum_{k=1}^{4} \varepsilon_k c_{ki},$$

where each $\varepsilon_k$ is a complex number of modulus 1 and each $c_{ki}$ is a real number with $0 \leq c_{ki} \leq |c_i|$. By Proposition 1.2 there is a space $F_1 \in \mathcal{F}$ and there are vectors $x_1 \in X$, $y_1 \in F_1$ with $\max(\| x_1 \|, \| y_1 \|) \leq 2$ and

$$x_1 \otimes y_1 = \sum_{i=1}^{r} c_{1i} E_{\lambda_i},$$

$$x_1 \otimes z = 0 \quad (z \in F), \qquad \| a_i \otimes y_1 \| < \varepsilon/4 \quad (i = 1, \ldots, s).$$

A repeated application of Proposition 1.2 allows us to choose spaces $(F_k)_{1 \leq k \leq 4}$ and pairs $(x_k, y_k)_{1 \leq k \leq 4}$ such that $F_k \in \mathcal{F}$, $y_k \in F_k$, $\max(\| x_k \|, \| y_k \|) \leq 2$ and

$$x_k \otimes y_k = \sum_{i=1}^{r} c_{ki} E_{\lambda_i}$$

holds for each k and such that

$$x_k \otimes z = 0 \quad (z \in F), \qquad \| a_i \otimes y_k \| < \varepsilon/4 \quad (i = 1, \ldots, s),$$

$$x_k \otimes y_i = 0, \qquad \| x_i \otimes y_k \| < \epsilon/6 \quad (i = 1, \ldots, k-1).$$

Then it suffices to define $x = \sum_{k=1}^{4} \epsilon_k x_k$, $y = \sum_{k=1}^{4} y_k$ and to observe that $y$ is contained in a suitable space $G \in F$.

To solve higher dimensional factorization problems we follow a scheme developed in [4]. First, we fix some notations.

For an arbitrary set $E$ and an arbitrary integer $N \geq 1$ we define

$$E^N = \{(x_1, \ldots, x_N) \, ; \, x_1, \ldots, x_N \in E\},$$

$$M(N,E) = \{(x_{jk}) \, ; \, x_{jk} \in E \text{ for } j, k = 1, \ldots, N\}.$$

We write $E^{\mathbf{N}}$, respectively $M(\mathbf{N},E)$, for the set of all infinite sequences $(x_i)_{i \in \mathbf{N}}$, respectively matrices $(x_{jk})_{j,k \in \mathbf{N}}$, with coefficients in $E$. If $E$ is a vector space, then $E^N$, $E^{\mathbf{N}}$, $M(N,E)$, $M(\mathbf{N},E)$ are vector spaces in a natural way. If $E$ is a normed space, then we shall consider the norms

$$\| (x_1, \ldots, x_N) \| = \max_{1 \leq i \leq N} \| x_i \| \, , \quad \| (x_{jk}) \| = \max_{1 \leq j,k \leq N} \| x_{jk} \|$$

on $E^N$, respectively $M(N,E)$.

If $x = (x_1, \ldots, x_N) \in X^N$ and $y = (y_1, \ldots, y_N) \in Y^N$ (resp. $y \in Y$), then we define $x \otimes y = (x_j \otimes y_k) \in M(N,Q(\Omega))$ (resp. $x \otimes y = (x_j \otimes y) \in Q(\Omega)^N$). The corresponding notation will also be used in the infinite case.

For $\gamma > 0$ and $N \in \mathbf{N}$ we shall denote by $B(\gamma,N)$ the closed ball of radius $\gamma$ in $M(N,Q(\Omega))$. Finally, if $N \geq 1$ is an integer, then we write $Q(N,\Omega)$ to denote the set of all matrices $L \in M(N,Q(\Omega))$ with the property that for given $a_1, \ldots, a_s \in X^N$, $F \in F$ and $\epsilon > 0$ there is a space $G \in F$ and there are $x \in X^N$, $y \in G^N$ with $\max(\| x \|, \| y \|) \leq 1$ and

$$\| L - x \otimes y \| < \epsilon, \quad x \otimes z = 0 \quad (z \in F^N), \qquad \max_{1 \leq i \leq s} \| a_i \otimes y \| < \epsilon .$$

**LEMMA 1.4.** *Define* $\gamma = 1/64$. *Then we have*

$$B(\gamma,1) \subset Q(1,\Omega).$$

PROOF. Assume that $L \in B(\gamma,1)$. Let $a_1, \ldots, a_s \in X$, $F \in F$ and $\epsilon > 0$ be given. Then there are pairwise distinct points $\lambda_1, \ldots, \lambda_r \in \Omega$ and complex numbers $c_1, \ldots, c_r$ with $\sum_{i=1}^{r} | c_i | \leq \gamma$ and

$$\left\| L - \sum_{i=1}^{r} c_i E_{\lambda_i} \right\| < \epsilon/2.$$

By Corollary 1.3 there is a space $G \in F$ and there are $x \in X$, $y \in G$ with $\mathbf{max}(\|x\|,$ $\|y\|) \leq 1$ and

$$\left\| x \otimes y - \sum_{i=1}^{r} c_i E_{\lambda_i} \right\| < \epsilon/2,$$

$$x \otimes z = 0 \quad (z \in F), \qquad \| a_i \otimes y \| < \epsilon \quad (i = 1, \ldots, s).$$

To prove the matrix version of Lemma 1.4 we need the following observation.

**PROPOSITION 1.5.** *For each integer $N \geq 1$ the inclusion*

$$M(N, Q(1, \Omega)) \subset N^2 \, Q(N, \Omega)$$

*holds.*

PROOF. Consider a matrix $L = (L_{jk}) \in M(N, Q(1, \Omega))$. Let $a_1, \ldots, a_s \in X^N$, $F \in F$ and $\epsilon > 0$ be arbitrary. Fix an enumeration $(i_n)_{1 \leq n \leq N^2}$ of the set $\{(j,k) \, ; \, 1 \leq j, k \leq N\}$ and define $L_n = L_{i_n}$ for $n = 1, \ldots, N^2$. By definition there is a space $F_1 \in F$ and there are $x_1 \in X$, $y_1 \in F_1$ with $\mathbf{max}(\|x_1\|, \|y_1\|) \leq 1$ and

$$\| L_1 - x_1 \otimes y_1 \| < \epsilon/N^2, \quad x_1 \otimes z = 0 \quad (z \in F), \qquad \max_{1 \leq i \leq s} \| a_i \otimes y_1 \| < \epsilon/N.$$

Inductively, one obtains a sequence $(F_k)_{1 \leq k \leq N^2}$ of spaces in $F$ and sequences $(x_k)_{1 \leq k \leq N^2}$ in $X$, $(y_k)_{1 \leq k \leq N^2}$ in $Y$ with $y_k \in F_k$, $\mathbf{max}(\|x_k\|, \|y_k\|) \leq 1$ and

$$\| L_k - x_k \otimes y_k \| < \epsilon/N^2, \quad x_k \otimes z = 0 \quad (z \in F + F_1 + \ldots + F_{k-1}),$$

$$\max_{1 \leq i < k} \| x_i \otimes y_k \| < \epsilon/N^2, \qquad \max_{1 \leq i \leq s} \| a_i \otimes y_k \| < \epsilon/N$$

for all $k = 1, \ldots, N^2$. Define $x_{i_k} = x_k$, $y_{i_k} = y_k$ for $k = 1, \ldots, N^2$ and set

$$x = \Big( \sum_{n=1}^{N} x_{jn} \Big)_{1 \leq j \leq N}, \qquad y = \Big( \sum_{m=1}^{N} y_{mk} \Big)_{1 \leq k \leq N}.$$

Then $y \in G^N$ for a suitable space $G \in F$, and for $j, k = 1, \ldots, N$ we have the estimate

$$\left\| L_{jk} - \sum_{n,m=1}^{N} x_{jn} \otimes y_{mk} \right\| = \left\| L_{jk} - x_{jk} \otimes y_{jk} - \sum_{\substack{n,m=1 \\ (n,m)\neq(k,j)}}^{N} x_{jn} \otimes y_{mk} \right\| <$$

$$< \epsilon/N^2 + (N^2 - 1)(\epsilon/N^2) = \epsilon.$$

By construction we have

$$x \otimes z = 0 \quad (z \in F^N), \qquad \max_{1 \leq i \leq s} \| a_i \otimes y \| < \epsilon, \qquad \max(\| x \|, \| y \|) \leq N.$$

Proposition 1.5 allows us to prove the announced generalization of Lemma 1.4.

**COROLLARY 1.6.** *Define* $\gamma = 1/64$. *For each integer* $N \geq 1$ *the inclusion*

$$B((\gamma/N^2),N) \subset Q(N,\Omega)$$

*holds.*

PROOF. If $L = (L_{jk}) \in B((\gamma/N^2),N)$, then

$$N^2 L \in M(N,Q(1,\Omega)) \subset N^2 Q(N,\Omega).$$

The next result contains the main idea of the Scott Brown technique.

**PROPOSITION 1.7.** *Let* $N \geq 1$ *be an integer. Assume that there is a real number* $\gamma > 0$ *with*

$$B(\gamma,N) \subset Q(N,\Omega).$$

*Then for* $L \in M(N,Q(\Omega))$, $F \in F$, $x_0 \in X^N$, $y_0 \in F^N$ *and* $\epsilon > 0$ *there is a space* $G \in F$ *and there are* $x \in X^N$, $y \in G^N$ *with*

$$\| L - x \otimes y \| \leq (\epsilon/\gamma) \| L - x_0 \otimes y_0 \|,$$

$$\max(\| x - x_0 \|, \| y - y_0 \|) \leq (1/\gamma)^{\frac{1}{2}} \| L - x_0 \otimes y_0 \|^{\frac{1}{2}}.$$

PROOF. Without loss of generality we may assume that $\eta = \| L - x_0 \otimes y_0 \| > 0$. By assumption $M = (\gamma/\eta)(L - x_0 \otimes y_0)$ belongs to $Q(N,\Omega)$. If $\delta > 0$ is arbitrary, then there is a space $H \in F$ and there are elements $a \in X^N$, $b \in H^N$ with $\max(\| a \|, \| b \|) \leq 1$ and

$$\| M - a \otimes b \| < \delta, \qquad a \otimes y_0 = 0, \qquad \| x_0 \otimes b \| < \delta.$$

In view of

$$\| L - (x_o + (\eta/\gamma)^{\frac{1}{2}} a) \otimes (y_o + (\eta/\gamma)^{\frac{1}{2}} b) \| \leq \| (\eta/\gamma)(M - a \otimes b) \| +$$

$$+ (\eta/\gamma)^{\frac{1}{2}} \| x_o \otimes b \| < (\eta/\gamma) \delta + (\eta/\gamma)^{\frac{1}{2}} \delta$$

it is clear that the vectors

$$x = x_o + (\eta/\gamma)^{\frac{1}{2}} a, \qquad y = y_o + (\eta/\gamma)^{\frac{1}{2}} b$$

satisfy all assertions, if $\delta$ is sufficiently small.

Now all details are gathered to prove our main factorization results.

**PROPOSITION 1.8.** *Let* $N \geq 1$ *be an integer, let* $\gamma > 0$ *be a real number with*

$$B(\gamma,N) \subset Q(N,\Omega)$$

*and let* $F \in \mathcal{F}$ *be arbitrary. Then for* $L \in M(N,Q(\Omega))$, $x_o \in X^N$, $y_o \in F^N$ *and* $0 < \epsilon < \gamma$ *there are* $x \in X^N$, $y \in Y^N$ *with*

$$L = x \otimes y,$$

$$\mathbf{max}(\| x - x_o \|, \| y - y_o \|) \leq (1/(\gamma^{\frac{1}{2}} - \epsilon^{\frac{1}{2}})) \| L - x_o \otimes y_o \|^{\frac{1}{2}}.$$

PROOF. By assumption we have $c = (\epsilon/\gamma) < 1$. Define $R = \| L - x_o \otimes y_o \|$. Proposition 1.7 shows that there is a space $F_1 \in \mathcal{F}$ and that there are $x_1 \in X^N$, $y_1 \in F_1^N$ satisfying

$$\| L - x_1 \otimes y_1 \| \leq cR,$$

$$\mathbf{max}(\| x_1 - x_o \|, \| y_1 - y_o \|) \leq (R/\gamma)^{\frac{1}{2}}.$$

By a repeated application of Proposition 1.7 we obtain sequences $(F_n)$ in $F$, $(x_n)$ in $X^N$ and $(y_n)$ in $Y^N$ such that $y_n \in F_n^N$ and such that

$$\| L - x_n \otimes y_n \| \leq c^n R,$$

$$\mathbf{max}(\| x_{n+1} - x_n \|, \| y_{n+1} - y_n \|) \leq (c^{\frac{1}{2}})^n (R/\gamma)^{\frac{1}{2}}$$

holds for all integers $n \geq 0$. But then the limits

$$x = \lim_{n \to \infty} x_n, \qquad y = \lim_{n \to \infty} y_n$$

exist and satisfy

$$L = x \otimes y,$$

$$\mathbf{max}(\| x - x_0 \|, \| y - y_0 \|) \leq (R/\gamma)^{\frac{1}{2}}(1/(1 - c^{\frac{1}{2}})) = (1/(\gamma^{\frac{1}{2}} - \epsilon^{\frac{1}{2}}))R^{\frac{1}{2}}.$$

Similarly, Proposition 1.7 can be used to show that Bergman operators on $L_a^p(\Omega)$, $1 \leq p < \infty$, satisfy an $\mathbf{A}_{\aleph_0}$-condition. To do this, we fix one more notation. If $L = (L_{jk}) \in M(\mathbf{N}, Q(\Omega))$ is an infinite matrix with coefficients in $Q(\Omega)$, then for $N \in \mathbf{N}$ we define $L_N = (L_{jk})_{1 \leq j,k \leq N}$. If $x \in X^N$ for some integer $N \geq 1$, then we shall not distinguish between $x$ and the vector $\tilde{x} \in X^\mathbf{N}$ defined by

$$\tilde{x}_j = x_j \quad \text{for } 1 \leq j \leq N, \quad \tilde{x}_j = 0 \quad \text{for } j > N.$$

Analogously we regard $Y^N$ as a subspace of $Y^\mathbf{N}$.

**PROPOSITION 1.9.** *For* $L = (L_{jk}) \in M(\mathbf{N}, Q(\Omega))$ *there are sequences* $(x_j)_{j \geq 1}$ *in* X, $(y_k)_{k \geq 1}$ *in* Y *with*

$$L_{jk} = x_j \otimes y_k \quad (j, k \geq 1).$$

PROOF. By Corollary 1.8 with $\gamma = 1/64$ the inclusion

$$B(\gamma/N^2, N) \subset Q(N, \Omega)$$

holds for each integer $N \geq 1$. Let $L = (L_{jk}) \in M(\mathbf{N}, Q(\Omega))$ be given. Define $c = 1/2$, choose a real number $a > \| L_{11} \|$ and a sequence $(a_n)_{n \geq 1}$ of real numbers with $0 < a_n < 1$ and

$$a_n \mathbf{max}(\{ \| L_{nk} \| \,; 1 \leq k \leq n\} \cup \{ \| L_{jn} \| \,; 1 \leq j \leq n\}) < c^{n-2}a$$

for all $n \geq 1$. Define $x_1 = 0$, $y_1 = 0$ and

$$M = (a_j a_k L_{jk})_{j,k \geq 1} \in M(\mathbf{N}, Q(\Omega)).$$

According to Proposition 1.7 there is a space $F_2 \in F$ and there are $x_2 \in X^2$, $y_2 \in F_2^2$ with

$$\| M_2 - x_2 \otimes y_2 \| < ca,$$

$$\mathbf{max}(\| x_2 - x_1 \|, \| y_2 - y_1 \|) \leq \tfrac{1}{4}a^{\frac{1}{2}}.$$

Again by Proposition 1.7 there is a space $F_3 \in F$ and there are $x_3 \in X^3$, $y_3 \in F_3^3$ with

$$\| M_3 - x_3 \otimes y_3 \| < c^2 a,$$

$$\mathbf{max}(\|x_3 - x_2\|, \|y_3 - y_2\|) \leq (3/8)(ca)^{\frac{1}{2}}.$$

Inductively, one obtains sequences $(x_N)_{N \geq 1}$, $(y_N)_{N \geq 1}$ with

(i) $x_N \in X^N$, $y_N \in Y^N$,

(ii) $\|M_N - x_N \otimes y_N\| < c^{N-1}a$

(iii) $\mathbf{max}(\|x_{n+1} - x_N\|, \|y_{N+1} - y_N\|) \leq ((N + 1)/8)(C^{N-1}a)^{\frac{1}{2}}$

for all $N \geq 1$. Because of condition (iii) the limits

$$x = \lim_{N \to \infty} x_N, \qquad y = \lim_{N \to \infty} y_N$$

exist in $\ell^\infty(N,X)$, respectively $\ell^\infty(N,Y)$. Let us denote by $x(j)$ and $y(j)$ the $j$-th component of $x$ and $y$, respectively. It follows from condition (ii) that

$$L_{jk} = (x(j)/a_j) \otimes (y(k)/a_k)$$

holds for all $j, k \geq 1$.

The solutions of the above factorization problems can be used in a standard way to show that the invariant subspace lattice $\mathbf{Lat}(B)$ of $B$ is extremely rich (cf. [4], [5]).

**THEOREM 1.10.** *Let $\Omega$ be a bounded open set in $\mathbf{C}^d$ and let $1 \leq p < \infty$. The invariant subspace lattice of the algebra*

$$B = \{M_f \; ; \; f \in H^\infty(\Omega)\}$$

*has the following properties:*

*(i) There is an infinite dimensional Banach space $Z$ such that $\mathbf{Lat}(B)$ contains a sublattice order isomorphic to $\mathbf{Lat}(Z)$.*

*(ii) There is a sequence $(M_n)_{n \in N}$ of non-trivial spaces in $\mathbf{Lat}(B)$ such that for each family $(F_\alpha)_{\alpha \in A}$ of subsets $F_\alpha$ of $N$ with the property that $\bigcap\limits_{\alpha \in A} F_\alpha = \emptyset$ and at least one of the sets $F_\alpha$ is finite, we have $\bigcap\limits_{\alpha \in A} ( \bigvee\limits_{n \in F_\alpha} M_n) = \{0\}.$*

PROOF. Fix an arbitrary point $\mu \in \Omega$. By Proposition 1.9 there are sequences $(x_j)_{j \geq 1}$ in $X$ and $(y_k)_{k \geq 1}$ in $Y$ with

$$\delta_{jk} E_\mu = x_j \otimes y_k \quad (j, k \geq 1).$$

Then the spaces $M, N \in \mathbf{Lat}(B)$ defined by

$$M = \overline{LH}\{fx_j \; ; \; f \in H^\infty(\Omega) \text{ and } j \geq 1\},$$

$$N = \overline{LH}\{fx_j \; ; \; f \in H^\infty(\Omega) \text{ with } f(\mu) = 0 \text{ and } j \geq 1\}$$

satisfy $N \subset M$ and $(g - g(\mu))M \subset N$ for all $g \in H^\infty(\Omega)$. Because of

$$\langle fx_j, y_k \rangle = x_j \otimes y_k(f) = \delta_{jk} f(\mu) \quad (f \in H^\infty(\Omega), \; j, k \geq 1)$$

the elements $y_k$ induce linear forms $\ell_k$ on $M/N$ with

$$\ell_k(x_j + N) = \langle x_j, y_k \rangle = \delta_{jk} \quad (j, k \geq 1).$$

Therefore $M/N$ is an infinite dimensional Banach space. If $q : M \twoheadrightarrow M/N$ denotes the quotient map, then one can define a lattice embedding by

$$\mathbf{Lat}(M/N) \to \mathbf{Lat}(B), \quad L \to q^{-1}(L).$$

To prove part (ii) choose an enumeration $(\mu_n)$ of a dense subset of $\Omega$ and apply Proposition 1.9 to find families $(x_j)_{j \geq 1}$ in $X$, $(y_{kn})_{k,n \geq 1}$ in $Y$ satisfying

$$x_j \otimes y_{kn} = \delta_{jk} E_{\mu_n} \quad (j, k, n \geq 1).$$

Obviously the spaces defined by

$$M_j = \{fx_j \; ; \; f \in H^\infty(\Omega)\}^- \quad (j \geq 1)$$

are non-zero and invariant under $B$. Let $(F_\alpha)_{\alpha \in A}$ be a family of subsets of $\mathbf{N}$ such that $\underset{\alpha \in A}{\cap} F_\alpha = \emptyset$ and $F_{\alpha_0} = \{k_1, \ldots, k_s\}$ is finite. Notice that $y_{kn} \in M_j^\perp$, whenever $j \neq k$. Fix an arbitrary vector $x$ in $\underset{\alpha \in A}{\cap} \underset{j \in F_\alpha}{(\vee M_j)}$. If $k \geq 1$ is an arbitrary integer, then there is an index $\alpha$ with $k \notin F_\alpha$. Hence $\langle x, y_{kn} \rangle = 0$ holds for all integers $k, n \geq 1$. As an element in $M_{k_1} \vee \ldots \vee M_{k_s}$ the vector $x$ can be represented in the form

$$x = \lim_{\nu \to \infty} \sum_{m=1}^s f_{m\nu} x_{km}$$

with suitable functions $f_{m\nu} \in H^\infty(\Omega)$. It follows that

$$0 = \langle x, y_{k_\ell n} \rangle = \lim_{\nu \to \infty} \sum_{m=1}^s \langle f_{m\nu} x_{k_m}, y_{k_\ell n} \rangle = \lim_{\nu \to \infty} f_{\ell \nu}(\mu_n)$$

holds for all $\ell = 1, \ldots, s$ and all $n \geq 1$. Since convergence in $L_a^p(\Omega)$ implies pointwise convergence, we conclude that $x(\mu_n) = 0$ for each $n$. This completes the proof, because the sequence $(\mu_n)$ is dense in $\Omega$.

## 2. REFLEXIVITY

Let E be a complex Banach space. A subalgebra $L$ of L(E) containing the identity operator is called reflexive if $L = \mathbf{Alg\,Lat}(L)$, where by definition the right hand side consists of all operators $T \in$ L(E) with the property that $\mathbf{Lat}(L) \subset \mathbf{Lat}(T)$. Obviously a reflexive algebra $L$ is closed in the weak operator topology (WOT).

A reflexive algebra $L$ is called super-reflexive (cf. [12] or [13]), if each WOT--closed subalgebra $L_o$ of $L$ containing the identity operator is reflexive. Since by definition an operator $T \in$ L(E) is reflexive if and only if W(T), i.e. the smallest WOT-closed subalgebra of L(E) containing T and the identity operator, is reflexive, each operator contained in a super-reflexive subalgebra of L(E) is reflexive.

Our aim in this section is to show that the algebra $B = \{M_f ; f \in H^{\infty}(\Omega)\}$ is super-reflexive. The following first observation is very elementary and very well known.

**LEMMA 2.1.** *Let E be a complex Banach space and let $L \subset$ L(E) be a subalgebra. Suppose that M, N $\in \mathbf{Lat}(L)$ with $N \subset M$ and A $\in \mathbf{Alg\,Lat}(L)$ are given. If $\tilde{L}$ is defined as the subalgebra of L(M/N) consisting of all operators*

$$\tilde{T} : M/N \longrightarrow M/N, \quad x + N \longrightarrow (Tx) + N \quad (T \in L)$$

*and if $\tilde{A} \in$ L(M/N) is defined analogously, then $\tilde{A} \in \mathbf{Alg\,Lat}(\tilde{L})$.*

PROOF. If $U \in \mathbf{Lat}(\tilde{L})$, then

$$\{x \in M ; x + N \in U\} \in \mathbf{Lat}(L) \subset \mathbf{Lat}(A).$$

This implies of course that $U \in \mathbf{Lat}(\tilde{A})$.

Let $\Omega$ be a bounded open set in $\mathbf{C}^d$ and let p be a real number with $1 \leq p < \infty$. As before we denote by $B$ the algebra $B = \{M_f ; f \in H^{\infty}(\Omega)\}$ of all multiplication operators on $L_a^p(\Omega)$ induced by a function in $H^{\infty}(\Omega)$. To prove the next result we combine ideas from [6] and [9].

**PROPOSITION 2.2.** *Let A $\in \mathbf{Alg\,Lat}(B)$. There is a unique function $g : \Omega \longrightarrow \mathbf{C}$ with the property that for every $\lambda \in \Omega$ the identity*

$$g(\lambda) = \langle Ax, y \rangle$$

*holds, whenever $x \in X$, $y \in Y$ are such that*

$$E_\lambda = x \otimes y$$

*and such that there are sequences $(x_n)$ in $X$, $(y_n)$ in $Y$ with*

   (i)  $x = \lim\limits_{n \to \infty} x_n$, $y = \lim\limits_{n \to \infty} y_n$,

   (ii)  *each $y_n$ is contained in a suitable space $F_n \in F$.*

*Moreover, this function $g$ belongs to $H^\infty(\Omega)$.*

   PROOF. Let $\lambda \in \Omega$ be arbitrary. Notice that by Proposition 1.8 and its proof there are converging sequences $(x_n)$ in $X$ and $(y_n)$ in $Y$ such that these sequences and their limits $x = \lim\limits_{n \to \infty} x_n$, $y = \lim\limits_{n \to \infty} y_n$ satisfy all required assumptions. We have to show that $\langle Ax,y \rangle$ is independent of the special choice of these sequences. To this end, assume that converging sequences $(x_{1n})_{n \geq 1}$, $(x_{2n})_{n \geq 1}$ in $X$, $(y_{1n})_{n \geq 1}$, $(y_{2n})_{n \geq 1}$ in $Y$ are given such that for each $n$ there is a space $F_n$ in $F$ with $y_{in} \in F_n$ for each $i = 1, 2$ and such that the limits

$$x_i = \lim\limits_{n \to \infty} x_{in}, \qquad y_i = \lim\limits_{n \to \infty} y_{in} \qquad (i = 1, 2)$$

satisfy $x_1 \otimes y_1 = x_2 \otimes y_2$.

   We shall prove that $\langle Ax_1, y_1 \rangle = \langle Ax_2, y_2 \rangle$. To begin with, we choose a sequence $(\alpha_n)_{n \geq 1}$ of positive real numbers which converges to zero and satisfies

$$\| x_{1n} \otimes y_{1n} - x_{2n} \otimes y_{2n} \| < \alpha_n / 2$$

for $n \geq 1$. Let us fix for the moment an arbitrary integer $n \geq 1$. For each pair $(i,j)$ of integers satisfying $1 \leq i, j \leq 2$, $(i,j) \neq (2,2)$, there are points $\lambda_{ij}(1), \ldots, \lambda_{ij}(r)$ in $\Omega$ and complex numbers $\alpha_{ij}(1), \ldots, \alpha_{ij}(r)$ with

$$\| x_{in} \otimes y_{jn} - \sum_{k=1}^{r} \alpha_{ij}(k) E_{\lambda_{ij}(k)} \| < \alpha_n / 2.$$

The reader should keep in mind that the length $r$ of the above sum as well as the points $\alpha_{ij}(k)$, $\lambda_{ij}(k)$ depend on the fixed natural number $n$. For $(i,j)$ as above we define

$$L_{ij} = \sum_{k=1}^{r} \alpha_{ij}(k) E_{\lambda_{ij}(k)}.$$

Moreover, we set $L_{22} = L_{11}$ and $\alpha_{22}(k) = \alpha_{11}(k)$, $\lambda_{22}(k) = \lambda_{11}(k)$ for all $k = 1, \ldots, r$. We denote by $\Lambda$ the finite set of all points $\lambda_{ij}(k)$ with $1 \leq i, j \leq 2$ and $1 \leq k \leq r$. We may of course assume that for each $i = 1, \ldots, d$ the $i$-th components of the points $\lambda \in \Lambda$ are pairwise distinct. For all pairs $(i,j)$ with $1 \leq i, j \leq 2$, $(i,j) \neq (2,2)$, we have

$$\| x_{in} \otimes y_{jn} - L_{ij} \| < \alpha_n/2.$$

For $(i,j) = (2,2)$ we obtain

$$\| x_{2n} \otimes y_{2n} - L_{22} \| \leq \| x_{2n} \otimes y_{2n} - x_{1n} \times y_{1n} \| + \| x_{1n} \otimes y_{1n} - L_{11} \| < \alpha_n.$$

If we define $L = (L_{ij}) \in M(2, Q(\Omega))$, $x_n = (x_{in})_{i=1,2}$, $y_n = (y_{in})_{i=1,2}$, then the estimate

$$\| L - x_n \otimes y_n \| < \alpha_n$$

follows. By Proposition 1.8 there are $x'_n = (x'_{in})_{i=1,2} \in X^2$, $y'_n = (y'_{in})_{i=1,2} \in Y^2$ satisfying

$$L = x'_n \times y'_n, \quad \max(\| x'_n - x_n \|, \| y'_n - y_n \|) \leq \tfrac{1}{4}\alpha_n^{\frac{1}{2}}.$$

We claim that

$$(*) \qquad \langle Ax'_{1n}, y'_{1n} \rangle = \langle Ax'_{2n}, y'_{2n} \rangle.$$

Notice first that by the above construction the identity

$$x'_{1n} \otimes y'_{1n} = L_{11} = L_{22} = x'_{2n} \otimes y'_{2n}$$

holds. We define invariant subspaces for $B$, respectively $B^* = \{(M_f)^* ; f \in H^\infty(\Omega)\}$, by

$$M = \overline{LH}\{fx'_{in} ; f \in H^\infty(\Omega) \text{ and } i = 1, 2\},$$

$$M^* = \overline{LH}\{fy'_{in} ; f \in H^\infty(\Omega) \text{ and } i = 1, 2\}$$

and notice that $N = M \cap {}^{\perp}M_* \in \mathbf{Lat}(B)$. By Lemma 2.1 we obtain that $\tilde{A} = (A| M)/N \in \mathbf{Alg\,Lat}(\tilde{B})$.

Let $S = (S_1, \ldots, S_d)$ be the tuple consisting of the Bergman operators

$$S_i : L_a^p(\Omega) \to L_a^p(\Omega), \quad f \to z_i f \quad (1 \leq i \leq d)$$

and write $\tilde{S} = (\tilde{S}, \ldots, \tilde{S}_d)$ for the induced tuple on $M/N$. We claim that $\tilde{S}$ is annihilated by each polynomial $p$ in $d$ variables which vanishes on $\Lambda$. To verify this, notice that for $1 \leq i, j \leq 2$ the identity

$$\langle p(S)fx'_{in}, gy'_{jn} \rangle = \langle (pfg)x'_{in}, y'_{jn} \rangle = x'_{in} \otimes y'_{jn}(pfg) = \sum_{k=1}^{r} \alpha_{ij}(k) E_{\lambda_{ij}(k)}(pfg)$$

holds for all functions $f, g \in H^\infty(\Omega)$. Since for each $z \in \mathbf{C}^d \setminus \Lambda$ there is a polynomial p in d variables which vanishes on $\Lambda$, but not in z, the spectral mapping theorem for the Taylor spectrum ([15], [16]) implies that $\sigma(\tilde{S}, M/N) \subset \Lambda$.

If f is a function in $H^\infty(\Omega)$, then there is an open neighbourhood U of $\Lambda$ and a sequence $(p_k)$ of polynomials in d variables that converges to f uniformly on U. Using the continuity of the analytic functional calculus one obtains

$$\langle f(\tilde{S})[gx'_{in}], hy'_{jn}\rangle = \lim_{k\to\infty} \langle p_k gx'_{in}, hy'_{jn}\rangle = L_{ij}(fgh) = \langle \tilde{M}_f[gx'_{in}], hy'_{jn}\rangle$$

for $1 \leq i, j \leq 2$ and all functions $g, h \in H^\infty(\Omega)$. We conclude that $f(\tilde{S}) = \tilde{M}_f$. Since $f \in H^\infty(\Omega)$ was arbitrary, we deduce that $\mathbf{Lat}(\tilde{B}) = \mathbf{Lat}(\tilde{S})$.

Notice that there is a finite direct sum decomposition

$$M/N = \bigoplus_{\lambda\in\Lambda} X_\lambda$$

of M/N into spaces $X_\lambda \in \mathbf{Lat}(\tilde{S})$ such that $\sigma(\tilde{S}\,|\,X_\lambda) \subset \{\lambda\}$ for all $\lambda$ ([16]). Since $\tilde{S}$ is annihilated by each of the polynomials

$$p_i(z) = \prod_{\lambda\in\Lambda} (z_i - \lambda_i) \quad (1 \leq i \leq d)$$

and since for each i the i-th components of the points in $\Lambda$ are pairwise distinct, it follows that the inclusion

$$X_\lambda \subset \bigcap_{i=1}^{d} \mathbf{Ker}(\lambda_i - \tilde{S}_i)$$

holds for each $\lambda$ in $\Lambda$. Because of $\tilde{A} \in \mathbf{Alg\,Lat}(\tilde{S})$ there is for each $\lambda$ in $\Lambda$ a complex number $a_\lambda$ such that $\tilde{A}\,|\,X_\lambda = a_\lambda I_{X_\lambda}$. Choose a polynomial in d variables with $p(\lambda) = a_\lambda$ for all $\lambda$ in $\Lambda$. If $x \in M/N$ has the representation $x = \sum_{\lambda\in\Lambda} x_\lambda$ with $x_\lambda \in X_\lambda$ for all $\lambda$, then

$$\tilde{A}x = \sum_{\lambda\in\Lambda} p(\lambda)x_\lambda = \sum_{\lambda\in\Lambda} p(\tilde{S})x_\lambda = p(\tilde{S})x.$$

Thus we have shown that $\tilde{A} = p(\tilde{S})$. This information allows us to prove the assertion $(*)$

$$\langle Ax'_{1n}, y'_{1n}\rangle = \langle p(\tilde{S})(x'_{1n} + N), y'_{1n}\rangle = \langle p(S)x'_{1n}, y'_{1n}\rangle =$$

$$= \langle p(S)x'_{2n}, y'_{2n}\rangle = \langle Ax'_{2n}, y'_{2n}\rangle.$$

Since this relation is true for each n, we obtain

$$\langle Ax_1, y_1\rangle - \langle Ax_2, y_2\rangle = \lim_{n\to\infty} (\langle Ax_{1n}, y_{1n} - y'_{1n}\rangle +$$

$$+ \langle A(x_{1n} - x'_{1n}), y'_{1n}\rangle + \langle A(x'_{2n} - x_{2n}), y'_{2n}\rangle + \langle Ax_{2n}, y'_{2n} - y_{2n}\rangle) = 0.$$

Thus we have shown that Proposition 2.2 yields a well defined function $g : \Omega \to \mathbf{C}$. It remains to prove that $g \in H^\infty(\Omega)$. The function $g$ can be used to define a linear form on $E = LH\{E_\lambda ; \lambda \in \Omega\}$ in an obvious way by

$$\psi : E \to \mathbf{C}, \qquad \sum_{i=1}^{r} \alpha_i E_{\lambda_i} \longrightarrow \sum_{i=1}^{r} \alpha_i g(\lambda_i).$$

Since the point evaluations $E_\lambda$ $(\lambda \in \Omega)$ form a linearly independent set, this functional is well defined. Our next aim is to prove that $\psi$ is continuous. To this end, consider an element $L = \sum_{i=1}^{r} \alpha_i E_{\lambda_i} \neq 0$ in $E$ with $(\lambda_i)_{1 \leq i \leq r}$ pairwise distinct and $\alpha_i \neq 0$ for all $i$. By Proposition 1.8 and its proof there are converging sequences $(x_n)$ in $X$ and $(y_n)$ in $Y$ such that the limits $x = \lim\limits_{n \to \infty} x_n$, $y = \lim\limits_{n \to \infty} y_n$ satisfy

$$L = x \otimes y, \qquad \max(\| x \|, \| y \|) < (1/4)\| L \|^{\frac{1}{2}}$$

and such that each $y_n$ is contained in a suitable space $F_n \in F$. Choosing for each $i = 1, \ldots, r$ a polynomial $h_i$ with

$$h_i(\lambda_j) = \delta_{ij}/\alpha_i \qquad (j = 1, \ldots, r)$$

and setting $y_{in} = h_i y_n$ $(n \in \mathbf{N})$ we obtain for each $i$ a converging sequence $(y_{in})_{n \geq 1}$ with $y_{in} \in F_n$ for $n \in \mathbf{N}$ and such that the limits $y_i = h_i y$ satisfy

$$x \otimes y_i(f) = \langle fh_i x, y\rangle = x \otimes y(fh_i) = f(\lambda_i)$$

for all $f \in H^\infty(\Omega)$. Since

$$x \otimes y = L = x \otimes \sum_{i=1}^{r} \alpha_i y_i,$$

the first part of the proof implies the relation

$$\langle Ax, y\rangle = \langle Ax, \sum_{i=1}^{r} \alpha_i y_i\rangle.$$

Applying the definitions of $\psi$ and $g$ we obtain

$$\psi(L) = \sum_{i=1}^{r} \alpha_i \langle Ax, y_i\rangle = \langle Ax, y\rangle.$$

But then the continuity of $\psi$ is obvious

$$|\psi(L)| \leq \|A\| \ \|x\| \ \|y\| \leq (\|A\|/16)\|L\| .$$

Since $E$ is dense in $Q(\Omega)$, there is a function $\tilde{g} \in H^{\infty}(\Omega)$ with $\psi(L) = \langle L,\tilde{g}\rangle$ for all $L \in E$. The proof is completed by the observation that

$$\tilde{g}(\lambda) = \psi(E_\lambda) = g(\lambda)$$

holds for all $\lambda \in \Omega$.

Our next aim is to prove that in the situation of Proposition 2.2 the operator A is the multiplication operator induced by g.

**THEOREM 2.3.** *The algebra B is reflexive.*

PROOF. Let $A \in \mathbf{Alg\ Lat}(B)$ and let $g \in H^{\infty}(\Omega)$ be defined as in Proposition 2.2. We first claim that the linear space

$$Y_0 = \cup (F \ ; F \in \mathcal{F})$$

separates the points of X. To see this, notice that for each $\lambda \in \Omega$ the space $N_0(\{\lambda\})$ is the one-dimensional space spanned by the point evaluation $X \to \mathbf{C}$, $x \to x(\lambda)$.

Let $x \in X$ and $y \in Y_0$ be arbitrary. Choose a sequence $(L_n)$ in $E$ with limit $x \otimes y$ and $L_n \neq x \otimes y$ for all n. By Proposition 1.8 and the proof of Proposition 2.2 there are sequences $(x_n)$ in X and $(y_n)$ in Y with

$$L_n = x_n \otimes y_n, \quad \mathbf{max}(\| x_n - x\|, \| y_n - y\|) < (1/4)\| L_n - x \otimes y\|, \quad \langle L_n,g\rangle = \langle Ax_n,y_n\rangle$$

for all natural numbers n. From

$$\langle Ax,y\rangle = \lim_{n \to \infty} \langle Ax_n,y_n\rangle = \lim_{n \to \infty} \langle L_n,g\rangle = \langle gx,y\rangle$$

we deduce that $Ax = gx$. Therefore we obtain that $A = M_g \in B$ and the proof is complete.

To prove that the algebra $B = \{M_f \ ; f \in H^{\infty}(\Omega)\}$ is even super-reflexive it suffices by Theorem 2.3 of [13] to check that for each linear functional $\phi$ on $B$, which is continuous with respect to the (relative) weak operator topology, there are elements $x \in X$, $y \in Y$ with

$$\phi(A) = \langle Ax,y\rangle \quad (A \in B).$$

Therorem 2.3 in [13] is formulated in a Hilbert space context, but its proof remains valid in the framework of arbitrary Banach spaces. If $\phi$ is as above, then in view of the sequential continuity of the map

$$(H^{\infty}(\Omega), w^*) \to (B, WOT), \qquad f \to M_f,$$

the induced functional $L : H^{\infty}(\Omega) \to \mathbf{C}$, $f \to \phi(M_f)$, belongs to $Q(\Omega)$. Hence by Proposition 1.8 there are elements $x \in X$, $y \in Y$ with

$$\phi(M_f) = x \otimes y(f) = \langle M_f x, y \rangle$$

for all $f \in H^{\infty}(\Omega)$.

**COROLLARY 2.4.** *Let* $\Omega \subset \mathbf{C}^d$ *be a bounded open set and let* $1 \le p < \infty$. *Then the algebra* $B = \{M_f ; f \in H^{\infty}(\Omega)\}$ *is super-reflexive. In particular, each multiplication operator*

$$M_f : L_a^p(\Omega) \to L_a^p(\Omega), \qquad g \to fg,$$

*is reflexive.*


### REFERENCES

1.  **Albrecht, E. ; Chevreau, B.** : Invariant subspaces for $\ell^p$-operators having Bishop's property $(\beta)$ on a large part of their spectrum, *J. Operator Theory* **18**(1987), 339-372.

2.  **Apostol, C.** : Functional calculus and invariant subspaces, *J. Operator Theory* **4**(1980), 159-190.

3.  **Apostol, C.** : The spectral flavour of Scott Brown's techniques, *J. Operator Theory* **6**(1981), 3-12.

4.  **Apostol, C. ; Bercovici, H. ; Foiaş, C. ; Pearcy, C.** Invariant subspaces, dilation theory, and the structure of the predual of a dual algebra, *J. Funct. Anal.* **63**(1985), 369-404.

5.  **Bercovici, H.** : The algebra of multiplication operators on Bergman spaces, *Arch. Math.* **48**(1987), 165-174.

6.  **Bercovici, H. ; Foiaş, C. ; Langsam, J. ; Pearcy, C.** : (BCP)-operators are reflexive, *Michigan Math. J.* **29**(1982), 371-379.

7.  **Bonsall, F.F. ; Duncan, J.** : *Numerical ranges. II*, Cambridge University Press, 1973.

8.  **Brown, S.** : Some invariant subspaces for subnormal operators, *Integral Equations Operator Theory* **1**(1978), 310-333.

9.  **Brown, S. ; Chevreau, B.** : Toute contraction a calcul fonctionnel isométrique est réflexive, *C. R. Acad. Sci. Paris*, to appear.

10. **Eschmeier, J.** : Operators with rich invariant subspace lattices, *J. Reine Angew. Math.* **396**(1989), 41-69.

11.    **Eschmeier, J.** : Invariant subspaces for operators with Bishop's property (β) and large spectrum, preprint Münster, 1989.

12.    **Hadwin, D. ; Nordgren, E.** : Subalgebras of reflexive algebras, *J. Operator Theory* 7(1982), 3-23.

13.    **Loginov, A. ; Sulman, V.** : Hereditary and intermediate reflexivity of $W^*$-algebras (Russian), *Izv. Akad. Nauk. SSSR Ser. Mat.* **39**(1975), 1260-1273; *Math. USSR Izv.* **9**(1975), 1189-1201.

14.    **Range, R.M.** : *Holomorphic functions and integral representations in several complex variables*, New York – Berlin – Heidelberg – Tokyo, Springer Verlag, 1986.

15.    **Taylor, J.L.** : A joint spectrum for several commuting operators, *J. Funct. Anal.* **6**(1970), 172-191.

16.    **Taylor, J.L.** : The analytic functional calculus for several commuting operators, *Acta Math.* **125**(1970), 1-38.

**Jörg Eschmeier**

Mathematisches Institut
Universität Münster
Einsteinstrasse 62, D-4400 Münster,
Federal Republic of Germany.

Operator Theory:
Advances and Applications, Vol. 43
© 1990 Birkhäuser Verlag Basel



# UNIFORM BOUNDEDNESS OF THE ENERGY FOR
# TIME-PERIODIC POTENTIALS

Vladimir S. Georgiev[1]

## INTRODUCTION

The key questions in the scattering theory for time-periodic potentials are as follows:

1) the uniform boundedness of the global energy;

2) the local energy decay.

In this article we study the following perturbation of the wave equation

$$(1.1) \qquad u_{tt} - \Delta_x u = q(t,x)u, \quad t \in \mathbf{R}, \ x \in \mathbf{R}^n,$$

where $n \geq 3$ is odd. The energy Hilbert space $H_0$ associated with this equation is the closure of $C_0^\infty(\mathbf{R}^n) \times C_0^\infty(\mathbf{R}^n)$ with respect to the energy norm

$$\| f \|^2 = \int_{\mathbf{R}^n} | \nabla_x f_1(x) |^2 + | f_2(x) |^2 dx, \quad f = (f_1, f_2).$$

The potential $q(t,x)$ is assumed to satisfy the properties (see [1], [2])

$$(H_1) \quad \begin{cases} \text{a) } q(t + T,x) = q(t,x) \text{ for some positive } T, \\ \text{b) there exists } \rho > 0 \text{ such that } q(t,x) = 0 \text{ for } |x| \geq \rho, \\ \text{c) } q \in C^0(\mathbf{R}_t, L^\infty(\mathbf{R}_x^n)). \end{cases}$$

For the case of potentials independent of time the answer to the first question is positive. Moreover, due to the results of R. Phillips [12] one can construct a suitable Hilbert space on which the local energy decay holds. For the case of time-periodic potentials the results in [1], [2] say us that the local energy decays on a suitable Hilbert subspace of $H_0$ if the global energy is uniformly bounded. Therefore, it remains open the question to define the maximal Hilbert subspace of $H_0$, where the global energy is uniformly bounded.

---

[1] The author was partially supported by Bulgarian Ministry of Culture, Science and Education under Contract 52/87.

The main goal of this work is to solve this problem. More precisely, denote by $U(t,s)$ the two parameter group, such that the solution to (1.1) with initial data

$$u(s,x) = f_1(x), \quad u_t(s,x) = f_2(x)$$

can be represented by

$$(u(t,\cdot),\, u_t(t,\cdot)) = U(t,s)f, \quad f = (f_1, f_2).$$

Let

$$V = U(T,0)$$

denote the monodromy operator. The local energy decay is closely connected with the properties of the operator

$$Z_\rho = P_+^\rho V P_-^\rho,$$

where $P_+^\rho$ (resp. $P_-^\rho$) denotes the orthogonal projection onto the orthogonal complement of the outgoing space $D_+^\rho$ (resp. incoming space $D_+^\rho$). The spaces $D_\pm^\rho$ together with the operator $Z_\rho$ have been introduced by Lax and Phillips for the stationary scattering theory and used by Cooper and Strauss for moving obstacles. We refer here to [7], [5], [8], [10], where the properties of the outgoing and incoming spaces are discussed.

Consider the Hilbert space $F_b^\rho$ spanned by the generalized eigenvectors of the operator $Z_\rho^*$ with eigenvalues $|\lambda| \geq 1$, i.e.

$$F_b^\rho = \{g \in H_o \,;\, (Z_\rho^* - \lambda)^k g = 0 \text{ for some } k \in \mathbf{Z}_+,\, |\lambda| \geq 1\}.$$

Then $F_b^\rho$ is a finite dimensional subspace of $H_o$, since $(Z_\rho)^m$ is a compact operator for some large integer m according to the results in [1], [2].

**DEFINITION 1.** The *space of scattering data* is the Hilbert space $H_{sc}^\rho$ of $f \in H_o$ such that the orbit $\{V^m f \,;\, m \in \mathbf{Z}_+\}$ is orthogonal to $F_b^\rho$, i.e. $f \in H_{sc}^\rho$ iff

$$(V^m f, g) = 0 \quad \text{for any } m \in \mathbf{Z}_+ \text{ and } g \in F_b^\rho.$$

The above definition allows us to define $H_{sc}^\rho$ by the equality

$$H_{sc}^\rho = H_o \ominus \overline{\{(V^*)^m g \,;\, g \in F_b^\rho\}}.$$

**REMARK.** We use the orthogonal complement of the orbits $\{(V^*)^m g \,;\, m \in \mathbf{Z}_+\}$ for $g \in F_b^\rho$ instead of the orthogonal complement $(F_b^\rho)^\perp$, since the space $(F_b^\rho)^\perp$ might be noninvariant by V.

The main result is

**THEOREM 1.** *Suppose* $n \geq 3$ *is odd. Then the Hilbert space* $H^\rho_{sc}$ *satisfies the properties*

a) $H^\rho_{sc}$ *is invariant by* $V$,

b) *the operator* $V|\, H^\rho_{sc}$ *is power bounded*,

c) $\lim\limits_{t \to +\infty} \|\, \phi U(t,0)f \,\| = 0$ *for* $f \in H^\rho_{sc}$ *and any* $\phi(x) \in C^\infty_o(\mathbf{R}^n)$.

## 2. UNIFORM BOUNDEDNESS OF THE ENERGY

The main goal of this section is to prove the uniform boundedness of the energy on the space of scattering data $H^\rho_{sc}$.

**THEOREM 2. 1.** *The Hilbert space* $H^\rho_{sc}$ *satisfies the properties*

a) $H^\rho_{sc}$ *is invariant by* $V$,

b) *the operator* $V|\, H^\rho_{sc}$ *is power bounded.*

PROOF. The property a) follows straightway from the definition of the space of scattering data.

Set

$$K^\rho = H_o \ominus (D^\rho_+ \oplus D^\rho_-).$$

The first step in the proof is

**LEMMA 2. 2.** *The Hilbert space* $H^\rho_{sc} \cap K^\rho$ *is invariant by* $Z_\rho$ *and there exists* $\beta$, $0 < \beta < 1$ *such that*

$$\|\, (Z_\rho)^m f \,\| \leq \beta^m \|\, f \,\| \quad for\ f \in H^\rho_{sc} \cap K^\rho.$$

PROOF OF LEMMA 2.2. Denote by $U_o(t)$ the unitary group associated with the free wave equation and set $U_o = U_o(T)$. Decompose $H_o$ as

$$H_o = F^\rho_b \oplus (F^\rho_b)^\perp.$$

It suffices to show that

(2.1) $$H^\rho_{sc} \cap K^\rho = (F^\rho_b)^\perp \cap K^\rho,$$

since the space in the right hand side of the equality is invariant by $Z_\rho$ and the restriction of the operator $Z_\rho$ on this space has no eigenvalues on $\{\,|\lambda| \geq 1\}$.

The inclusion

$$H^{\rho}_{sc} \cap K^{\rho} \subseteq (F^{\rho}_{b})^{\perp} \cap K^{\rho}$$

follows from the definition of the space $H^{\rho}_{sc}$.

To prove (2.1) take $f \in (F^{\rho}_{b})^{\perp} \cap K^{\rho}$ and let us suppose $g \in F^{\rho}_{b}$ is a generalized eigenvector of $(Z_{\rho})^{*}$ of multiplicity k so that

$$(Z^{*}_{\rho} - \lambda)^{k} g = 0, \quad (Z^{*}_{\rho} - \lambda)^{k-1} g \neq 0.$$

Set

$$(2.2) \quad \begin{cases} g_{j} = (Z^{*}_{\rho} - \lambda)^{j} g, \\[2mm] d_{-,j} = (V^{*} - \lambda) g_{j} - g_{j+1}, \quad d_{-,k} = 0 \end{cases}$$

for $j = 0, 1, \ldots, k - 1$. Then $d_{-,j} \in D^{\rho}_{-}$ and we derive inductively

$$(2.3) \qquad (V^{*})^{N} g_{j} = \sum_{s=0}^{k-j-1} [\binom{N}{s} \lambda^{N-s} g_{j+s} + \sum_{r=0}^{N-s-1} c^{j,N}_{r,s} \lambda^{r} U^{N-s-r-1}_{o} d_{-,j+s}],$$

where $c^{j,N}_{r,s}$ are constants. First choose $j = k - 1$. Then (2.3) becomes

$$(V^{*})^{N} g_{k-1} = \lambda^{N} g_{k-1} + \sum_{r=0}^{N-1} c^{k-1,N}_{r,s} \lambda^{r} U^{N-r-1}_{o} d_{-,k-1}.$$

Multiplying by f we get

$$(f, (V^{*})^{N} g_{k-1}) = 0.$$

Similarly from (2.3) we obtain inductively with respect to $j = 0, 1, \ldots$ that

$$(f, (V^{*})^{N} g_{k-1-j}) = 0, \quad g_{-m} = 0 \quad \text{for } m \in \mathbf{Z}_{+}.$$

This shows $f \in H^{\rho}_{sc}$ and completes the proof of the lemma. $\qquad \square$

**LEMMA 2.3.** *The operator V is power bounded on* $H^{\rho}_{sc} \cap (D^{\rho}_{-})^{\perp}$ *and local energy decay*

$$\lim_{t \to +\infty} \| \phi U(t,0) f \| = 0 \quad \text{for } \phi(x) \in C^{\infty}_{o}(\mathbf{R}^{n})$$

*is fulfilled for any* $f \in H^{\rho}_{sc} \cap K^{\rho}$.

PROOF OF LEMMA 2.3. The key of the proof is the identity (see [2])

$$(2.4) \qquad U(t,0)f = U_{o}(t)f + \int_{0}^{t} U_{o}(t - \sigma) Q(\sigma) U(\sigma,0) f \, d\sigma, \quad Q(\sigma) = \begin{bmatrix} 0 & 0 \\ q(\sigma, \cdot) & 0 \end{bmatrix}.$$

Picking $f \in H^\rho_{sc} \cap K_\rho$ we quote Lemma 2.2 and derive

$$\| Q(\sigma)U(\sigma,0)f \| \le C \exp(-\epsilon\sigma) \| f \|.$$

Having in view the unitarity of $U_o(t)$ we obtain

$$\| V^m f \| \le C \| f \|$$

with some constant $C$ independent of $m$. On the other hand, for $f \in D^\rho_+$ we have $\| V^m f \| = \| U^m_o f \| = \| f \|$. This proves the operator $V$ is power bounded on $H^\rho_{sc} \cap (D^\rho_-)^\perp$. If $\phi(x) \in C^\infty_o(\mathbf{R}^n)$ we can use the identity (2.4) and get

$$\| \phi V^m f \| \le C \| Z^m f \| \le C\beta^m \| f \|, \qquad 0 < \beta < 1.$$

This completes the proof.                                                                          $\square$

Our next goal is to strenghten the uniform boundedness of the energy onto the space $H^\rho_{sc} \cap (D^R_-)^\perp$ where $R > \rho$. In fact we have the property

$$f \in (D^R_-)^\perp \Longrightarrow U(t,0)f \in (D^\rho_-)^\perp \quad \text{for } t > R - \rho.$$

Now choosing the integer $p$ so that $pT > R - \rho$ we conclude that

$$V^p f \in H^\rho_{sc} \cap (D^\rho_-)^\perp \quad \text{provided} \quad f \in H^\rho_{sc} \cap (D^R_-)^\perp.$$

Then the equality $V^N f = V^{N-p}V^p f$ and Lemma 2.3 lead to the following

**LEMMA 2.4.** *The operator $V$ is power bounded on $H^\rho_{sc} \cap (D^R_-)^\perp$ for $R > \rho$ and local energy decay*

$$\lim_{t \to +\infty} \| \phi U(t,0)f \| = 0 \quad \text{for } \phi(x) \in C^\infty_o(\mathbf{R}^n)$$

*is fulfilled for $f \in H^\rho_{sc} \cap (D^R_-)^\perp$.*

**COROLLARY 2.5.** *The operator $Z_R = P^R_+ V P^R_-$ has no eigenvalues with $|\lambda| \ge 1$ on $H^\rho_{sc} \cap K^R$.*

Turning to the proof of Theorem 2.1 we shall use the translation representation for the free wave equation. Recall that this representation is a unitary operator $R_n : H_o \longrightarrow L^2(\mathbf{R} \times \mathbf{S}^{n-1})$ defined by (see [7], [8])

$$R_n f = d_n D_s^{(n-1)/2}(\partial_s \tilde{f}_1 - \tilde{f}_2), \qquad d_n = 2^{-n/2}(\pi)^{(1-n)/2}$$

for n odd. Here $\tilde{f}_j(s,\omega)$ is the Radon transform of $f_j$. We refer here to [7], [9] for the properties of the translation representation and the Radon transform.

Let $h \in H^\rho_{sc} \cap D^\rho_-$. Make the following partition of the unity on $\{s < -\rho\}$

$$1 = \sum_{r=0}^{\infty} \chi_r(s),$$

where $\chi_r$ is the characteristic function of the interval

$$\Delta_r^- = \{s \; ; \; -\rho - k(r+1)T \le s \le -\rho - krT\}, \quad r = 0, 1, 2\ldots,$$

where $k$ is the maximal multiplicity of the eigenvalues $\lambda$ of the operator $Z_\rho$ with $|\lambda| \ge 1$. Set

$$h_r = R_n^{-1} \chi_r R_n h, \quad r = 0, 1, 2, \ldots .$$

It is important to note that our choice of the partition of the unity yields

(2.5) $$h_r \in H^\rho_{sc}.$$

Indeed, let $g \in F_b$ be such that $((Z_\rho^*) - \lambda)^k g = 0$ and $((Z_\rho^*) - \lambda)^{k-1} g \ne 0$. Introduce the vectors $g_j$, $d_{-,j}$ according to (2.2). Furthermore, for $h \in D^\rho_-$ we have the property

(2.6) $$h \perp g_j \text{ for } j = 0,\ldots,k-1 \quad \text{iff} \quad h \perp d_{-,j} \text{ for } j = 0,\ldots,k-1.$$

The identity (2.3) shows that $\operatorname*{supp}_s R_n d_{-,j} \subset \{-\rho - kT \le s \le -\rho\}$. Exploiting the unitary of the translation representation we see that the property $h \perp d_{-,j}$ is equivalent to

$$(R_n h, R_n d_{-,j})_{L^2(\mathbf{R} \times \mathbf{S}^{n-1})} = 0.$$

Now the choice of $\chi_r$ guarantees that

$$(\chi_r R_n h, R_n d_{-,j})_{L^2(\mathbf{R} \times \mathbf{S}^{n-1})} = 0.$$

Using (2.6) again we obtain the desired property (2.5).

Similarly, we make a partition of the unity on $\{s > \rho\}$

$$1 = \sum_{q=0}^{\infty} \theta_q(s),$$

where $\theta_q$ is the characteristic function of the interval

$$\Delta_q^+ = \{s \; ; \; \rho + kqT \le s \le \rho + k(q+1)T\}, \quad q = 0, 1, 2, \ldots .$$

**LEMMA 2.6.** *Suppose* $f \perp D_+^\rho$. *Then for any integers* $N, q \geq 0$ *we have*

$$V^N f = \sum_{q=0}^{M} U_o^{N-kq} R_n^{-1} \theta_o R_n V^{kq} f,$$

*where* M *is any integer with* $M \geq N/k + \rho/kT$.

PROOF. We lose no generality assuming $f \in S(\mathbf{R}^n) \times S(\mathbf{R}^n)$. Then the function

$$k(s,\omega,t) = R_n(U(t,0)f)(s,\omega)$$

is smooth and satisfies the equation (see [7])

$$(\partial_t + \partial_s)k = 0 \quad \text{for } s > \rho.$$

Hence $k(s,\omega,t) = k(s-\tau,\omega,t-\tau)$ for $s > \rho$, $0 < \tau < s - \rho$. Thus we derive

$$\theta_q(s)k(s,\omega,t) = \theta_o k(s-kqT,\omega,t-kqT).$$

Choosing $t = NT$ and taking advantage of the property

$$R_n(k(s-\tau,\omega)) = U_o(\tau)R_n(k(s,\omega)),$$

we get

$$R_n^{-1}\theta_q R_n V^N f = U_o^{N-kq} R_n^{-1}\theta_o R_n V^{kq} f.$$

The assumption $f \perp D_+^\rho$ implies that

$$R_n^{-1}\theta_q R_n V^N f = 0$$

for $q > N/k + \rho/kT$. Thus we derive

$$V^N f = \sum_{q=0}^{M} U_o^{N-kq} R_n \theta_o R_n^{-1} V^{kq} f.$$

This proves the lemma.                                                                            $\square$

The following identity is essential in the proof of Theorem 2.1

(2.7)
$$V^N h = \sum_{r=0}^{M} V^{N-kr} V^{kr} h_r + U_o^N \left( \sum_{r=M+1}^{\infty} h_r \right),$$

where M is any integer with $M \geq N/k + \rho/kT$. To verify (2.7) we take advantage of the equality

$$U_o^N h_r = V^N h_r$$

holding in view of our choice of $h_r$.

The norm of

$$U_o^N \left( \sum_{r=M+1}^{\infty} h_r \right)$$

is uniformly bounded, since

$$\sum \| h_r \|^2 \leq \| h \|^2.$$

Note that

$$V^{kr} h_r = U_o^{kr} h_r$$

and

$$\operatorname*{supp}_s R_n V^{kr} h_r \subset \{ -\rho - kT \leq s \leq -\rho \}.$$

Hence $P_+^R P_-^R V^{kr} h_r = V^{kr} h_r$ for $R = \rho + kT$. Thus it remains to estimate

$$\left\| \sum_{r=0}^{M} V^N h_r \right\|^2 = \sum_{r,r'} (V^N h_r, V^N h_{r'}).$$

Pick the integers $r, r'$ with $r - r' \geq 0$, quote Lemma 2.6 and find

$$[V^N h_r, V^N h_{r'}] = \sum_{q=0}^{M-r} \sum_{q'=0}^{M-r'} (U_o^{N-k(q+r)} w_{q,r}, U_o^{N-k(q'+r')} w_{q',r'}),$$

where

$$w_{q,r} = R_n^{-1} \theta_o R_n V^{k(q+r)} h_r.$$

On the other hand, we have the equality

$$(U_o^{N-k(q+r)} f, U_o^{N-k(q'+r')} f') = 0$$

provided $q + r \neq q' + r'$ and $\operatorname*{supp}_s f \cup \operatorname*{supp}_s f' \subset [\rho, \rho + kT]$. Hence, we obtain the equality

$$(2.9) \qquad (V^N h_r, V^N h_{r'}) = \sum_{q=0}^{M-r} (w_{q,r}, w_{q+r-r',r'}).$$

Furthermore we have

$$w_{q,r} = R_n^{-1} \theta_o R_n V^{kq} V^{kr} h_r = R_n^{-1} \theta_o R_n Z_{\rho+kT}^{kq} U_o^{kr} h_r.$$

Since $U_o^{kr} h_r \in H_{sc}^{\rho} \cap (D_-^R)^{\perp}$, $R = \rho + kT$, and $Z_{\rho+kT}$ has no eigenvalues $\lambda$ on $H_{sc}^{\rho} \cap (D_-^R)^{\perp}$ with $|\lambda| \geq 1$ according to the Corollary 2.5, we conclude that

$$\| w_{q,r} \| \leq c \beta^{kq} \| h_r \|$$

for some $\beta$ with $0 < \beta < 1$. Then the identity (2.9) leads to the estimates

$$| (V^N h_r , V^N h_{r'})| \le C \sum_{q=0}^{M-r} \beta^{2kq} \beta^{k(r-r')} \| h_r \| \| h_{r'} \| \le C'\beta^{k(r-r')} \| h_r \| \| h_{r'} \| .$$

From the Cauchy-Schwartz inequality and the estimate

$$\sum_{r'} \| h_{r'} \|^2 \le \| h \|^2$$

we get

$$\sum_{r'} \sum_{r \ge r'} \beta^{k(r-r')} \| h_r \| \| h_{r'} \| \le (\sum_{r'} ( \sum_{r \ge r'} \beta^{2k(r-r')} \| h_r \| )^2)^{\frac{1}{2}} \| h \| .$$

Now the Young inequality yields

$$(\sum_{r'} ( \sum_{r \ge r'} \beta^{2k(r-r')} \| h_r \| )^2)^{\frac{1}{2}} \le C' \| h \| .$$

Thus we get

$$\| \sum_{r=0}^{M} V^N h_r \| \le C \| h \| .$$

Turning back to the key identity (2.7) we obatin

$$\| V^N h \| \le C \| h \| .$$

This completes the proof of Theorem 2.1.                                                    $\square$

## 3. LOCAL ENERGY DECAY ON $H_{sc}^{\rho}$

**THEOREM 3.3.** *For any* $f \in H_{sc}^{\rho}$ *and any* $\phi(x) \in C_o^{\infty}(\mathbf{R}^n)$ *we have*

$$\lim_{N \to \infty} \| \phi V^N f \| = 0.$$

PROOF. The uniform boundedness of the energy on $H_{sc}^{\rho}$ shows it suffices to prove the local energy decay for f varying in a dense subset of $H_{sc}^{\rho}$. Thus, we loose no generality assuming

$$f \in H_{sc}^{\rho} \cap (D_-^R)^{\perp}$$

for some sufficiently large R > 0. Then Lemma 2.4 completes the proof of the theorem.    $\square$


**REFERENCES**

1.    **Bachelot, A. ; Petkov, V.** : Existence de l'opérateur de diffusion pour l'équation des ondes avec un potentiel périodique en temps, *C. R. Acad. Sci. Paris, Série I* **303**(1986), 671-673.

2.    **Bachelot, A. ; Petkov, V.** : Existence des opérateurs d'ondes pour les systèmes hyperboliques avec un potentiel périodique en temps, *Ann. Inst. H. Poincaré Sect. A (N.S.)* **47**(1987), 383-428.

3.    **Cooper, J. ; Strauss, W.** : Energy boundedness and local energy decay of waves reflecting of a moving obstacle, *Indiana Univ. Math. J.* **25**(1976), 671-690.

4.    **Cooper, J. ; Strauss, W.** : Scattering of waves by periodically moving bodies, *J. Funct. Anal.* **47**(1982), 180-229.

5.    **Cooper, J. ; Strauss, W.** : Abstract scattering theory for time periodic systems with applications to electromagnetism, *Indiana Univ. Math. J.* **34**(1985), 33-83.

6.    **Georgiev V. ; Petkov, V.** : Théorème de type RAGE pour des opérateurs à puissances bornées, *C. R. Acad. Sci. Paris, Série I*, **303**(1986), 605-608.

7.    **Lax, P. ; Phillips, R.** : *Scattering theory*, Academic Press, New York, 1967.

8.    **Melrose, R.** : Singularities and energy decay in acoustical scattering, *Duke Math. J.* **46**(1979), 43-59.

9.    **Petkov, V.** : *Scattering theory for hyperbolic operators*, Notas de Curso, No.24, Departemento de Matematica, Universidade Federal de Pernambuco, Recife, 1987.

10.   **Petkov, V. ; Georgiev, V.** : RAGE theorem for power bounded operators and local energy decay for periodically moving obstacles, preprint.

11.   **Petkov, V. ; Rangelov, Tz.** : Leading singularity of the scattering kernel for moving obstacles, preprint. *C. R. Acad. Bulg. Sci.* **40**(1987), No.12, 5-8.

12.   **Phillips, R.** : Scattering theory for the wave equation with a short range perturbation, *Indiana Univ. Math. J.* **31**(1982), 602-639.

13.   **Popov, G. ; Rangelov, Tz.** : On the exponential growth of the local energy for periodically moving obstacles, *C. R. Acad. Bulg. Sci.* **40**(1987), No.8.

14.   **Stefanov, P.** : Unicité du problème inverse de diffusion pour l'équation des ondes avec un potential dépendant du temps, *C. R. Acad. Sci. Paris, Serie I* **305**(1987), 411-413.

**Vladimir S. Georgiev**

Bulgarian Academy of Sciences
Institute of Mathematics
Laboratory of Differential Equations, 1090 Sofia
**Bulgaria.**

Operator Theory:
Advances and Applications, Vol. 43
© 1990 Birkhäuser Verlag Basel



# HOMOLOGY IN BANACH AND POLYNORMED ALGEBRAS:
# SOME RESULTS AND PROBLEMS

**A.Ya. Helemskii**

The purpose of this paper is to review the current state of some key questions of the area indicated in the title. Recently, in Romania and elsewhere, some mathematicians began to call this area "topological homology". We shall try to get rather fresh results as soon as possible. But from the very beginning we must, naturally, recall some old definitions and theorems.

Throughout the text we denote the category of Banach left (respectively, right, bi-) modules over a Banach algebra A, and of their continuous morphisms, by A-**mod** (**mod**-A, A-**mod**-A). Sometimes - it will be clear from the context - we shall use the same denotions for the category of polynormed modules of respective type over a polynormed algebra. (As a rule, all underlying spaces are complete and all inner and exterior multiplications are jointly continuous.) All these categories are additive (but not Abelian). We denote by $A_+$ the unitization of A, by $A^{op}$ the algebra opposite to A, by $A^{env}$ the envelopping algebra of A (that is $A_+ \widehat{\otimes} A_+^{op}$). We recall that thanks to these algebras any module of each type can be identified with some left module (which is also unital): a Banach A-bimodule with a Banach left $A^{env}$-module, etc. For the details see, e.g., [B, no.1][1], [B, no.174], [B], [12].

## 1. COHOMOLOGY GROUPS AND SOME RELATED RESULTS

The oldest and till now the most popular concept of topological homology is that of continuous (or, because of later modifications, "ordinary") cohomology groups. Its definition which is carrying over to the "algebras from analysis" the classical concept of Hochschild [B, no.206] is due to Kamowitz in his pioneering work [B, no.107]. In slightly generalized form (Guichardet, [B, no. 84]) it is as follows.

Let A be a fixed Banach algebra, $X \in$ A-**mod**-A. We denote by $C^n(A, X)$, $n = 1, 2, \ldots$, the space of continuous n-linear operators from A to X (in this context we

---

[1] To avoid a formidable list of literature, only the works of the last five years will be cited "personally". As to others, reference [B, no. N] means the item number N in the bibliography of the book [B].

call them *n-dimensional cochains*), and we put $C^o(A, X) = X$. Let us consider the so-called *standard cohomological complex*

$$0 \to C^o(A, X) \xrightarrow{\delta^o} \ldots \to C^n(A, X) \xrightarrow{\delta^n} C^{n+1}(A, X) \to \ldots \quad (St(A, X))$$

where "coboundary operator" $\delta^n$ is defined by $\delta^n(a_1, \ldots, a_{n+1}) = a_1 f(a_2, \ldots, a_{n+1}) +$

$+ \sum_{k=1}^{n} (-1)^k f(a_1, \ldots, a_k a_{k+1}, \ldots, a_{n+1}) + (-1)^{n+1} f(a_1, \ldots, a_n) \cdot a_{n+1}$  (in  particular,

$\delta^o x(a) = a \cdot x - x \cdot a$). The n-th cohomology of this complex, denoted by $H^n(A, X)$, is called the *continuous*, or ordinary, *n-dimensional cohomology group of Banach algebra A with coefficients in the Banach A-bimodule X*.

One can easily extend this definition to various classes of polynormed algebras and polynormed coefficients; see, e.g., [B, no. 174], [B]. (In this process, however, we can loose the natural structure of a complete prenormed space in $H^n(A, X)$, but it is not so dramatic.)

We are devoid of the possibility to dwell on numerous applications and interpretations of the continuous cohomology; see, e.g., [B, nos. 1, 3, 155], [B], [13], [19] and their bibliography. Let us recall only that Kamowitz in his cited paper aimed to apply 2-dimensional groups to the extension theory of Banach algebras, especially to the problem of Wedderburn decomposition. Later Kadison and Ringrose [B, no. 127, 128] had the principal interest in 1-dimensional cohomology, which describes the derivations "modulo inner ones" (with the bridge to automorphisms and their one-parameter groups which are important to mathematical apparatus of quantum mechanics [B, no. 118]). At last, we mention the problem of the stability of Banach algebras, or of some of their properties, after a small perturbation (Johnson [B, no. 97], Raeburn-Taylor [B, no. 158]): it involves for the most part the 2- and the 3-dimensional cohomology.

Now we can state the oldest concrete result and the oldest open question in the area.

**THEOREM 1.** (Kamowitz [B, no. 107]). *Let $\Omega$ be a compact set, X be a symmetric (that means $a \cdot x = x \cdot a$) Banach bimodule over $C(\Omega)$. Then $H^1(C(\Omega), X) = H^2(C(\Omega), X) = 0$.*

(Afterwards Johnson [B, no. 3] has managed to extend this result from $C(\Omega)$ to all amenable algebras which will be discussed later.)

**QUESTION 1.** Is it true, in the assumptions of Theorem 1, that $H^n(C(\Omega), X) = 0$ for every, or at least for some, $n \geq 3$ ?

Nevertheless, the other natural question - can one dispence with the symmetry condition? - was long ago answered in the negative. More precisely, the following general theorem is true.

**THEOREM 2.** [B, nos. 194, 200, 202]. *Let* A *be a commutative Banach algebra. Then*

(I) (as in pure algebra) $H^1(A, X) = 0$ *for all* $X \in$ A-**mod**-A *iff* $A = \mathbf{C}^m$ *(with coordinatewise multiplication) for some* m.

(II) (in contrast to pure algebra) *If* $H^2(A, X) = 0$ *for all* X, *then the spectrum of* A *is finite (hence if, in addition,* A *is semisimple, then* $A = \mathbf{C}^m$*).*

The second part of the theorem, which is more complicated, reflects some "pathological" properties of Banach structures such as the possible absence of Banach complements to subspaces. But now we shall present a rather fresh result. It shows that if one replaces Banach spaces by the nuclear ones, the cohomology behavior of an algebra obtains an enviable geometrical transparency.

**THEOREM 3.** (Ogneva [16]). *Let* $C^\infty(M)$ *be a Fréchet algebra of infinitely smooth functions on an infinitely smooth* m-*dimensional manifold* M. *Then for every* $k = 1, \ldots, m$ *there exists a Fréchet* $C^\infty(M)$-*bimodule* X *with* $H^k(C^\infty(M), X) \neq 0$. *However, if* $k > m$, *then* $H^k(C^\infty(M), X) = 0$ *for every Fréchet* $C^\infty(M)$-*bimodule* X.

Quite recently it was shown that replacing the infinite smoothness by the finite one, we obtain homologically hopeless algebras (Banach and polynormed):

**THEOREM 4.** (Kleshchiov, diploma work [20]). *Let* $C^r(M)$ *be an algebra of* r- *-smooth functions on an* r-*smooth manifold* M *(of arbitrary topological dimension),* $x \in$ $\in M$, $\mathbf{C} = \mathbf{C}(x)$ *be the complex plane equipped with* $f \cdot z = z \cdot f = f(x)z$; $f \in C^r(M)$, $z \in \mathbf{C}$. *Then* $H^k(C^r(M), \mathbf{C}) \neq 0$ *for any* $k > 0$.

*

*    *

Among the Banach (bi)modules – in what follows "Banach" is essential – one can distinguish a class which is much more tractable than in general case [B, nos. 3, 127]. Namely, a Banach left (right, bi-) module X over A is called *dual* if it has a predual space (it is dual as a Banach space to some another space), and, besides, operators $x \mapsto a \cdot x$ or/and $x \mapsto x \cdot a$ are weak* continuous for every $a \in A$. Equivalent definition: there exists a Banach right (respectively, left, bi-) module $X_*$ over A which is predual

as a space to our X and such that the operations in X can be defined by $\langle a \cdot x, x_* \rangle =$
$= \langle x, x_* \cdot a \rangle$ or/and $\langle x \cdot a, x_* \rangle = \langle x, a \cdot x_* \rangle$; $a \in A$, $x \in X$, $x_* \in X_*$.

We shall discuss "dual cohomology" ( = continuous cohomology groups with dual coefficients) later. Now we proceed to another variant of cohomology groups, besides the ordinary ones – to "normal" or "ultraweak" cohomology. These groups were introduced by Kadison and Ringrose [B, no. 127] only for operator algebras, and they reflect essentially the operator framework.

Let A be an operator $C^*$-algebra in a Hilbert space H ( = self-adjoint uniformly closed subalgebra of $B(H)$ where $B(\cdot)$ stands for all bounded operators), and let $X = (X_*)^*$ be a dual bimodule over such A. A cochain $F : A \times \ldots \times A \to X$ is called *normal* provided it is normal (that is, ultraweak-weak$^*$ continuous) separately in each variable. As to X itself, it is called *normal* provided the operators $A \to X : a \to a \cdot x$, $a \to x \cdot a$ are normal for every $x \in X$.

If X is normal, then in the standard complex $\mathbf{St}(A,X)$ there exists the subcomplex $\mathbf{St}_w(A,X)$ consisting of normal cochains (among them, by definition, are all 0-dimensional cochains). The n-th cohomology of $\mathbf{St}_w(A,X)$, denoted by $H^n_w(A,X)$, is called *normal n-dimensional cohomology group of* (operator $C^*$-) *algebra* A *with coefficients in* (normal) X.

Originally, the normal cohomology was introduced as an effective tool for computing the ordinary cohomology. The core of applications is

**THEOREM 5.** (Johnson-Kadison-Ringrose [B, no. 99]). *Let* A *be an operator* $C^*$-*algebra*, X *be a normal A-bimodule. Then* $H^n_w(A,X) \simeq H^n(A,X)$ *for every* $n \geq 0$.

To be precise, this theorem was proved by the three authors with one extra assumption on X. Quite recently, this assumption was ridded of, and the proof was simplified [11].

The following theorem, which uses Theorem 5, shows that the normal cohomology groups (or at least normal coefficients) are valuable in their own merit. The "if" part in it was obtained just in [B, no. 99], and "only if" part is due to Connes [B, no. 114] with the final stroke of Elliott [B, no. 222].

**THEOREM 6.** *Let* A *be a von Neumann algebra. Then* $H^n_w(A,X) = 0$ *for all normal* $X \in A$-**mod**-A *and* $n > 0$ *iff* A *is hyperfinite.*

Recently, still in terms of standard complexes, appeared new variants – from topological or algebraic point of view – of cohomology groups of algebras. We shall only mention the interesting results on so-called completely continuous cohomology

(Christensen-Effros-Sinclair [2]) and on the Banach modification of the cyclic cohomology of Tsygan and Connes (Christensen-Sinclair [3]).

## 2. PROJECTIVE, INJECTIVE, FLAT MODULES. EXPRESSION
## OF THE COHOMOLOGY BY MEANS OF Ext

The majority of results presented in Section 1 was obtained by methods which rely heavily on the definition of the cohomology by way of a respective standard complex - that is, which are based on the investigation of the equation $\delta^n g = f$. We call here these, sometimes very sophisticated, methods "direct".

Theorems 2 and 3, however, were proved with the help of another approach – the ideas and technics of "full" homology of Cartan, Eilenberg, MacLane. We want to recall only one achievement of their blossoming theory. Namely, they treat cohomology groups of various algebraic systems as particular cases of the unified notion of derived functor. In the computation this save us from being tied to standard complexes. Instead, we can use specially chosen resolutions and the machinery of long exact sequences.

The work of carrying over these ideas to algebras with the topology was done in [B, nos. 194-198] for Banach structures and independently by Taylor [B, no. 177] for polynormed structures of general type. (Taylor's striking results on multi-operator analytic calculus which were obtained by homological methods are stated in [B, nos. 7, 175--178]; for some further advances see Vasilescu [B, no. 15] and Putinar [B, nos. 150-152], [18]). Besides that, some concepts (e.g. **Tor**) were introduced by Kiehl. Douady and Verdier for some polynormed algebras from complex analysis (cf. [B, no. 62]).

To begin with, one must give the "right" definitions of the projectivity, injectivity (and then of the flatness as well). It was done with the help of the ideas of relative homology, originated to Hochschild [B, no. 209], as follows.

Consider the category A-**mod** (first in the Banach case). A complex in A-**mod** is said to be *admissible* if it splits (=has a contracting homotopy) as a complex in *Ban*, that is in category of Banach spaces and continuous operators. We call a given P (respectively, J) $\epsilon$ A-**mod** *projective* (*injective*) if the standard co(contra)variant functor of morphisms $_A h(P, ?)$ (respectively, $_A h(?, J)$) from A-**mod** to *Ban* sends every admissible complex to an exact complex. Equivalent definition: for any X $\epsilon$ A-**mod** every morphism P $\to$ X (respectively, X $\to$ J) in A-**mod** has a right (left) inverse in this category provided it has a right (left) inverse in *Ban*.

Further, we call an X $\epsilon$ A-**mod** *flat* if its dual $X^* \epsilon$ **mod**-A = $A^{op}$-**mod** is injective (for an equivalent definition in terms of tensor products see, e.g., [B]). Every projective module is flat but not conversely: non-trivial one-dimensional C[0,1]--modules are obviously not projective but, as we shall see later, every C[0,1]-module is flat.

Let X, Y be in A-**mod**. An admissible complex

$$0 \leftarrow X \leftarrow P_0 \leftarrow P_1 \leftarrow \ldots \qquad (0 \leftarrow X \leftarrow P)$$

$$(\text{respectively, } 0 \rightarrow Y \rightarrow J_0 \rightarrow J_1 \rightarrow \ldots \qquad (0 \rightarrow Y \rightarrow J))$$

with projective (injective) modules excepting, perhaps, X (Y) itself, is called a *projective (injective) resolution of* X (Y). Note that every Banach module certainly has pro- and injective resolutions.

Now we come to one of the principal facts of homological algebra. In our "Banach packing" it is as follows. Let $0 \leftarrow X \leftarrow P$ and $0 \rightarrow Y \rightarrow J$ be as before. Then *for every* $n \geq 0$ *the* n-*th cohomology of the complex*

$$0 \rightarrow {}_A h(P_0, Y) \rightarrow {}_A h(P_1, Y) \rightarrow \ldots \qquad ({}_A h(P, Y))$$

*coincides with the* n-*th cohomology of*

$$0 \rightarrow {}_A h(X, J_0) \rightarrow {}_A h(X, J_1) \rightarrow \ldots \qquad ({}_A h(X, J))$$

*and both of them do not depend on the particular choice of resolutions* P (J). This n-th cohomology has a special denotion $\mathbf{Ext}_A^n(X, Y)$. It is easy to observe that $\mathbf{Ext}_A^n(?, ?)$ is a co(contra)variant functor of first (second) argument.

By analogy, one can define **Ext** and preceding concepts for right modules – in this case we write $\mathbf{Ext}_{A^{op}}(\cdot, \cdot)$ – and for bimodules (we write $\mathbf{Ext}_{A^{env}}(\cdot, \cdot)$) or $\mathbf{Ext}_{A-A}(\cdot, \cdot)$). For the details and the general definition of the derived functor see, e.g., [B] or [12].

As to categories of more general polynormed (by)modules, the definitions (with the exception of that of flatness) are the same, but now one can not be sure about the existence of injective resolutions for a given module. (With the "projective" part of the theory everything is all right.) See [B, no. 177] for a detailed discussion. Here we only want to ask the rather concrete

**QUESTION 2.** Can one find a (non-normed) Fréchet algebra A which has no injective Fréchet left A-modules excepting O? In particular, what's about A = O(U); $U \subseteq \mathbf{C}^n$?

Now we proceed to the principal applications of **Ext**. Let A be a polynormed algebra; we shall consider A and $A_+$ as A-bimodules with obvious operations. Then by analogy with pure algebra we have

**THEOREM 7.** *Let* X *be a polynormed* A-*bimodule. Then* $H^n(A,X) = \mathbf{Ext}^n_{A-A}(A_+,X)$ *for every* $n \geq 0$.

In the particular case of Banach algebras — now it is essential — we have, in addition

**THEOREM 8.** (see, e.g., [B] or [12]). *Let* $X = (X_*)^*$ *be a dual Banach* A-*bimodule. Then* $H^n(A,X) = \mathbf{Ext}^n_{A-A}(X_*,A_+^*)$ *for every* $n \geq 0$.

Quite recently it turned out that the normal cohomology can also be described in terms of the "full" homology. And one is not obliged to seek for this aim any exotic "normal" **Ext**'s: the just defined standard Banach **Ext** is fitted for it, but with somewhat different arguments.

In what follows A is an operator $C^*$-algebra and $\overline{A}_*$ is the A-bimodule which is the predual of the ultraweak closure of A (in other words, $\overline{A}_*$ is the closed sub-A--bimodule of $A^*$ consisting of normal functionals).

**THEOREM 9.** [11]. *Let* $X = (X_*)^*$ *be a normal* A-*bimodule. Then* $H^n_w(A,X) =$ $= \mathbf{Ext}^n_{A-A}(X_*,\overline{A}_*)$ *for every* $n \geq 0$.

As a corollary we obtain a rather transparent proof of Theorem 5 in its slightly stronger form. In the language of **Ext**land this result sounds as follows.

**THEOREM 10.** *Let* $X = (X_*)^*$ *be a dual* A-*bimodule such that at least one of the operators* $A \rightarrow X : a \mapsto a \cdot x$ *and* $a \mapsto x \cdot a$ *is normal for all* $x \in X$. *Then* $\mathbf{Ext}^n_{A-A}(X_*,A_+^*) = \mathbf{Ext}^n_{A-A}(X_*,\overline{A}_*)$ *for every* $n \geq 0$.

The proof of both theorems relies heavily on the construction of the injective resolution for $\overline{A}_*$ from the following building material.

**THEOREM 11.** (the main lemma for Theorems 9 and 10). *Let* $A^{n*}_\alpha$ *be the space of continuous* n-*linear functionals on* A *which are normal in several (arbitrarily chosen) variables. Then* $A^{n*}_\alpha$ *is injective Banach* A-*bimodule.*

### 3. ALGEBRAS, DEFINABLE IN HOMOLOGICAL TERMS

A polynormed algebra A is called *contractible* if $H^1(A,X) = 0$ (or, equivalently, $H^n(A,X) = 0$; n = 1,2, ...) for every polynormed A-bimodule X. Note that in the case of Banach algebra we obtain the same definition if we restrict ourselves to only Banach bimodules. Further, a Banach algebra A is called *amenable* if $H^1(A,X) = 0$ (or, equivalently, $H^n(A,X) = 0$; n = 1, 2, ...) for every *dual* A-bimodule X. (Sometimes these are called *amenable-after-Johnson*, who has discovered them in [B, no. 3]). At last, an operator $C^*$-algebra A is called *amenable-after-Connes* (cf. [B, no. 113]), and also ultraweakly amenable [B, no. 130] or normally amenable [6] if $H^1(A,X) = 0$ for every *normal* A-bimodule X. (Again, like in the preceding definitions, it implies $H^n(A,X) = 0$; n = 1, 2, ... for every X of respective type, but the proof is more difficult).

In the language of the "full" homology these three classes of algebras have the following description. In parallel to the well known fact of the pure homological algebra one can prove

**THEOREM 12.** *A polynormed algebra A is contractible iff* $A_+$ *is a projective A-bimodule, or iff A is a projective A-bimodule and A has a unit.*

More specific is

**THEOREM 13.** *A Banach algebra A is amenable iff* $A_+^*$ *is injective* ( $= A_+$ *is flat*) *A-bimodule, or iff* $A^*$ *is injective* ( $= A$ *is flat*) *and A has a bounded approximate unit.*

As to the proof, it is a joint corollary of Theorem 8 and of the following general result.

**THEOREM 14.** [B, no. 205]. *Let A be a Banach algebra, let I be a closed left ideal in* $A_+$. *Then* $A_+/I \in A\text{-}\mathbf{mod}$ *is flat iff I has a right bounded approximate unit, provided that for "only if" part* $I^\perp$ *has a Banach complement in* $A_+^*$.

The following very recent result completes the picture.

**THEOREM 15.** [11]. *An operator* $C^*$-*algebra A is amenable-after-Connes iff* $A_*$ (cf. Theorem 9) *is an injective A-bimodule.*

The proof is less transparent than that of Theorems 12 and 13. It relies heavily on a deep lemma of Effros [5, Lemma 2.3] about so-called "binormal tensor squares" of certain von Neumann algebras.

Note that *every left and every right module over a contractible polynormed algebra* (respectively, *over an amenable Banach algebra*) *is projective* (respectively, *flat*). We do not know whether the converse to any of these two propositions is true. (In pure algebra the similar conjecture is false; see, e.g., Kaplansky [B, no. 109].)

*
*   *

If we weaken the contractibility condition in another direction, (compared with the amenability), we get the following rather interesting class of algebras. Namely, we call a polynormed algebra A *biprojective* if A (and not, generally speaking, $A_+$) is a projective A-bimodule. The structure of semi-simple biprojective Banach algebras with the approximation property was described by Selivanov [B, no. 166]: roughly speaking, these are topological direct sums of algebras of nuclear operators in Banach spaces. But the original stimulus to distinghuish such algebras was

**THEOREM 16.** (cf. [B, nos. 200, 202]). *Let* A *be a polynormed biprojective algebra. Then* $H^n(A,X) = 0$ *for every* $X \in$ A-**mod**-A *provided* $n \geq 3$.

(For n = 2 it is certainly not the case; see, e.g., the next section).

Let us define, into the bargain, *biflat* Banach algebras A as those with flat A $\in$ A-**mod**-A. Then the relations among all classes of algebras discussed above can be expressed by the following diagram.

(for operator $C^*$-algebras)

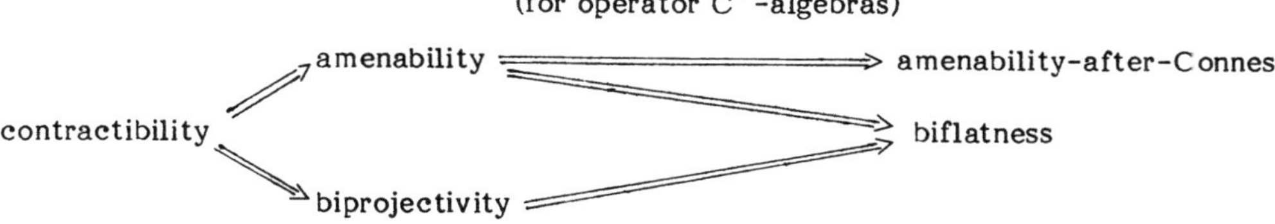


None of these arrows can change its direction, and one cannot add any arrow connecting amenability with biprojectivity.

*
*   *

Which functional-analytic properties do correspond to the amenability and other homological ones, when we discuss various concrete classes of "algebras from analysis"? One of the most profound result of this type is due to Connes ("only if", [B, no. 114]) and Haagerup ("if", [B, no. 187]):

**THEOREM 17.** *A* $C^*$-*algebra is amenable-after-Johnson iff it is a nuclear* $C^*$-*algebra ( = its envelopping von Neumann algebra is hyperfinite).*

The proof of the both parts uses essentially the normal cohomology and, in particular, Theorem 6 from above. The same theorem combined with Theorem 15 easily gives the following.

**THEOREM 18.** *An operator* $C^*$*-algebra is amenable-after-Connes iff its ultraweak closure is hyperfinite.*

Returning to abstract $C^*$-algebras we note that *such an algebra is contractible (respectively, biprojective) iff it is a direct* $C^*$*-sum of a finite (respectively, arbitrary) family of full matrix* $C^*$*-algebras* ([B, no. 199] *and* [B, no. 166]).

Let us now turn to group algebras. The appearance of the very term "amenable algebra" is connected with the following well known achievement of Johnson. Let G be a locally compact group.

**THEOREM 19.** ([B, no. 3], for alternative proofs see [B, no. 205], [B, no. 153]). *The Banach algebra* $L^1(G)$ *is amenable iff* G *is amenable as a locally compact group (that is,* G *has an invariant mean).*

At the same time $L^1(G)$ *is biprojective iff* G *is compact* [B, nos. 199, 204] and $L^1(G)$ is contractible iff G is finite.

Now we shall discuss in more details the strongest homological property – that of contractibility. We have just seen on the examples that it is indeed very rigid. However, in contrast to similarly defined separable algebras of the "pure" homology, the contractible polynormed algebras need not in general to be finite sums of matrix algebras:

**THEOREM 20.** (Taylor, [B, no. 177]; for the accurate proof see [10]). *Let A be a commutative locally multiplicatively-convex complete algebra. Then A is contractible iff every polynormed A-module is projective, or iff* $A = \mathbf{C}^M$ *(for some (arbitrary) set M) with the topology of pointwise convergence.*

Nevertheless it is open

**QUESTION 3.** Let A be a *Banach* algebra which is contractible or at least every left Banach A-module is projective. Does it imply that A is (up to an isomorphism) a finite sum of full matrix algebras?

The affirmative answer is known for a lot of algebras with "good geometry". The most useful is, perhaps, the result of Selivanov [B, no. 165]: *algebra with projective* left modules does possess the desired structure provided either every irreducible $X \in A\text{-}\mathbf{mod}$ *or* $A/\mathbf{Rad}\,A$ *has, as a space, approximate property.* Hence this is the case for $C^*$-algebras, all $L^1(G)$, commutative ones, ... . For other sufficient conditions see Taylor [B, no. 177], Curtis-Loy [4].

## 4. HOMOLOGICAL DIMENSIONS

If a given module is not projective, one can be interested by the measure to which it is worse than projective modules. This characteristic is so-called *homological dimension* of $X \in A\text{-}\mathbf{mod}$, denoted by $\mathbf{dh}_A X$. It is defined as the smallest n such that there exists a projective resolution $0 \leftarrow X \leftarrow P_0 \leftarrow P_1 \leftarrow \ldots$ of X with $P_k = 0$ for $k > n$. Equivalently, $\mathbf{dh}_A X = \{\min n : \mathbf{Ext}_A^k(X,Y) = 0 \text{ for all } k > n \text{ and } Y \in A\text{-}\mathbf{mod}\}$. If there is no such n, we put $\mathbf{dh}_A X = \infty$. The similarly defined homological dimension of a right module is denoted by $\mathbf{dh}_{A^{op}} X$, and that of a bimodule – by $\mathbf{dh}_{A\,\mathbf{env}} X$ or $\mathbf{dh}_{A-A} X$. Of course, $\mathbf{dh}_{(\cdot)} X = 0$ means just that X is projective.

Several characteristics of A itself – these are also called homological dimensions – show how nice (that is, close to projective) are some classes of A-modules. We shall discuss now only two of them. The (left) *global dimension of* A, denoted by $\mathbf{dg}A$, is $\sup\{\mathbf{dh}_A X : X \in A\text{-}\mathbf{mod}\}$. The *bidimension of* A, denoted by $\mathbf{db}A$, is $\mathbf{dh}_{A-A} A_+$. The same number (or $\infty$) – and from here one can see its importance – is the smallest n such that $H^n(A,X) = 0$ for all $k > n$ and $X \in A\text{-}\mathbf{mod}\text{-}A$.

It is known that $\mathbf{dg}A \leq \mathbf{db}A$ for every polynormed A [B], [12]. However, in contradistinction with the situation in pure algebra, there is still open

QUESTION 4. Does there exist a Banach algebra A with $\mathbf{dg}A < \mathbf{dg}A$?

What do we know about the behaviour of homological dimensions?

The oldest general result is the following (cf. Theorem 2): *one has* $\mathbf{dg}A, \mathbf{db}A \geq 2$ *for every commutative Banach algebra with infinite spectrum* [B, no. 200]. Later these estimates were extended to some other classes of algebras, for most part infinite-dimensional and semi-simple. Let us mention the recent result of Lykova [14] who has managed to establish them for CCR-algebras.

On the contrary, Theorem 16 implies $\mathbf{dg}A, \mathbf{db}A \leq 2$ for any biprojective A. Both of estimates – from below and from above – help to distinguish some family of algebras A with $\mathbf{dg}A = \mathbf{db}A = 2$. These include $C_0(M)$ with M discrete (or with its one-point compactification), $L^1(G)$ and $C^*(G)$ for a compact G, the algebra $N(H)$ of nuclear operators, ... . As to concrete modules, it turned out that $\mathbf{dh}_{\ell^1}\ell_\infty = \mathbf{dh}_{c_0}\ell_\infty = 2$

[B, no. 199] and, as non-commutative analog, $\mathbf{dh}_{N(H)}B(H) = \mathbf{dh}_{K(H)}B(H) = 2$. Note also that $\mathbf{dh}_{C[0,1]}L^p[0,1] = 1$ for $1 \leq p < \infty$ (Yakovlev, diploma work).

Group algebras of non-compact groups are much worse. Let G be a locally compact group, $I = \{f : \int_G f(s)ds = 0\}$ – the so-called augmenting ideal in $L^1(G)$. Then one has $\mathbf{dh}_{L^1(G)}L^1(G) = 0$ for all G, and also $\mathbf{dh}_{L^1(G)}I = 0$ for all compact G. But more difficult is

**THEOREM 21.** (Sheinberg, cf. [B, no. 213]). *Assume that G is non-compact and that it contains some closed infinite amenable subgroup (e.g., G is amenable itself).* *Then*

$$\mathbf{db}L^1(G) = \mathbf{dg}L^1(G) = \mathbf{dh}_{L^1(G)}I = \infty.$$

The first example of an algebra covered by this theorem is of course Wiener algebra $W = \ell_1(\mathbf{Z})$. Nobody knows whether the result holds for all non-compact groups, including that of "non-Burnside" type.

Generally speaking, the computation of homological dimensions is a rather delicate matter. Look at the following question which was mentioned by many authors (e.g., [B, nos. 95, 104], [B], [13]). Till now it presents a real challenge because of seemingly simplicity of participating algebras.

**QUESTION 5.** Which is the exact value of $\mathbf{dg}A$ and $\mathbf{db}A$ for $A = C[0,1]$ and for $K(H)$ ($K(\cdot)$ stands for compact operators)?

(To try to answer, one has to work with tensor projective powers of these "simple" algebras, and they are by no means simple ...).

Now let us pay attention to a circle of questions about numbers, which can be the values of both dimensions in various important classes of algebras. As to "most traditional" function ( = commutative and semi-simple) Banach algebras, it follows from above that the values 0, 2 and $\infty$ are permitted, but 1 is forbidden. Further, Selivanov [B, no. 161] and Krichevetz [B, no. 120] have constructed function algebras A with $\mathbf{dg}A = \mathbf{db}A = 2n$ for any given integer n.

**QUESTION 6.** Does there exist a function Banach algebra A with $\mathbf{dg}A$ and/or $\mathbf{db}A$ equal to 3? to any given odd integer?

Compared with the Banach case, the behavior of homological dimensions in certain classes of non-normed "algebras from analysis" can be much more intelligible.

For various algebras of test functions in the theory of distributions these numbers coincide, as a rule, with the number of variables [B, nos. 7, 143]. The most advanced of these "variants of Hilbert's Syzygy theorem" is the "Syzygy theorem for manifolds" due to Ogneva [16]: *for* $A = C^\infty(M)$ *one has* $\mathbf{dg}A = \mathbf{db}A = \dim M$ *where* $\dim$ *stands for topological dimension.* (As to bidimension, it is just Theorem 3 above).

The key lemma of Ogneva is as follows: *for any open set* $U \subseteq M$ *the (left) Fréchet* $C^\infty(M)$-*module* $C^\infty(U)$ *is projective.* Mihai Putinar (unpublished?) observed that if we replace a smooth manifold by Stein manifold and smooth functions by holomorphic ones, the corresponding "lemma" is false. It gives more interest to the

**QUESTION 7.** Which are the values of both dimensions for the Fréchet algebra $O(V)$ of holomorphic functions on a Stein manifold $V$ ?

(For the domains in $\mathbf{C}^n$ the answer is in affirmative: it is an old result of Taylor [B, no. 7]).

At last, we shall mention a very natural question which seems to be explicitly raised first in [7]. It concerns a possible "degree of non-amenability" of a given Banach algebra. To put it in the frame of "full" homology, let us define a *weak (or flat) homological dimension* $\mathbf{wdh}_A X$ of $X \in A$-**mod** as the smallest n such that there exists an injective resolution $0 \to X^* \to J_0 \to J_1 \to \dots$ of $X^*$ in **mod**-A with $J_k = 0$ for $k > n$. Then the *"weak bidimension of* A" $\mathbf{wdb}\,A$ which is defined as $\mathbf{wdh}_A \mathbf{env} A_+$, is just the smallest n such that $H^k(A,X) = 0$ for all $k > n$ and every *dual* $X \in A$-**mod**-A. (Thus A with $\mathbf{wdb}\,A = 0$ is just an amenable algebra).

**QUESTION 8.** Does there exist a semi-simple Banach algebra A with $\mathbf{wdb}\,A = 1$ ? with $\mathbf{wdb}\,A = n$ for a given n; $0 < n < \infty$ ?

(Extra assumptions like semi-simplicity are necessary: e.g., algebra of $2 \times 2$ matrixes with zeros in the second column provides an example of A with $\mathbf{db}\,A = \mathbf{wdb}\,A = 1$).

## REFERENCES

1.     **Bade, W.G. ; Curtis, P.C., Jr. ; Dales, H.G.** : Amenability and weak amenability for Beurling and Lipschitz algebras, *Proc. London Math. Soc.* **55**(1987), 359-377.

2.     **Christensen, E. ; Effros, E.G. ; Sinclair, A.** : Completely bounded multilinear maps and $C^*$-algebraic cohomology, *Invent. Math.* **90**(1987), 279-296.

3.     **Christensen, E. ; Sinclair, A.M.** : On the vanishing of $H^n(A,A^*)$ for certain $C^*$-*algebras*, Kobenhavns Univ., Mat. Inst., Preprint Ser. **19**(1987).

4.    **Curtis, P.C., Jr. ; Loy, R.J.** : The structure of amenable Banach algebras, *J. London Math. Soc.*, to appear.

5.    **Effros, E.G.** : Amenability and virtual diagonals for von Neumann algebras, *J. Func. Anal.* **78**(1988), 137-153.

6.    **Effros, E.G.** : Advances in quantized functional analysis, in *Proceedings of the ICM*, 1986.

7.    **Effros, E.G. ; Kishimoto, A.** : Module maps and Hochschild-Johnson cohomology, *Indiana Univ. Math. J.* **36**(1987).

8.    **Golovin, Yu.O.** : Homological properties of Hilbert modules over nest operator algebras (Russian), *Mat. Zametki* **41**:6(1987).

9.    **Groenbaek, N.** : A characterization of weakly anemable Banach algebras, *Studia Math.*, to appear.

10.   **Helemskii, A.Ya.** : Description of commutative Arens-Michael algebras of homological dimension zero, in *Theoretical and applied problems of mathematics*, Tartu State Univ., 1985, Part 1, pp. 200-201.

B.    **Helemskii, A.Ya.** : *Homology in Banach and topological algebras* (Russian), Moscow State Univ., 1986; to appear in English at Reidel, Dordrecht, 1989.

11.   **Helemskii, A.Ya.** : Homological algebra background of the "amenability after A. Connes": injectivity of the predual bimodule (Russian), *Mat. Sb.* **7**(1989).

12.   **Helemskii, A.Ya.** : *Banach and polynormed algebras. General theory. Representations. Homology* (Russian), Moscow, Nauka, to appear in 1989; English translation, Oxford University Press, 1990.

13.   **Johnson, B.E.** : Low dimensional cohomology of Banach algebras, in *Proc. Symp. Pure Math.*, **38**, Providence, 1982, Part 2, pp. 253-259.

14.   **Lykova, Z.A.** : An estimate from below of the global homological dimension of infinite-dimensional CCR-algebras (Russian), *Uspekhi Mat. Nauk* **41**:1(247), (1986), 197-198.

15.   **Krichevetz, A.N.** : On the homological dimension of $C(\Omega)$, (Russian), *VINITI*, dep. no. 9012-B86, 1986.

16.   **Ogneva, O.S.** : The coincidence of homological dimensions of the Frechet algebra of smooth functions on a manifold with the dimension of the manifold (Russian), *Funct. Anal. i Pril.* **20**(1986), 92-93.

17.   **Pugach, L.I.** : Homological properties of function algebras and analytic polydiscs in their maximal ideals spaces (Russian), *Rev. Roumaine Math. Pure Appl.* **31**(1986), 347-356.

18.   **Putinar, M.** : On analytic modules: softness and quasicoherence, in *Proc. 3-rd Conf. Complex Analysis*, Varna, 1985.

19.   **Ringrose, J.R.** : Cohomology theory for operator algebras, in *Proc. Symp. Pure Math.* **38**(1982), Providence, Part 2, pp. 229-252.

20.   **Kleshchiov, A.S.** : Homological dimension of Banach algebras of smooth functions in equal to infinite (Russian), *Vestnik Moscow Univ.*, ser. 1. mat., mech. **6**(1988), 57-60.

**A. Ya. Helemskii**

Chair of Function Theory and Functional Analysis
Faculty of Mechanics and Mathematics
Moscow State University, Moscow 119899 GSP
U.S.S.R.

Operator Theory:
Advances and Applications, Vol. 43
© 1990 Birkhäuser Verlag Basel



# LARGE ANALYTIC FUNCTIONS

**H. Helson**

## INTRODUCTION

Let f be a complex summable function on the circle **T**. f is called *analytic* if these equivalent conditions hold: the Fourier coefficients

$$
(1) \qquad a_n = (2\pi)^{-1} \int f(x) e^{-nix} dx
$$

all vanish for $n < 0$; and there is a holomorphic function F in the open disc $\Delta$ satisfying

$$
(2) \qquad \int | F(re^{ix}) | dx \le K < \infty, \quad 0 < r < 1,
$$

whose boundary function is f. The space $H^1$ is, ambiguously, the set of all such functions f or F.

If f is not summable the first definition is not available, but the second one can be generalized provided the growth condition on F implies that F has a boundary function in some sense. The spaces $H^p$ for $0 < p < 1$ are defined in this way. To obtain still larger classes of analytic functions one can form quotients F/G where F, G are analytic in $\Delta$ and G has no zeros. The class of functions so obtained depends on conditions put upon F and G.

If f is in $H^1$ and is real a.e., then f is constant; if f is in $H^{\frac{1}{2}}$ and non-negative a.e., then f is constant [4]. Easy examples show that both exponents are best possible. This note aims to investigate real and non-negative analytic functions, necessarily outside the classes $H^1$ and $H^{\frac{1}{2}}$, respectively. It will use known results and ideas about inner and outer functions, which can be found for example in [1]. The subject is closely related to that of [2], [3], [5], [6], [8]. The author is grateful to Don Sarason and Eric Hayashi for conversations about this subject.

We say *invariant subspace* to mean a closed subspace of $H^2$ invariant under multiplication by z. Beurling's theorem on the form of invariant subspaces is an important tool.

Here is our definition of analyticity for large functions defined on **T**: f is *analytic* if f is null, or if $\log |f|$ is summable and fg is in $H^2$ for every g in $H^2$ such that the product is in $L^2$. Here the class of such functions will be called $\mathfrak{A}$ , although the same class is known by the name $N^+$.

**THEOREM 1.** $\mathfrak{A}$ *is the set of all quotients g/h where g, h are in* $H^\infty$ *and h is outer. If g and h belong to any spaces* $H^p$ *with h outer, then g/h is in* $\mathfrak{A}$ .

It follows that $\mathfrak{A}$ is an algebra.

The proof, using conventional means, is omitted.

A function f in $\mathfrak{A}$ is called *outer* if it is not divisible in $\mathfrak{A}$ by a non-constant inner function. For this it is necessary and sufficient that the function g of Theorem 1 be outer. Thus an analytic function is outer if and only if its reciprocal is analytic, and then the reciprocal is also outer. In other words, the outer functions are the units of $\mathfrak{A}$ .

**COROLLARY.** *Every non-null function in* $\mathfrak{A}$ *is outer or is the product of an inner function and an outer analytic function.*

We see also that the inner factor is unique (up to constant factors).

**THEOREM 2.** *The function f in* $\mathfrak{A}$ *is outer if and only if the set of products fg, where g and fg are in* $H^2$, *is dense in* $H^2$.

Let f be analytic and non-null. Denote by $\mathfrak{M}_f$ the closure of the set of all such products fg. This is an invariant subspace of $H^2$, which by Beurling's theorem is equal to $q \cdot H^2$ for some inner q. If q is constant it is easy to see that f cannot have a non-constant inner factor, so that f is outer. Otherwise q divides f in the space of analytic functions. The statement of the theorem follows.

## 2. THE MULTIPLICATION OPERATOR

Denote of $\mathfrak{A}_r$ the class of functions f in $\mathfrak{A}$ that are real a.e., and by $\mathfrak{A}_p$ those that are non-negative.

**LEMMA.** *If f is in* $\mathfrak{A}_p$, *outer and bounded from 0 on* **T**, *then f is constant.*

The set of products fg, where g and fg are in $H^2$, is closed and therefore is all of $H^2$. Let $g = f^{-1}$, a bounded function. Then multiplication by g carries $H^2$ into itself; therefore g is in $H^\infty$. But g is non-negative, so it is constant. Thus f is constant.

Here are some examples of functions in these classes. They will be referred to again by name. The function $\phi$ defined by $\phi(z) = -i(z - 1)/(z + 1)$ is a Riemann map of $\Delta$ onto the upper half-plane. Obviously $\phi$ is in each $H^p$ for $p < 1$, so that $\phi$ belongs to $\mathfrak{A}_r$.

For any inner function q, also $\phi \circ q = -i(q - 1)/( + 1)$ is in $\mathfrak{A}_r$, and is outer.

$\phi^2$ is a Riemann map of $\Delta$ onto the complement of the interval $[0, \infty)$; this function is in $H^p$ for $p < \frac{1}{2}$, and belongs to $\mathfrak{A}_p$.

$(\phi^2 + 1)/4$ is the Koebe function $k(z) = z/(z + 1)^2$, which is a Riemann map of $\Delta$ onto the complement of $[\frac{1}{4}, \infty)$. For q inner, $k \circ q = q/(q + 1)^2$ is a function in $\mathfrak{A}_p$ whose inner factor is exactly q, because $(q + 1)^2$ is an outer function. (To see this, note that $q + 1$ takes values in the right half-plane and therefore is outer; and a product of outer functions is outer.) If f is any function in $\mathfrak{A}_p$ with inner factor q, then $g = f(q + 1)^2/q$ is in $\mathfrak{A}_p$ and is outer. This remark gives us an option in studying a given f in $\mathfrak{A}_p$ : $f + 1$ is bounded from 0 and has a bounded inverse, but cannot be outer; whereas the function g just defined is outer, but cannot be bounded from 0.

Choose f in $\mathfrak{A}_r$ and non-constant. Define $\mathfrak{D}$ to be the set of g in $H^2$ such that fg is in $L^2$ (and therefore in $H^2$). For g in $\mathfrak{D}$ let $Sg = fg$. Then S is an unbounded operator with domain $\mathfrak{D}$.

**LEMMA.** $\mathfrak{D}$ *is dense in* $H^2$.

The closure of $\mathfrak{D}$ is an invariant subspace of $H^2$, which is $q \cdot H^2$ for some inner q if it is not all of $H^2$. But if g is in $H^2$ and qg in $\mathfrak{D}$, then g itself is in $\mathfrak{D}$. Therefore the second alternative occurs.

S is a closed symmetric linear operator on $\mathfrak{D}$. Hence S − w has closed range and is invertible on its range for any non-real w. We define $\eta(w)$ to be the codimension of this range. Then $\eta$ is constant in each half-plane. If f is non-negative, $\eta$ takes the same value in both half-planes, and S has a selfadjoint extension. In this case $\eta(w)$ has the same value for real negative w as for non-real w.

S would be selfadjoint if $\eta = 0$, but this cannot happen (f being non-constant). For then multiplication by $f + i$ would map $\mathfrak{D}$ onto all of $H^2$, so that (as before) $(f + i)^{-1}$ would be a bounded analytic function; similarly $(f - i)^{-1}$ would be bounded and analytic, hence also $(f^2 + 1)^{-1}$, and f would be constant.

**LEMMA.** *Let* f *belong to* $\mathfrak{A}_r$ *and let* w *be a non-real number.* $\eta(w)$ *is finite if and only if the inner factor of* $f - w$ *is a finite Blaschke product.*

The range of S − w is $q \cdot H^2$, where q is the inner divisor of $f - w$. The codimension of this subspace equals the number of zeros in q.

**THEOREM 3.** *The defect numbers of* f *in* $\mathfrak{A}_r$ *are both finite if and only if* f *is a rational function.*

Suppose that f belongs to $\mathfrak{A}_r$ and is rational. The range of S - w is $q \cdot H^2$, where q is the inner divisor of f - w. For a rational function f this can only be a finite Blaschke product.

Next suppose that f is in $\mathfrak{A}_p$ with finite defect $\eta$. By hypothesis, the inner divisor q of f + 1 is a finite Blaschke product. Write f + 1 = qg, so that g is outer and bounded from 0. Then $k = (f + 1)^{-1} = h/q$ where h = 1/g is outer and bounded. Denote the meromorphic extension of k to $\Delta$ by K.

For $|z| > 1$ define $K(z) = \overline{K}(1/\bar{z})$. This function is analytic outside the disc and has the same boundary values as K (because k is real). Standard arguments from function theory show that K is analytic across the circle. Thus K is analytic everywhere except at zeros of q in $\Delta$ and at points symmetric to these poles outside $\Delta$. The point at infinity is also regular or a pole. Therefore K is rational, and it follows that f is rational.

Now let f be in $\mathfrak{A}_r$ with finite defect numbers. Then the inner divisor of $(f + i)(f - i) = f^2 + 1$ is a finite Blaschke product. By what has just been proved, $f^2$ is rational. Let F be the holomorphic extension of f to $\Delta$. Then $G(z) = \overline{F}(1/\bar{z})$ is holomorphic outside the circle, with the same boundary function f. Since $f^2$ is rational, F and G are bounded near the circle **T** except near the poles of $f^2$. Therefore F and G continue each other everywhere in the plane except at the poles of $f^2$, where there can only be poles. Therefore f is rational.

The mapping function $\phi$ defined above has defect numbers (1, 0), because the inner factor of $\phi$ - w is a single Blaschke factor when w lies in the upper half-plane, and $\phi$ - w is outer when w is in the lower half-plane. The function $\phi \circ q$ has defect numbers (n, 0) if q is inner with n zeros; the numbers are ($\infty$, 0) if q is inner but not a finite Blaschke product. Similarly, $k \circ q$ has defect numbers (n, n) or ($\infty$, $\infty$) if q is a Blaschke product with n zeros or infinitely many zeros, respectively. Other possibilities are less easy to realize.

**THEOREM 4.** *For m and n any non-negative integers there is a function f in* $\mathfrak{A}_r$ *such that the associated multiplication operator has defect numbers (m, n).*

To prove the theorem we construct a class of functions in $\mathfrak{A}_r$. Let $\nu$ be a positive singular measure on **T**. The function

$$(3) \qquad F(z) = i \int \frac{e^{it} + z}{e^{it} - z} d\nu(t)$$

maps $\Delta$ to the upper half-plane, and has real boundary values a.e. Since F takes values in a half-plane it belongs to $H^p$ for every $p < 1$, and thus to $\mathfrak{A}$. If $\nu$ is formed from a finite set of point masses, then F is rational. If $\nu$ is a *real* singular measure, F still belongs to $\mathfrak{A}_r$ and to $H^p$ for $p < 1$.

Now let $\nu$ consist of m point masses that are positive and n that are negative. If t is a point carrying positive mass of $\nu$, the rational function F sends $e^{it}$ to $\infty$ and maps a neighborhood of $e^{it}$ in $\Delta$ to a neighborhood of $\infty$ in the upper half-plane. Thus F on $\Delta$ covers a neighborhood of $\infty$ in the upper half-plane m times (at least). If $n = 0$, then F has exactly m poles and so restricted to $\Delta$ covers a neighborhood of $\infty$ at most m times. Thus F has defects (m, 0). If $n \neq 0$, the same argument shows that F has defects (m, n).

We can obtain defect $(\infty, n)$ by taking for $\nu$ a measure with infinitely many positive point masses, and n negative ones.

Since the function F defined by (3), where $\nu$ is a real singular measure, all belong to $H^p$ for each $p < 1$, they form only a small subset of $\mathfrak{A}_r$. However if f in $\mathfrak{A}_r$ has second defect equal to 0, that is if F takes values in the upper half-plane, then F has such a representation with $\nu$ positive (up to an additive real constant). For $-iF$ has positive real part, equal to the Poisson integral of a positive measure $\nu$, which is singular because $-iF$ has imaginary values a.e.

It is also obvious that each f in $\mathfrak{A}_r$ with second defect 0 has the form $\phi \circ q$ for some inner q, because $\phi^{-1} \circ F$ maps the disc into itself and is inner.

A different kind of parametric representation can be given for the elements of $\mathfrak{A}_p$. Let u be any summable real function on $\mathbf{T}$ whose conjugate function v takes values in the set $2\pi\mathbf{Z} + c$ for some real number c. ($\mathbf{Z}$ is the set of integers.) Then $f = C\exp(u + iv) = C\exp u$, where $C = \exp(-ic)$, belongs to $\mathfrak{A}_p$, and every *outer* member of $\mathfrak{A}_p$ is obtained in this way. If v is bounded, then f belongs to some $H^p$. We can start with v taking values in $2\pi\mathbf{Z}$, take for $-u$ the conjugate of v, and thus construct complicated outer function in $\mathfrak{A}_p$. However we cannot describe the whole class because for given v it is not easy to tell whether u is summable.

### 3. VALUE DISTRIBUTION

If f is in $\mathfrak{A}_p$ with $\eta = \infty$ the inner factor of $f - w$ is not a finite Blaschke product, for all w outside $[0, \infty)$. If this inner factor contains an infinite Blaschke product, then F assumes the value w infinitely often in $\Delta$. If the inner factor is purely singular, on the other hand, then F does not take the value w, and this can happen at least for some values of w. However, such w are limits of f along certain radii of the disc. All this brings to mind the Nevanlinna theory of value distribution. The obvious question is whether the set of w assumed by F only finitely often in $\Delta$ can be large. The next theorem shows that it is not. Denote by $\zeta(w)$ the number of $z$ in $\Delta$ (counting multiplicity) where $F(z) = w$.

**THEOREM 5.** *For $f$ in $\mathfrak{A}_p$ with $\eta = \infty$, we have $\zeta(w) = \infty$ for all $w$ in a dense $G_\delta$ of the complex plane.*

The theorem is a consequence (by Baire's theorem) of this

**LEMMA.** *For each positive integer $n$, the set $E$ of $w$ such that $\zeta(w) \geq n$ is open and dense in the complex plane.*

Suppose that $F(z) = w$ for $z = z_1, \ldots, z_n$ in $\Delta$ (where the $z_j$ may be repeated). Let $\Delta_j$ be an open disc centered at $z_j$ whose closure is contained in $\Delta$. (Choose the same disc $\Delta_j$ for repeated roots.) Let the discs be so small that they do not intersect. F carries each $\Delta_j$ onto a neighborhood of $w$. If the $z_j$ are distinct then each $w'$ in a neighborhood of $w$ is the image of at least $n$ distinct points of $\Delta$, one from each $\Delta_j$. If $z_j$ is a repeated root, $w'$ is the image of a corresponding number of points from $\Delta_j$. This proves that the set $E$ is open.

Suppose that $E$ is not dense. Find a disc $\Delta'$ not touching the real axis in which $\zeta < n$. Let $w$ be a point in $\Delta'$ such that $\zeta(s) \leq \zeta(w) = k$ for all $s$ in $\Delta'$. Let $z_1, \ldots, z_k$ be the points of $\Delta$ (counted with multiplicity) where $F = w$. For all $s$ near $w$, $F^{-1}(s)$ contains $k$ points close to the $z_j$; there are no more in $\Delta$ because $\zeta(s) \leq k$. It follows that $F(z) - w$ is bounded from 0 outside the union of $k$ discs centered at the points $z_j$ whose closures are contained in $\Delta$.

Hence the inner factor of $F(z) - w$ is a finite Blaschke product. This implies that $\eta(w)$ is finite, contrary to assumption. Thus the lemma is proved.

A theorem of Rudin [7], generalizing a classical theorem of Frostman, asserts that the inner factor of $F - w$ (where $F$ is an analytic function in $\Delta$ satisfying hypotheses that hold for functions in $\mathfrak{A}$) is a Blaschke product for all $w$ not belonging to a set of capacity 0. This result improves the conclusion of our theorem by showing that the exceptional set has capacity 0 in the plane, and furthermore that the inner factor of $f - w$ has no singular factor when $w$ is outside the exceptional set. It does not by itself, however, show that the Blaschke product is infinite; this depends on the constancy of the defect number.

## 4. STRONGLY OUTER FUNCTIONS

A function $f$ in $H^1$ is *strongly outer* if whenever $g$ is in $H^1$ and $f/g \geq 0$ a.e., necessarily g is a constant multiple of f. For f of norm 1, this means that f is an exposed point of the unit ball of $H^1$. (We are following T. Nakazi in this definition; a different definition of strongly outer functions is given in [5].) A strongly outer function is outer, for if it has the inner factor q, then $f(1 + q^2)/q = g$ is in $H^1$ with the same argument as f. Obviously the factor $(1 + q^2)/q = 1/(k \circ q)$ is not constant unless q is.

Let f be an outer function in $H^1$. Set $w = |f|$, and form the spaces $L^p(w)$ of functions on **T** based on the measure $w\,dx$. It is known [2], [3], [6] that the property of being strongly outer is related to a prediction problem in the space $L^p(w)$. Here we point out that these questions are related to the structure of $\mathfrak{A}$.

In $L^p(w)$, let $\mathfrak{F}^p$ be the closure of the set of analytic trigonometric polynomials.

**THEOREM 6.** *A function f in $H^1$ is strongly outer if and only if these equivalent conditions hold: $\mathfrak{F}^1$ and $\mathfrak{A}_p$ have only constant functions in common; $\mathfrak{F}^2$ and $\mathfrak{A}_r$ have only constant functions in common.*

The main fact needed in the proof is this

**LEMMA.** *For any finite p, the intersection of $\mathfrak{A}$ with $L^p(w)$ is $\mathfrak{F}^p$.*

The lemma shows that the functions of $\mathfrak{A}$ are exactly the limits in $L^1(w)$ of analytic trigonometric polynomials, where $w = |f|$ and f varies over the non-null elements of $H^1$.

Beyond pointing out the connection between the class $\mathfrak{A}$ and this prediction problem we cannot add anything to the results of Hayashi and Nakazi.


## REFERENCES


1.      **Duren, P.L.** : *Theory of $H^p$ spaces*, Academic Press, 1970.

2.      **Bloomfield, P.B. ; Jewell, N.P. ; Hayashi, E.** : Characterizations of completely nondeterministic stochastic processes, *Pacific J. Math.* **107** (1983), 307–317.

3.      **Hayashi, E.** : Completely nondeterministic processes and exposed points in $H^1$, preprint.

4.      **Helson, H. ; Sarason, D.** : Past and future, *Math. Scand.* **21**(1967), 5–16.

5.      **deLeeuw, K. ; Rudin, W.** : Extreme points and extremum problems in $H_1$, *Pacific J. Math.* **8**(1958), 467–485.

6.      **Nakazi, T.** : Exposed points and extremal problems in $H^1$, *J. Func. Anal.* **53** (1983), 224–230.

7.      **Rudin, W.** : A generalization of a theorem of Frostman, *Math. Scand.* **21** (1967), 136–143.

8.     **Yabuta, K.** : Some uniqueness theorems for $H^p(U^n)$ functions, *Tôhoku Math. J.* **24** (1972), 353-357.

**Henry Helson**
Mathematics Department
University of California
Berkeley, California 94720
U.S.A.

Operator Theory:
Advances and Applications, Vol. 43
© 1990 Birkhäuser Verlag Basel



# TOEPLITZ OPERATORS ON THE SEGAL – BARGMANN SPACE
# OF INFINITELY MANY VARIABLES

**J. Janas** and **K. Rudol**

## INTRODUCTION

The importance of Toeplitz operators acting on the Segal – Bargmann space $B_N$ (also referred to as the Fock space) became clear when V. Bargmann obtained in [1], [2] a model for certain quantum mechanics operators. Such fundamental concepts as the creation, or annihilation operators (of a Bose gas particle at state $e_j$) can be represented as Toeplitz operators on $B_N$, the symbols being the coordinate functions $z_j$ on $\mathbf{C}^N$, or their complex conjugates $\bar{z}_j$.

The integer N corresponds to the number of degrees of freedom, hence in some physically significant cases we have $N = \infty$. Our standing assumption is the separability of spaces and $\infty$ means "countable infinity".

The "finite-dimensional" case, $N < \infty$, investigated by Coburn and Berger in [3], [4] has motivated our study suggesting many important questions. Our aim here is to consider the possibilities of extending their results to the case $N = \infty$. This means that the underlying space $\mathbf{C}^N$ is replaced by a separable Hilbert space H. As we shall see, certain properties, true for $N < \infty$, now fail. The main source of difficulties is the behaviour of the Gaussian measures $\mu_N$ as $N \to \infty$ : the obtained formal limit, a cylindric measure on $H = \ell^2$, fails to be $\sigma$-additive. Only its cylindric extension $\mu$ (our $\mu_\infty$) to some larger space $H_-$ has this property. Since now $\mu(H) = 0$, even the definition of the Bargmann space $B_N$ needs a cautious generalization, if one wants to define Toeplitz operators on $B \overset{\text{def}}{=} B_\infty$ in the same way as for $N < \infty$.

Despite the mentioned difficulties, a number of positive results, mainly generalizations of [3], [4] hold in our setting. In § 2 we construct a canonical isomorphism between the Fock space $\exp(H)$ and $B$. This provides a "justification" for our definition of $B$ as a subspace of $L^2(\mu)$. Next, we present a few basic properties of Toeplitz operators on $B$. In § 3 we study the Weyl system over H. As in [3], [4], this is related to certain Toeplitz operators via the *Basic Identities* that generalize to the case $N = \infty$ and yield a form of the CCR relations. Section 4 contains examples pointing out a feature that has no counterpart for $N < \infty$ — the lack of compact Toeplitz operators.

There also exists another approach to analytic models of the Fock space and the Canonical Commutation Relations as $N = \infty$. Segal uses in [10] a form of inductive limit techniques. The model space, however, becomes more abstract.

## 1. BACKGROUND

In this section we recall two basic concepts related to the Bargmann spaces. Let us begin with the *Gaussian measure* $\mu = \mu_\infty$ on a separable Hilbert space H. It generalizes the measures $\mu_N$ defined for $N < \infty$ on $\mathbf{C}^N$ by the density $\mathbf{exp}(-|z|^2/2)$ w.r. to the *Lebesgue measure* (volume element) $dV(z)$. The setting for our $\mu$ will be a triple of (complex) spaces: $E_+ \subseteq H \subseteq E_-$, the inclusion mappings being continuous and with dense ranges, where $E_-$ is a Fréchet space, $E_+ = (E_-)^*$ and for $\phi \in E_+$, $z \in H$ we have $\phi(z) = \langle z, \phi \rangle$. Here $\langle \cdot, \cdot \rangle$ denotes the inner product of H and from now on it will be used also for the pairing: $E_- \times E_+ \to \mathbf{C}$. About the norm in H, denoted by $|\cdot|$ we assume that

(1.1)   the mapping: $z \to |z|$ is continous on $E_+$ in its nuclear (Sazonov) topology.

This is the case e.g. if $E_+$ is nuclear, its inclusion in H being continuous, or if $E_+$ is a Hilbert space such that

(1.1a)   the inclusion: $E_+ \subseteq H$ is a Hilbert – Schmidt operator.

The Milnos – Sazonov theorem gives under (1.1) a probabilistic measure $\mu$ on $E_-$, whose Fourier transform $\hat{\mu}$ satisfies $\hat{\mu}(x) = \mathbf{exp}(-|x|^2/2)$ for $x \in E_+$   [8]. The assumption (1.1a) appears more restrictive than it really is: recall that in the important case $E_+ = \{x \in \ell^\alpha : x_n = 0 \text{ for n large enough}\}$, $H = \ell^2$, the measure $\mu$ is actually carried by a Hilbert space containing H [7].

We shall use throughout this paper a coordinate system $(z_j)$ arising from a fixed orthogonal basis $(k_j)$ $(j = 1, 2, \ldots)$ of H.

**1.2. FACT.** Assume that $k_j \in E_+$ $(\forall j)$. For $z \in E_-$ put $z_j(z) = \langle z, k_j \rangle$. For *cylindric* $L^1$-*functions*, i.e. the functions f of the form $f(z) = \psi_N(z_1, \ldots, z_N)$ where $N < \infty$, $\psi_N \in L^1(\mu_N)$ we have $f \in L^1(\mu)$ and

$$\int f d\mu = \int \psi_N d\mu_N = \int \psi_N(z) \, \mathbf{exp}(-|z|^2/2) d\mu_N(z).$$

This formula is the only prerequisite concerning the Gaussian measures. We also need the notion of analyticity on a Fréchet space $E$ (usually set in local terms) [5], [6, II.2].

**1.3. DEFINITION.** Given a function $f : E \to \mathbf{C}$ we say that

(i) $f \in H_G(E)$ (f is *G-holomorphic* on $E$) iff its "affine restrictions": $\lambda \to f(a + \lambda w)$ ($\forall a, w \in E$) are analytic functions of $\lambda \in \mathbf{C}$;

(ii) $f \in H(E)$ (f is *holomorphic* on $E$, or *entire*) iff $f \in H_G(E)$ and f is continuous (equivalently: locally bounded) on $E$.

Any function $f \in H_G(E)$ has a unique representation as a pointwise convergent series $\sum f_m$ of homogeneous polynomials of degree m (m-homogeneous, for short) given at points $z \in E$ by

$$(1.4) \qquad f_m(z) = (2\pi i)^{-1} \int_{|\lambda| = r} f(\lambda z) \lambda^{-m-1} d\lambda = (d_o^m f)(z, \ldots, z)/m!$$

(m-th derivative at 0). If, moreover, $f \in H(E)$, these polynomials are continuous [6, II.2.9].

The following notation for certain spaces will be useful

$$P_n(E) = (\text{continuous, n-homogeneous complex polynomials on } E)$$

$$E_a^{\odot n} = \text{the n-th algebraic (i.e. not completed) symmetric power of } E.$$

In the case of Hilbert spaces we consider also $H^{\odot n}$ – the completed n-th symmetric power. This space is the closure of $H_a^{\odot n}$ in $H^{\otimes n} = H \otimes \ldots \otimes H$, the (completed) n-fold tensor product. The inner product and the norm of both $H^{\odot n}$ and $H^{\otimes n}$ will be denoted by $\langle \cdot, \cdot \rangle_n$, $\| \cdot \|_n$.

## 2. THE SPACE B

In the "finite-dimensional" case $N < \infty$, the Segal – Bargmann space $B_N$ is defined as $H(\mathbf{C}^N) \cap L^2(\mu_N)$. For $N = \infty$ the analogous space $B$ was defined as a subspace of $H(H)$ described by certain $\ell^2$-condition on the coefficients appearing in the Taylor series expansion at 0. Keeping the notations of §1 we formulate the following (as will turn out – equivalent) definition.

**2.0. DEFINITION.** *The Bargmann space B* (over H) *is the closure in* $L^2(\mu)$ *of the set of continuous on* $E_-$ (complex) *polynomials.*

As for $N < \infty$, the space $B$ is a canonical image of the Fock space $\mathbf{exp}(H)$. Recall that for a Hilbert space H, $\mathbf{exp}(H)$ is the (completed) orthogonal sum over $n = 0, 1, 2, \ldots$ of $H^{\odot n}$, i.e. the set $\{ \oplus e_n ; e_n \in H^{\odot n}, \sum \| e_n \|_n^2 < \infty \}$. Note first that the dual product

$\langle \cdot, \cdot \rangle : E_- \times E_+ \to \mathbf{C}$ extends to a pairing of n-th symmetric powers, $\langle \cdot, \cdot \rangle_n :$
$: (E_-)_a^{\odot n} \times (E_+)_a^{\odot n} \to \mathbf{C}$, coinciding on $H_a^{\odot n} \times (E_+)_a^{\odot n}$ with the inner product of $H^{\odot n}$.
Next, given $g \in (E_+)_a^{\odot n}$, $z \in E_-$ put

$$(Jg)(z) = (n!2^n)^{-\frac{1}{2}} \langle z^{\otimes n}, g \rangle_n .$$

Then J maps $(E_+)_a^{\odot n}$ **onto** the set $P_n(E_-)$. Moreover, we have

**2.1. PROPOSITION.** *The above mapping J extends (by linearity and by continuity) to an isometry* $J : \mathbf{exp}(H) \to L^2(\mu)$ *satisfying* $J(\mathbf{exp}(H)) = B$.

Before proving this result, we formulate one of its consequences that provides a more satisfactory description of the space $B$.

**2.2. THEOREM.** (i) *B is equal to the set of functions* $f \in L^2(\mu)$ *that are representable as norm-convergent series* $f = \sum f_n$, *where each* $f_n$ *satisfies the following condition.*

(*)    *There exists a sequence* $(f_{nk})$ *in* $P_n(E_-)$ *convergent to* $f_n$ *in the norm of* $L^2(\mu)$ *as* $k \to \infty$.

(ii) *Moreover, any such* $(f_{nk})$ *converges uniformly on bounded subsets of H (i.e. "locally uniformly on H"). If* $\mathbf{f}_n(z)$ *denotes its limit at* $z \in H$, *then the series* $\mathbf{f}(z) = \sum \mathbf{f}_n(z)$ *converges locally uniformly on H, thus defining an analytic function* $\mathbf{f}$ *on H.*

(iii) *One can choose the* $(f_{nk})$ *satisfying* (*) *to be also convergent (as* $k \to \infty$) *pointwise on some linear subspace* $M_{\mathbf{f}}$ *of* $E_-$, *having the full measure* $\mu$ *i.e., such that* $\mu(E_- \setminus M_{\mathbf{f}}) = 0$.

**2.3. REMARKS.** (I) It will be apparent from the proof that the functions $f_n$ are determined uniquely by (*) and mutually orthogonal. Hence the norm-summability in (i) means $\sum \| f_n \|^2 < \infty$.

(II) Since $\mu(H) = 0$ and any $f \in B$ is an "equivalence class modulo the equality a.e. $[\mu]$", there is no direct meaning for the value of f on H. The function $\mathbf{f}$ constructed in 2.2(ii) belongs to the space defined by Bargmann in [2]. The mapping $f \to \mathbf{f}$ is an isomorphism between $B$ and that space.

(III) Although continuous on H, the (representatives of the equivalence class) $\mathbf{f}_n$ can behave badly on the space $E_-$.

**EXAMPLE.** Take $H = \ell^2$, $E_- = \{(\lambda_n) : \lambda_n \in \mathbf{C}, \sum |n^{-1}\lambda_n|^2 < \infty\}$. Then for $f(\lambda) =$

$= \sum n^{-1}\lambda_n$ we have $f \in B$, $M_f = \{\lambda \in E_- ; \sum |n^{-1}\lambda_n| < \infty\}$. Note that here $f_n = 0$ for $n \neq 1$ and $f_1 = f$.

(IV) In view of 2.2 (iii), $B$ is a space of "almost everywhere analytic functions": the limits of $f_{nk}$ ($k \to \infty$) are n-homogeneous polynomials on $M_f$. (To speak of polynomials, one must have a linear structure on the domain, hence our condition on $M_f$.)

(V) Because of (1.2), we can use the tensor product notations: for a cylindric function $f(z) = \phi_N(z_1, \ldots, z_N)$, write $f = \phi_N \otimes 1_{\infty(N)}$, also $\mu = \mu_N \otimes \mu_{\infty(N)}$. Then $B = B_N \otimes \otimes B_{\infty(N)}$, where "$\infty(N)$" informally indicates the dependence on coordinates $z_{N+1}$, $z_{N+2}, \ldots$ only.

PROOF OF 2.1. Since $(E_-)^* = E_+$, it follows by the universal diagram (factorization) property that $J((E_+)_a^{\odot n}) = P_n(E_-)$. The dense subset of $\exp(H)$ consisting of sums $\sum g_n$, where $g_n \in (E_+)_a^{\odot n}$ are equal to zero for all but finitely integers $n \geq 0$, is mapped **onto** the (dense in $B$) set of continuous polynomials on $E_-$. Therefore it remains to show that $J$ maps this dense set isometrically into $L^2(\mu)$. Since the summands $g_n$ are mutually orthogonal and since $P_n(E_-)$ is orthogonal to $P_m(E_-)$ ($\forall m \neq n$) by (1.2) and by the corresponding property for $N < \infty$, it suffices to show that

$$(2.4) \qquad \int \| (Jg_n)(z) \|^2 d\mu(z) = \| g_n \|_n^2.$$

A proof of this equality can be based on the following fact. If $(k_j) \subseteq E_+$ is an orthogonal basis of $H$ as in (1.2), then the multi-indexed vectors $K_\alpha$, $\alpha$ ranging over the set

$$A_n = \{\alpha ; \exists p < \infty, \exists \text{ integers } \alpha_j \geq 0; \alpha = (\alpha_1, \ldots, \alpha_p, 0, 0, \ldots), \sum \alpha_j = n\}$$

form an orthogonal basis of $H^{\odot n}$, where for $\alpha$ as above, $K_\alpha$ is the orthogonal projection onto $H^{\odot n}$ (=symmetrization) of the vector

$$K_\alpha^* = (n!/\alpha_1! \ldots \alpha_n!)^{\frac{1}{2}} k_1^{\otimes \alpha_1} \otimes \ldots \otimes k_p^{\otimes \alpha_p}.$$

(We write $h^{\otimes m}$ for the m-fold product $h \otimes \ldots \otimes h$, so that $K_\alpha^* \in H^n$.) For the vectors $g_n$ having only finitely many non-zero coefficients $c_{\alpha,n}$ in the above basis $\{K_\alpha ; \alpha \in A_n\}$, if $z_j(z) = \langle z, k_j \rangle$, then we have

$$(Jg_n)(z) = \sum_{|\alpha|=n} c_{\alpha,n} \alpha! 2^{|\alpha|} z_1^{\alpha_1} \ldots z_p^{\alpha_p},$$

where $\alpha! = \alpha_1! \ldots \alpha_p!$, and $p$ is such that ($c_{\alpha,r} = 0 \; \forall r > p$). Using (1.2) and the case $N < \infty$ we conclude that distinct multiindices give orthogonal in $L^2(\mu)$ polynomials and that the equality (2.4) holds.

A simple calculation based on the similar ideas gives for $z \in H$, $g = \sum g_n$, with $g_n \in (E_+)^{\odot n}$ the following useful estimate

$$(**) \qquad \sum |(Jg_n)(z)| \leq \sum \|g_n\|_n |z|^n (n! 2^n)^{-\frac{1}{2}} \leq \|g\|_{\exp(H)} e^{|z|^2/4}.$$

This estimate holds also for arbitrary $g = \sum g_n \in \exp(H)$, if for $z \in H$ the evaluation $(Jg_n)(z)$ is taken in the sense of 2.2(ii).

PROOF OF 2.2. Clearly, the vectors of the form $\sum f_n$ satisfying (*) belong to $B$. Conversely, let $f \in B$. By (2.1) we can find the vectors $e_m \in H^{\odot m}$ such that $e = \oplus e_m$ belongs to $\exp(H)$ and satisfies $Je = f$. Since $H^{\odot m}$ are direct orthogonal summands of $\exp(H)$, we obtain the desired convergence of $\sum f_m$, where $f_m = J(e_m)$. By the density of $(E_+)_a^{\odot m}$ in $H^{\odot m}$ one can find for each integer $k > 0$ vectors $e_{m,k} \in (E_+)_a^{\odot m}$ satisfying $\sum \|e_{m,k} - e_m\|^2 \to 0$ as $k \to \infty$. Then $f_{m,k} = J(e_{m,k}) \in P_m(E_-)$ satisfy (*), since $J$ is an isometry. Part (ii) of (2.2) follows now from the estimates (**) and from a normal family argument applied to $f_{m,k}$.

To prove (iii) note first that a subsequence of $(f_{m,k})$, call it again $(f_{m,k})$, converges almost everywhere on $E_-$. This means that the set $N_m = \{z \in E_- ; \lim_{k \to \infty} f_{m,k}(z)$ exists$\}$ is of full measure i.e., that $\mu(E_- \setminus N_m) = 0$. The set $M_f = \cap N_m$ is also of full measure and $cM_f \subseteq M_f$ ($\forall c \in \mathbf{C}$). The problem, as to whether $M_f + M_f \subseteq M_f$ seems open, even when $f = f_m$. (It suffices to consider here only this case – for a fixed m.) However, if as $f_{m,k}$ we take the (finite) partial sums of the expansion $f_m = \sum c_\alpha K_\alpha$ ($K_\alpha$ as in the proof of (2.4), the sum over $\alpha \in A_m$) then the values $f_{m,k}(z)$ will be partial sums of the corresponding function series. This series will converge absolutely on some set $M = M_m$ of full measure, by a classical result on $L^p$-norms. That $M + M \subseteq M$ follows now from a multi-index version of Minkowski's inequality proved by J.B. McLeod [9].

**2.7. NOTE.** The reproducing property known for $B_N$, $N < \infty$ appears in $B$ only in a rudimentary form: at points $\lambda \in E_+$ the functions $\varepsilon_\lambda$ defind on $E_-$ by $\varepsilon_\lambda(z) = \exp(\langle z, \lambda/2 \rangle)$ belong to $B$ and satisfy $f(\lambda) = \langle f, \varepsilon_\lambda \rangle$ for any $f \in B$. For cylindric functions this follows from (1.2) and then extends by continuity to arbitrary f's. The set $\{\varepsilon_\lambda ; \lambda \in E_+\}$ is linearly dense in $B$.

**2.8. DEFINITION.** Let $P : L^2(\mu) \to B$ be the orthogonal projection. The Toeplitz operator $T_\phi : D(T_\phi) \to B$ with a $[\mu]$-measurable symbol $\phi$ ($\phi$ defined a.e. $[\mu]$ on $E_-$) is given by

$$T_\phi f = P(\phi \cdot f) \quad \text{for} \quad f \in D(T_\phi) \stackrel{\text{def}}{=} \{f \in B \; ; \; \phi \cdot f \in L^2(\mu)\}.$$

The following properties are easy to check.

**2.9. PROPOSITION.** (i) $T_\phi$ *is always a closed operator.*

(ii) *The adjoint operator satisfies* $(T_\phi)^* \supset T_{\bar\phi}$ *whenever* $D(T)_\phi$ *is dense in B* (e.g. *when $\phi$ is cylindric and of polynomial growth*).

(iii) *If* $\phi \in L^\infty(\mu)$ *then* $T_\phi$ *is bounded and* $\|T_\phi\| \leq \|\phi\|_\infty$. *(The latter inequality can be sharp, even for the infinite right hand side.)*

(iv) *For cylindric functions* $\phi(z) = \Phi(z_1, \ldots, z_N)$, $f(z) = F(z_1, \ldots, z_N)$, $F \in B_N$ *if* $F \in D(T_\Phi)$, *then* $f \in D(T_\phi)$ *and we have*

$$(T_\phi f)(z) \equiv (T_\Phi F)(z_1, \ldots, z_N) \quad \text{a.e. } [\mu].$$

(v) $T_\phi = 0$ *implies* $\phi \equiv 0$ a.e. $[\mu]$.

The proof of (v) proceeds as that of Theorem 4 in [3], via the equalities $\langle T_\phi f, g \rangle = \langle \phi f, g \rangle = \langle \phi, \bar{f}g \rangle$ valid for $f \in D(T_\phi)$, $g \in B$. Indeed, since the cylinder $L^2$-functions form a dense set in $L^2(\mu)$, so do the polynomials in $z_j$ and $\bar{z}_j$.

### 3. WEYL SYSTEM ON H

Following Bargmann [1] we define the Weyl system $(V_\lambda)_{\lambda \in H}$ of unitary operators on $B$. First, for $\lambda \in E_+$, $z \in E_-$ and $f \in B$ put

$$(3.1) \qquad\qquad (V_\lambda f)(z) = \exp(\langle z, \lambda/2 \rangle - |\lambda/2|^2) f(z - \lambda).$$

**3.2. PROPOSITION.** *The mapping* $\lambda \to V_\lambda$ *is unitary-valued and strongly continuous on* $E_+$ *in the norm topology induced there from H. Therefore it extends to a strongly continuous mapping that assigns to each $\lambda \in H$ a unitary operator on B, denoted also by* $V_\lambda$.

PROOF. To show the isometry of $V_\lambda$ it suffices to check that $\|V_\lambda f\| = \|f\|$ for any finite sum $f = \sum_s a_s \epsilon_{\lambda_s}$, where $\lambda_s \in E_+$, since the reproducing kernels form a total set in $B$ (cf. 2.7). Using the reproducing property twice, we calculate:

$$\|V_\lambda f\|^2 = \int \Big| \sum_s a_s \epsilon_{\lambda_s} \Big|^2 d\mu = \int \Big| \sum_s a_s e^{\langle z, \lambda/2 \rangle - |\lambda/2|^2} e^{\langle z - \lambda, \lambda_s/2 \rangle} \Big|^2 d\mu(z) =$$

$$= \sum_{s,t} a_s \bar{a}_t \langle \epsilon_{\lambda + \lambda_s}, \epsilon_{\lambda + \lambda_t} \rangle \, e^{-|\lambda|^2/2 - \langle \lambda, \lambda_s/2 \rangle - \langle \lambda, \lambda_t/2 \rangle} =$$

$$= \sum_{s,t} a_s \bar{a}_t e^{\langle \lambda_s, \lambda_t \rangle /2} = \sum_s a_s \bar{a}_t \varepsilon_{\lambda_s}(\lambda_t) = \langle \sum_s a_s \varepsilon_{\lambda_s}, \sum_t a_t \varepsilon_{\lambda_t} \rangle = \| f \|^2.$$

Hence $\| V_\lambda f \| = \| f \|$ for all $f \in B$ and to prove the strong continuity it suffices to show that $\langle V_\lambda f, g \rangle$ converges to $\langle f, g \rangle$ as $|\lambda| \to 0$, $\lambda \in E_+$ for any cylindric functions f, g from B. Assume that in a coordinate system $(z_j)$, $z_j = \langle \cdot, k_j \rangle$ satisfying (1.2) we have $f(z) = F(z_1, \ldots, z_N)$, $g(z) = G(z_1, \ldots, z_N)$. As in (2.3.V), we can use the decomposition onto the spaces corresponding to the first N and to the remaining set of coordinates (or vectors $k_j$ – for our $\lambda$). Accordingly, $\mu = \mu_N \otimes \mu_{\infty(N)}$ and using Fubini's theorem we reduce our problem to the case treated in [1] and [3], since we have

$$\langle V_\lambda f, g \rangle = \int e^{\langle w, \theta/2 \rangle - |\theta/2|^2} d\mu_{\infty(N)}(w) \langle V_\gamma^{(N)} F, G \rangle.$$

Here the last scalar product was taken in $L^2(\mu_N)$, $\lambda = \gamma \oplus \theta$ is the decomposition of $\lambda$. (We make it feasible by a proper choice of the $k_j$'s, proving the convergence (sequential continuity) first on dense subspaces algebraically generated by $(k_j)$.) $V_\gamma^{(N)}$ is the corsponding Weyl operator on $B_N$.

This second part of the proof can be also carried out by a direct, but tedious estimate of $\| V_\lambda \varepsilon_y - \varepsilon_y \|$. Finally, to see that the $V_\lambda$ are unitary it suffices to note that $V_{-\lambda} V_\lambda f = f$ ($\forall f \in B$).

Like in the "finite-dimensional" case it turns out that the $V_\lambda$ can be expressed in terms of Toeplitz operators.

**3.3. PROPOSITION.** *For any* $\lambda \in E_+$ *we have*

$$\exp(iT_{Re\langle \cdot, \lambda \rangle}) = V_{-i\lambda}$$

Here i stands for the imaginary unit and $\langle \cdot, \lambda \rangle$ is the mapping: $z \in E_- \to \langle z, \lambda \rangle \in \mathbf{C}$. Since by (2.8), $T_{Re\langle \cdot, \lambda \rangle}$ is densely defined and self-adjoint, the compared operators are both unitary on B.

PROOF. Like the continuity in (3.2), we reduce the proof to the "finite" case. Since $\lambda$ is now fixed, the basis $(k_j)$ can be chosen to satisfy $\lambda = |\lambda| k_1$. By the continuity of both sides, it suffices to compare their values at vectors $f \in B$ of the form $f(z) = F(z_1, \ldots, z_N)$. Both sides of the desired equality are then equal to the tensor products of the corresponding operators on $B_N$ evaluated at F with the constant function **1** (from $B_{\infty(N)}$). The result reduces to the corresponding one in [3].

An alternative proof can be based as in [1] on finding the infinitesimal generator of the strongly continuous unitary group $\{V_{t\lambda}\}_{t\in R}$. Simple calculations yield the equality

(3.4)        $$V_{\lambda} V_{\gamma} = V_{\lambda+\gamma} \exp(\langle \lambda/2, \gamma/2 \rangle - \langle \gamma/2, \lambda/2 \rangle) \quad \text{for } \lambda, \gamma \in H.$$

The last proposition can be used to obtain in our context the so called ([3], [4]) *Basic Identities*:

**3.5. PROPOSITION.** *For any* $v, w \in E_+$ *we have*

(a)        $$T_{\exp(i\,\mathbf{Re}\langle \cdot, w \rangle)} = e^{-|w|^2/4}\, e^{iT\,\mathbf{Re}\langle \cdot, w \rangle}$$

(b)        $$T_{\exp(i\,\mathbf{Re}\langle \cdot, w \rangle)}\, T_{\exp(i\,\mathbf{Re}\langle \cdot, v \rangle)} = e^{\langle v, w \rangle/2}\, T_{\exp(i\,\mathbf{Re}\langle \cdot, v+w \rangle)}.$$

PROOF. Let L and R denote the left (resp. right) hand side of the equality (a). Since the reproducing kernels $\varepsilon_{\lambda}$ form a total set in $B$ (cf.2.7), to prove this equality it suffices to show that $\langle L\varepsilon_{\lambda}, \varepsilon_{\gamma} \rangle = \langle R\varepsilon_{\lambda}, \varepsilon_{\gamma} \rangle$ for all $\lambda, \gamma \in E_+$. This is done via Proposition 3.3, exactly as in [3]. To prove (b) we apply (a) to each of the left hand side factors. The new factors are containing the exponents of $iT_{\mathbf{Re}\langle \cdot, \lambda \rangle}$ for $\lambda = v$ and $\lambda = w$. These exponents are "converted" into the operators $V_{-i\lambda}$ by (3.3). Now the multiplication rule (3.4) applied to these Weyl operators and the application of (3.3) and (a) - now in the reverse directions - returns us to the right hand side of (b).

By *trigonometric polynomial* on $E_-$ we shall understand a finite sum $\sum c_k e^{i\,\mathbf{Re}\langle \cdot, \lambda_k \rangle}$ of functions on $E_-$, where $c_k \in \mathbf{C}$, $\lambda_k \in E_+$. Let *TP* denote the set of all trygonometric polynomials on $E_-$. The results of this section show that $\tau\{TP\}$, the $C^*$-algebra generated by $\{T_{\phi}; \phi \in TP\}$ coincides with the Weyl algebra of Canonical Commutation Relations, generated by $\{V_{\lambda}; \lambda \in H\}$ or, equivalently, by $\{e^{iT\,\mathbf{Re}\langle \cdot, w \rangle};$ $w \in E_-\}$. The closure of *TP* in $L^{\infty}(\mathbf{C}^N, \mu_N)$ was the algebra $AP(\mathbf{C}^N)$ of *almost periodic functions* on $\mathbf{C}^N$ and $\tau\{TP\} = \tau\{AP(\mathbf{C}^N)\}$. Since for $N = \infty$, $H$ is no longer an LCA-group, the term AP becomes inadequate.

**PROBLEM.** Describe in function-theoretic terms the closure of *TP* in $L^{\infty}(E_-, \mu)$.


### 4. NONCOMPACTNESS

In the case $N < \infty$ compactly supported functions $\phi$ generated Toeplitz operators $T_{\phi}$ belonging to the ideal $K(B_N)$ of compact operators. As N becomes infinite, the situation changes. But let us begin with a lemma valid for any $N \leq \infty$ stated here for

$N = \infty$. The function $\phi$ is, by assumption, measurable and **nonnegative**.

**4.1. LEMMA.** $T_\phi \in K(B)$ *iff the multiplication operator* $M_\phi : B \to L^2(\mu)$, $M_\phi f = \phi f$ *is compact.*

PROOF. One implication is obviuos since $T_\phi = PM_\phi$. For the converse implication, assume $T_\phi \in K(B)$ and put $\omega = \phi^{\frac{1}{2}}$. Then for any weakly convergent to zero sequence $(f_n)$ in $B$ we have $\| M_\omega f_n \|^2 = \int \phi |f_n|^2 d\mu = \langle T_\phi f_n, f_n \rangle$ going to zero as $n \to \infty$. Hence $M_\omega$ is a compact operator and so is $M_\phi = MM_\omega$, where $Mg = \omega g$, $g \in L^2(\mu)$.

**4.2. EXAMPLE.** Let $r_n \in R$, $r_n \geq n$ be such that $\sum \mu_1 \{w \in \mathbf{C}; \ |w| > r_n\} < \infty$ and $\int_{|w| < r_n} |w|^2 d\mu_1 \geq 1/4$. For $H = \ell^2$ let $E_- = \{(w_n) ; (w_n/(n^2 r_n)) \in \ell^2\}$. Then the set $\beta = \{(w_n) \in E_- ; \ |w_n| \leq r_n\}$ is compact in $E_-$, since the sums over $n > k$ of $(n^2 r_n)^{-1}$ converge uniformly on $\beta$ to zero as $k \to \infty$. Although the coordinate functions $z_n$ converge weakly to zero in $B$, for $\phi$ equal to the characteristic function of $\beta$ we have

$$\| M_\phi z_n \|^2 = \int_\beta |z_n|^2 d\mu(z) =$$

$$= \int_{|z_n| \leq r_n} |z_n|^2 d\mu_1(z_n) \cdot \prod_{j \neq n} \mu_1 \{w \in \mathbf{C} ; \ |w| \leq r_j\}.$$

Now, in view of 4.1, this $T_\phi$ cannot be compact, since the latter sequence fails to converge to zero. Let $\phi$ be measurable $[\mu]$. Even more discouraging is the following.

**4.3. EXAMPLE.** If $\phi \geq 0$ a.e. $[\mu]$, but $T_\phi \neq 0$ then $T_\phi \notin K(B)$.

PROOF. Again by (4.1), it suffices to show that $M_\phi$ is not compact. Since $T_\phi \neq 0$, we can find (by the tightness of $\mu$) a compact set $K$ in $E_-$ with $\delta = \mu(K) > 0$ such that $\phi > 0$ on $K$. We may even assume, without loss of generality, that $\phi \geq 1$ on $K$ (multiply, if necessary, $\phi$ by a constant). Take $r > 0$ so small that

$$(+) \qquad\qquad \mu_1(\{\xi \in \mathbf{C} ; \ |\xi| < r\}) < \delta/4.$$

Let the coordinates system $(z_n)$, $z_n = \langle \cdot, k_n \rangle$ be chosen to satisfy (1.2). Now we are going to use the fact that the "metric structure" of $\mu$ is "*homogeneous*", the same at any coordinate $z_n$. Precisely, let us fix one coordinate, say for $n = 1$. As in (2.3.V), we write $\mu = \mu_1 \otimes \mu_{\infty(1)}$ and, abbreviating $\mu_{\infty(1)}$ as $\mu^\sim$, we apply Fubinni's theorem to obtain $\mu(K) = \int \mu^\sim(K_y) \, d\mu_1(y)$. Here, according to the decomposition onto the first and the remaining coordinates, $K_y$ is the section over $y \in \mathbf{C}$ of $K$ by the hyperplane

$\{z \in E_- ; z_1 = y\}$.

(This notation is evident especially for $E_-$, a sequential space like in (2.3 III), or after (1.1a).) The mentioned homogenity is to mean that if, instead of $z_1$, we distinguish any other coordinate $z_n$, the corresponding tensor product decomposition: $\mu = \mu^{(n)} \otimes \mu^{\sim n}$ has the factors $\mu^{(n)}$ and $\mu^{\sim n}$ metrically identical with $\mu_1$ and $\mu^\sim$, respectively. Only the coordinate sections of $K$ are essentially different, but we need here only the estimate

$$(++) \qquad\qquad 0 \leq \mu^\sim(K_y) \leq 1.$$

Hence the following estimate of $\| M_\phi z_n \|^2$, carried here for $n = 1$, applies to any other n. We again use the Fubini's theorem. Writing $\xi$ for $z_1$, to simplify the notation, we calculate

$$\| \xi\phi \|^2 \geq \int_K |\xi|^2 d\mu = \int_{\mathbf{C}} (|\xi|^2 \mu^\sim(K_\xi)) d\mu_1(\xi) = \left\{ \int_{|\xi|>r} + \int_{|\xi|<r} \right\}(\dots) \geq$$

$$\geq \int_{|\xi|>r} r^2 \mu^\sim(K_\xi) d\mu_1(\xi) - \int_{|\xi|<r} r^2 \mu^\sim(K_\xi) d\mu_1(\xi) =$$

$$= \int_{\mathbf{C}} r^2 \mu^\sim(K_\xi) d\mu_1(\xi) - 2 \int_{|\xi|<r} r^2 \mu^\sim(K_\xi) d\mu_1(\xi) = r^2 \left\{ \mu(K) - 2 \int_{|\xi|<r} \mu^\sim(K_\xi) d\mu_1(\xi) \right\} \geq$$

$$\geq r^2 \{ \delta - 2\mu_1(\{\lambda \in \mathbf{C} ; |\lambda| < r\}) \} \geq r^2 \{ \delta - \delta/2 \}.$$

We were using only inequalities independent on n, like (+) and (++), hence everything works for any n, not just for $n = 1$. In particular, $\| M_\phi z_n \| \geq r^2 \delta/2$, independently of n and since $(z_n)$ converges weakly to zero as $n \to \infty$, $M_\phi$ cannot be compact.


### REFERENCES

1. **Bargmann, V.** : On a Hilbert space of analytic functions and an associated integral transform, *Comm. Pure Appl. Math.* **14**(1961), 187-241.

2. **Bargmann, V.** : Remarks on a Hilbert space of analytic functions, *Proc. Nat. Acad. Sci. U.S.A.* **18**(1962), 199-204.

3. **Berger, C.A. ; Coburn, L.A.** : Toeplitz operators and quantum mechanics, *J. Funct. Anal.* **68**(1986), 273-299.

4. **Berger, C.A. ; Coburn, L.A.** : Toeplitz operators on the Segal–Bargmann space, *Trans. Amer. Math. Soc.* **301**(1987), 813-829.

5. **Bochnak, J. ; Siciak, J.** : Analytic functions in topological vector spaces, *Studia Math.* **39**(1971), 77-112.

6. **Dinnen, S.** : *Complex analysis in locally convex spaces*, North Holland, Math. Stud. **57**, 1981.

7. **Hida, T.** : *Brownian motion*, Springer–Verlag, 1980.

8.      **Kuo, H.-H.** : *Gaussian measures in Banach spaces*, Lect. Notes in Math. **463**, Springer-Verlag, 1975.

9.      **McLeod, J.B.** : On four inequalities in symmetric functions, *Proc. Edinburgh Math. Soc.* **11**(1959), 211–219.

10.     **Segal, I.E.** : The complex wave representation of the free boson field, *Adv. in Math., Suppl. Studies*, **3**(1978), 321–343.

**J. Janas** and **K. Rudol**
Instytut Matematyczny PAN
ul. Solskiego 30,
31-027 Kraków,
Poland.

Operator Theory:
Advances and Applications, Vol. 43
© 1990 Birkhäuser Verlag Basel



# A NOTE ON PERTURBATIONS OF SELFADJOINT OPERATORS IN KREIN SPACES

Peter Jonas

Let $H$ be a separable Krein space, A a definitizable selfadjoint operator in $H$ and B a selfadjoint operator in $H$ with nonempty resolvent set. Let the difference of the resolvents of A and B belong to some Schatten-von Neumann ideal $S_p$, $1 \leq p < \infty$, of compact operators in $H$ ([3]). If, in addition, A is fundamentally reducible ([1]), then B possesses a spectral function with singularities and the set S of the spectral singularities of B has no more than a finite number of accumulation points. More precisely, the set S' of the accumulation points of $S \cup (\sigma(B) \setminus R)$ is contained in the union of the set of critical points and the nonreal spectrum of A. For any interval [a,b] with a,b $\notin$ S and [a,b] $\cap$ S' = $\emptyset$, the restriction of B to the spectral subspace corresponding to [a,b] is a definitizable operator. In [5] such an operator B was called definitizable over $\bar{R} \setminus S'$ (see the definition given below). For bounded operators this result was proved by H. Langer in [7] (even for the Macaev ideal $S_\omega$ instead of $S_p$), for its generalization to unbounded operators see [5].

For a given definitizable selfadjoint differential operator A in a Krein function space it is often difficult to decide whether A is fundamentally reducible or not. Therefore it would be useful to have a criterion for B to be definitizable over some open set $\Delta \subset R$, a criterion which does not contain the assumption that A is fundamentally reducible. The aim of this note is to prove a criterion of this kind.

We shall require instead of the fundamental reducibility of A that certain real points are no accumulation points of $\sigma(B) \setminus R$. From the viewpoint of the perturbation theory of differential operators this is a natural assumption (see e.g. [6]).

We restrict ourselves to the case when $c_\infty(A) \cap \Delta = \emptyset$ (for the definition of $c_\infty(A)$ see Section 1). The opposite case will be considered in a subsequent paper.

We shall admit that A is definitizable only over some open subset of $\bar{R}$ and that A and B are selfadjoint with respect to different Krein spaces with the same underlying linear space.

The proof of our criterion is based on the result mentioned in the beginning and on a functional calculus studied in [5].

## 1. NOTATIONS AND DEFINITIONS

Let $(H,[\cdot,\cdot])$ be a separable Krein space. All topological notions are understood with respect to some Hilbert norm $\|\cdot\|$ on $H$ such that $[\cdot,\cdot]$ is $\|\cdot\|$-continuous. Any two such norms are equivalent. If $K$ is a Krein subspace of $H$, we denote the least upper bound $(\leq\infty)$ of the dimensions of the positive (negative) definite subspaces of $K$ by $\kappa_+(K)$ (resp. $\kappa_-(K)$).

Let A be a densely defined selfadjoint operator in $H$. Then $\sigma(A)$ is symmetric with respect to $\mathbf{R}$. We put $\sigma_0(A) := \sigma(A)\setminus\mathbf{R}$ and $R(z;A) := (A - zI)^{-1}$ for every z in the resolvent set $\rho(A)$ of A. An eigenvalue $\lambda$ of A is said to be *normal* if the root space corresponding to $\lambda$ is finite-dimensional and has a closed complemetary subspace $N_\lambda$ which is invariant for A such that $\lambda \in \rho(A|N_\lambda)$. The set of normal eigenvalues of A is denoted by $\sigma_{p,\,norm}(A)$.

The operator A is called *nonnegative* if $\rho(A) \neq \emptyset$ and $[Ax,x] \geq 0$ holds for all x in the domain $D(A)$ of A. A is called *definitizable* if $\rho(A) \neq \emptyset$ and there exists a polynomial p such that $[p(A)x,x] \geq 0$ holds for all $x \in D(p(A))$. If A is definitizable, then $\sigma_0(A)$ is contained in the set of zeros of every polynomial p with the above property.

In this note we consider a class of selfadjoint operators in $H$ containing the definitizable selfadjoint operators which was introduced in [5]. For convenience we restrict ourselves to selfadjoint operators which fulfil the following condition:

$(f_0)$   $\sigma_0(A)$ has no more than a finite number of nonreal accumulation points.

Let $\Delta$ be an open subset of $\overline{\mathbf{R}}$. Here and in the following $\overline{\mathbf{R}}$ denotes the closure of R in the complex sphere $\overline{\mathbf{C}}$ .

**DEFINITION.** ([5; Definition 2.3]). A selfadjoint operator A in $H$ satisfying the condition $(f_0)$ is called *definitizable over* $\Delta$ if the following holds:

(i)   No point of $\Delta$ is an accumulation point of $\sigma_0(A)$.

(ii)   For every closed subset $\delta$ of $\Delta$ there exist $m \geq 1$ and $M > 0$ such that

$$\| R(z;A)\| \leq M(1 + |z|)^{2m-2}|\operatorname{Im} z|^{-m}$$

for all z in a neighbourhood of $\delta$ (in $\overline{\mathbf{C}}$) with $z \neq \infty$ and $\operatorname{Im} z \neq 0$.

(iii)   For every $t \in \Delta \setminus \{\infty\}$ there exist open intervals of definite type of the form $(t',t)$ and $(t,t'')$, $t' < t < t''$. Here an open set $\delta \subset \Delta$ is said to be of definite type if for every nonnegative $f \in C^\infty(\overline{\mathbf{R}})$ with $\operatorname{supp} f \subset \delta$ the operator f(A) (defined, in view of (ii), by extension of the Riesz-Dunford-Taylor functional calculus) is nonnegative or

-f(A) is nonnegative for every such f. If $\infty \in \Delta$, then there exist intervals of definite type of the form $(t',\infty)$ and $(-\infty,t'')$.

For an arbitrary selfadjoint operator A in $H$ the union of all open subsets $\Delta$ of $\bar{R}$ such that A is definitizable over $\Delta$ is denoted by $\Delta(A)$. By $c(A)$ we denote the set of all $t \in \Delta(A)$ such that there is no open subset $\delta \subset \Delta(A)$ of definite type containing t. The elements of $c(A)$ are called *critical points* of A. For the spectral function $E(\cdot\,;A)$ of a selfadjoint operator A definitizable over an open subset of $\bar{R}$ and its properties we refer to [5]. We mention only that for every subinterval $\delta$ of $\Delta(A)$ whose endpoints do not belong to $c(A)$, $E(\delta\,;A)$ is defined and this projection is selfadjoint in $H$.

By $c_\infty(A)$ we denote the set of all $t \in c(A)$ such that $\kappa_+(E(\delta\,;A)H) = \kappa_-(E(\delta\,;A)H) = \infty$ for every connected open set $\delta \subset \Delta(A)$ with $t \in \delta$ such that $E(\delta\,;A)$ is defined. A critical point is called *regular* if there exists an open deleted neighbourhood $\delta_o$ (in $\Delta(A)$) of t such that the set of the projections $E(\delta)$, where $\delta$ runs through all intervals with $\bar{\delta} \subset \delta_o$, is bounded.

By $\psi$ and $\phi$ we denote the linear fractional transformations defined by

(1)                    $$\psi(z) = -(z - i)/(z + i), \qquad \phi(\zeta) = i(1 - \zeta)/(1 + \zeta).$$

We have $\phi \circ \psi = \mathbf{id}$ and $\psi(\bar{R}) = \mathbf{T} := \{z : |z| = 1\}$. If A is a selfadjoint operator in $H$ with i, $-i \in \rho(A)$, then the Cayley transform $\psi(A)$ of A is a unitary operator in $H$.


## 2.

Let $H' = (H, [\cdot\,,\cdot\,]')$ and $H'' = (H, [\cdot\,,\cdot\,]'')$ be Krein spaces with the same underlying linear space $H$. Assume that there exists a positive definite scalar product $(\cdot\,,\cdot\,)$ on $H$ such that $(H,(\cdot\,,\cdot\,))$ is a separable Hilbert space and $[\cdot\,,\cdot\,]'$ and $[\cdot\,,\cdot\,]''$ are continuous with respect to $(\cdot\,,\cdot\,)$. We define bounded selfadjoint operators G' and G'' in $(H,(\cdot\,,\cdot\,))$ by

$$(G'x,y) = [x,y]', \qquad (G''x,y) = [x,y]'', \qquad x,y \in H.$$

**THEOREM.** *Assume that* $G' - G'' \in S_p$ *for some* $p \in [1,\infty)$. *Let* A *be a selfadjoint operator in* $H'$ *such that* $\sigma_o(A) \setminus \sigma_{p,\,norm}(A)$ *is empty or a finite set,* A *is definitizable over some connected open subset* $\Delta$ *of* $\bar{R}$ *and* $\Delta \cap c_\infty(A) = \emptyset$.

*Let* B *be a selfadjoint operator in* $H''$ *with* $\rho(B) \neq \emptyset$ *such that the following holds:*

(i) *For some (and hence all)* $z \in \rho(A) \cap \rho(B)$ *we have*

$$R(z\,;A) - R(z\,;B) \in S_p.$$

(ii) *The endpoints* $t_1$ *and* $t_2$ *of* $\Delta$ *(in* $\bar{\mathbb{R}}$*) are no accumulation points of* $\sigma_o(B)$.
*Then* B *is definitizable over* $\Delta$. *Moreover, if we have* $\kappa_-(E(\Delta';A)H') < \infty$ *for every connected set* $\Delta'$, $\bar{\Delta}' \subset \Delta$, *such that* $E(\Delta';A)$ *is defined, then the same holds for* A *and* H' *replaced by* B *and* H'', *respectively.*

PROOF. 1. By our assumptions on A and condition (i) the operators A and B fulfil the condition $(f_o)$.

We may assume that $\Delta$ is bounded, $\Delta = (t_1, t_2)$, and that i, $-i \in \rho(A) \cap \rho(B)$. This is no restriction. Further, it is sufficient to prove the theorem under the additional assumption that for every connected subset $\Delta'$ of $\Delta$ such that $E(\Delta';A)$ is defined,

$$(2) \qquad \kappa_-(E(\Delta';A)H') < \infty$$

holds. In the general case the result can be obtained by a covering of $\Delta$.

Assign $\epsilon > 0$. Then there exists an open interval $\delta_1'$, $\overline{\delta_1'} \subset (t_1, t_1 + \epsilon)$, and a positive number c such that

$$\sup\{\|R(t + i\eta; A)\| : t \in \delta_1'\} \leq c \, |\eta|^{-1}$$

for sufficiently small $|\eta| > 0$, and $\overline{\delta_1'}$ contains no accumulation points of $\sigma_o(B)$. Then by a local version of an estimation in [2; p.168] (see [5; proof of Theorem 3.6]) there exist an open interval $\delta_1$, $\overline{\delta_1} \subset \delta_1' \subset (t_1, t_1 + \epsilon)$, and positive numbers $C_1$ and $L_1$ such that

$$(3) \qquad \sup\{\|R(t + i\eta; B)\| : t \in \overline{\delta_1}\} \leq C_1 \exp\{L_1 |\eta|^{-p-1}\}$$

for sufficiently small $|\eta| > 0$. Similarly, we find an open interval $\delta_2$, $\overline{\delta_2} \subset (t_2 - \epsilon, t_2)$, and positive numbers $C_2$ and $L_2$ such that $\overline{\delta_2}$ contains no accumulation points of $\sigma_o(B)$ and

$$(4) \qquad \sup\{\|R(t + i\eta; B)\| : t \in \overline{\delta_2}\} \leq C_2 \exp\{L_2 |\eta|^{-p-1}\}$$

for sufficiently small $|\eta| > 0$.

2. On account of the condition (i) there exists a compact subset K of $\bar{\mathbb{C}}$ symmetric with respect to $\mathbb{R}$ such that the following holds:

(a) $\mathbb{C}^+ \setminus K$ and $\mathbb{C}^- \setminus K$ are simply connected domains, i, $-i \notin K$;

(b) $\bar{\mathbb{R}} \setminus K = \delta_1 \cup \delta_2$;

(c) $\sigma(A) \cup \sigma(B) \subset \mathbb{R} \cup K$.

Then the relations (3) and (4) imply that the Riesz – Dunford – Taylor functional calculus of B can be extended by continuity to the function space $A_K^{(s)}(\mathbb{R})$,

$s := 1 + (1 + p)^{-1}$ (see [5; § 1.1]).

Let $u_1'$, $u_1'' \in \delta_1$, $u_2'$, $u_2'' \in \delta_2$, $u_1' < u_1'' < u_2'' < u_2'$, and let $\chi$ be a function belonging to $A_K^{(s)}(\overline{\mathbf{R}})$ which satisfies the following conditions:

    ($\alpha$) $\chi(\overline{\mathbf{R}}) = \mathbf{T}$.

    ($\beta$) $(d/dt)(\arg \chi)(t) \geq 0$, $\quad t \in \mathbf{R}$.

    ($\gamma$) $\chi(t) = \psi(t)$ for $t \in (u_1'', u_2'')$ (see (1)).

    ($\delta$) $\chi(t) = -1$ for $t \in \overline{\mathbf{R}} \setminus (u_1', u_2')$.

Then $\chi(A)$ is unitary in $H'$, $\chi(B)$ is unitary in $H''$. We have

$$(5) \qquad\qquad \chi(A) - \chi(B) \in S_p.$$

Indeed, expressing the resolvents of A and B by the resolvents of the Cayley transforms $\psi(A)$ and $\psi(B)$, which fulfil the relation $\psi(A) - \psi(B) \in S_p$, we easily see that there exist positive numbers $C_3$ and $L_3$ such that

$$(6) \qquad \sup\{\| R(t + i\eta ; A) - R(t + i\eta ; B)\|_{S_p} : t \in \overline{\delta_1} \cup \overline{\delta_2}\} \leq C_3 \exp\{L_3 |\eta|^{-p-1}\}$$

for sufficiently small $|\eta| > 0$. Here $\| \cdot \|_{S_p}$ denotes the norm defined by $\| T \|_{S_p} = (\sum_j s_j^p(T))^{1/p}$ where $s_j(T)$ are the s-numbers of $T \in S_p$ with respect to the scalar product $(\cdot,\cdot)$. We consider the continuous linear functionals $f \mapsto \mathrm{Tr}\{F(f(A) - f(B))\}$ on $A_K^{(s)}(\overline{\mathbf{R}})$ where F runs through all operators of finite rank with $\| F \|_{S_q} \leq 1$, $q^{-1} + p^{-1} = 1$. From (6) and [5; Proposition 1.1] it follows that these functionals are equicontinuous in $A_K^{(s)}(\mathbf{R})$, which implies (5).

By the assumptions on A and $\chi$ the unitary operator $\chi(A)$ in $H'$ is definitizable (see e.g. [4]). Moreover, if $-1$ is a critical point of $\chi(A)$, it is a regular one and we have $\kappa_-(E(\mathbf{T} \setminus \{-1\}; \chi(A))H') < \infty$. Then by (5) and [4; Theorems 2.5 and 3.1] the unitary operator $\chi(B)$ in $H''$ is definitizable on $\mathbf{T} \setminus \{-1\}$ and, in view of (2),

$$(7) \qquad\qquad \kappa_-(E(\gamma ; \chi(B))H'') < \infty$$

for every closed arc $\gamma \subseteq \mathbf{T} \setminus \{-1\}$ such that $E(\gamma ; \chi(B))$ is defined. In particular, no point in $\mathbf{T}$ different from $-1$ is an accumulation point of $\sigma(\chi(B)) \setminus \mathbf{T}$.

3. We claim that $(u_1'', u_2'')$ contains no accumulation points of $\sigma_0(B)$:

$$(8) \qquad\qquad (u_1'', u_2'') \cap \overline{\sigma_0(B)} = \emptyset.$$

Indeed, let $K_o$ be the bounded connected component of K. Then for all $\lambda$ in a neighbourhood of $K_o$ the function $g_\lambda$,

$$g_\lambda(z) := (\chi(z) - \chi(\lambda))(\psi(z) - \psi(\lambda))^{-1}$$

belongs to $A_K^{(s)}(\mathbf{R})$. By the functional calculus it follows that for all $\lambda \in K_o \setminus \mathbf{R}$ with $\chi(\lambda) \notin \sigma(\chi(B))$, i.e. for all $\lambda$ in $K_o \setminus \mathbf{R}$ with the exception of no more than a finite number of points, the operator $(\chi(B) - \chi(\lambda)I)^{-1}g_\lambda(B)$ is the inverse of $\psi(B) - \psi(\lambda)I$. This proves (8).

4. Now we see as above that the resolvent of B satisfies a relation similar to (3) or (4) with $\overline{\delta_1}$ or $\overline{\delta_2}$ replaced by $\overline{\delta_1} \cup (u_1'', u_2'') \cup \overline{\delta_2}$. We may assume that $\sigma(A) \cup \sigma(B) \subset$ $\subset \mathbf{R} \cup K_\infty$, where $K_\infty$ is the unbounded connected component of K. Then the Riesz––Dunford–Taylor functional calculus of B can be extended by continuity to the function space $A_{K_\infty}^{(s)}(\mathbf{R})$, $s = 1 + (1 + p)^{-1}$. Moreover, if $f \in A_{K_\infty}^{(s)}(\mathbf{R})$ and **supp** $f \subset (u_1'', u_2'')$, then $f(B) = (f \circ \phi)(\chi(B))$ (see (1)). By this relation the special properties of the functional calculus of $\chi(B)$ and the spectral function of $\chi(B)$ (regarded as an extension by continuity of the functional calculus) can be carried over to B. From the fact that $\chi(B)$ is definitizable over $\mathbf{T} \setminus \{1\}$ it follows that B is definitizable over $(u_1'', u_2'')$. Let $\delta$ be a closed subinterval of $(u_1'', u_2'')$ such that $E(\delta ; B)$ is defined. Then according to (7) we have $\kappa_-(E(\delta ; B)H'') < \infty$. Since $\epsilon$ was arbitrary, this proves the theorem.

**REMARK.** Replace in the above theorem $S_p$ by $S_\infty$ and assume, in addition, that $\Delta$ is bounded, the endpoints of $\Delta$ belong to $\rho(A) \cap \rho(B)$ and $\kappa_-(E(\Delta ; A)H') < \infty$ holds. In this case the theorem has a simple proof. Indeed, $E(\Delta ; A)$ can be written as a Riesz––Dunford integral

$$E(\Delta ; A) = -(2\pi i)^{-1} \int_C R(\lambda ; A)d\lambda$$

with a curve C contained in $\rho(A) \cap \rho(B)$. Then setting $E_B := -(2\pi i)^{-1}\int_C R(\lambda ; B)d\lambda$ we obtain $E(\Delta ; A) - E_B \in S_\infty$. Therefore $\kappa_-(E_B H'') < \infty$. This implies that B is definitizable over $\Delta$ and $\kappa_-(E(\Delta ; B)H'') < \infty$. Similarly for an unbounded $\Delta$.

## REFERENCES


1.  **Bognár J.** : *Indefinite inner product spaces*, Springer-Verlag, Berlin--Heidelberg-New York, 1974.

2.  **Colojoară, I. ; Foiaş, C.** : *Theory of generalized spectral operators*, Gordon and Breach, New York-London-Paris, 1968.

3.  **Gohberg, I.C. ; Krein, M.G.** : *Introduction to the theory of linear nonselfadjoint operators* (Russian), Nauka, Moscow, 1965.

4.      **Jonas, P.** : On a class of unitary operators in Krein space, in *Advances in invariant subspaces and other results of operator theory,* Birkhäuser Verlag, Basel-Boston-Stuttgart, 1986, pp. 151-172.

5.      **Jonas, P.** : On a class of selfadjoint operators in Krein space and their compact perturbations,*Integral Equations Operator Theory* **11**(1988), 351-384.

6.      **Kuroda, S.T.** : Scattering theory for differential operators, *J. Math. Soc. Japan,* I: 25(1973), 75-104; II: **25**(1973), 222-234.

7.      **Langer. H.** : Spektralfunktionen einer Klasse J-selbstadjungierter Operatoren, *Math. Nachr.* **33**(1967), 107-120.

**Peter Jonas**

Karl-Weierstrass-Institut für Mathematik
Akademie der Wissenschaften der DDR
Mohrenstrasse 39, Berlin 1086
GDR

Operator Theory:
Advances and Applications, Vol. 43
© 1990 Birkhäuser Verlag Basel



# PROJECTIVE REPRESENTATIONS OF COMPACT GROUPS
# IN $C^*$-ALGEBRAS

**Mircea Martin**

## INTRODUCTION

In a recent paper [9], Norberto Salinas proved that the closed subset $P$ of all hermitian idempotents of a unital $C^*$-algebra $A$ is a real analytic manifold, and studied some of its geometrical properties. This space, called the Grassmann manifold of $A$, has a simple alternative description. There is a bijective correspondence from $P$ to the set of all unitary representations of the cyclic group $\mathbb{Z}/2$ in $A$. More precisely, each hermitian idempotent e in $A$ corresponds to a representation $\alpha$ of $\mathbb{Z}/2 = \{0,1\}$ in $A$, uniquely defined by $\alpha(1) = 1 - 2e$.

Our aim in the present paper is to study a more general space of this type, namely, the set of all unitary representations of a compact group G in $A$. Actually, in order to include a larger area of possible applications, it seems useful to consider projective representations with a prescribed multiplier. Although different objects have to be studied in different concrete cases, our hope is to provide tools for a unified treatment.

Our approach to the theory of representations is motivated only in part by the example of the Grassmann manifold. Other reasons arise in connection with some questions in differential geometry and topology. Roughly speaking, when $A$ is the algebra of bounded linear operators on a complex separable Hilbert space, the space of representations of a given group G in $A$ is, as in the case of the Grassmann manifold, a classifying space for an appropriate group, an object related to the study of certain types of vector bundles with additional structures (see, for instance, [5] and [6]). In spite of the fact that these aspects will not appear in the present paper, they lie on the basis of our work.

Section 1 contains a few preliminaries. We introduce the space $R$ of projective reprsentations with a prescribed multiplier, and discuss some algebraic properties of these representations as elements of a convolution algebra.

In Section 2 we describe the Zariski tangent spaces to the "algebraic set" $R$.

Section 3 begins with the proof that $R$ has a real analytic structure. Then we introduce the canonical connection on $R$ and obtain an explicit form of the exponential map corresponding to this connection.

Section 4 is devoted to an example which is the straightforward generalization of the Grassmann manifold mentioned above.

The fact that the space $R$ of all projective representations of $G$ in $A$ has a real analytic structure and is equipped with a natural connection represents only a first step in its understanding. Other significant geometrical and topological properties of this space, as well as some applications, will be presented in a subsequent paper.

## 1. CONVOLUTION ALGEBRAS AND PROJECTIVE REPRESENTATIONS

Throughout this paper G will denote a compact, Hausdorff, topological group, satisfying the second axiom of countability, and $A$ will be a separable unital $C^*$-algebra.

In the next paragraphs we summarize, in a suitable form, a few general facts regarding the projective representations. Some elementary properties are presented without proofs. The reader who wants to get a deeper insight into the standard theory should consult, for example, [1], or, if is interested in some relevant applications, [10].

**1.1.** We begin by recalling the notion of two-cocycle. Let $\mathbf{T}$ be the multiplicative group of complex numbers of modulus 1. By a two-cocycle for G we shall mean a continuous map $\sigma$ of $G \times G$ into $\mathbf{T}$, such that

$$(1.1) \qquad \sigma(g,h)\sigma(gh,k) = \sigma(g,hk)\sigma(h,k),$$

$$(1.2) \qquad \sigma(e,g) = \sigma(g,e) = 1,$$

for all g, h, k $\epsilon$ G, where e denotes the identity of G.

With pointwise multiplication, the set $Z^2(G)$ of all two-cocycles for G is an abelian group. Its identity is the two-cocycle $\sigma_0$ defined by $\sigma_0(g,h) = 1$ (g, h $\epsilon$ G). It will be called the trivial two-cocycle for G.

Given the elements $\sigma$ and $\sigma'$ in $Z^2(G)$, they are said to be cohomologous if there exists a continuous map $\tau$ of G into $\mathbf{T}$ such that

$$(1.3) \qquad \sigma'(g,h) = \sigma(g,h)\tau(g)\tau(gh)^{-1}\tau(h) \qquad (g, h \ \epsilon \ G).$$

**1.2.** Assume now that $\sigma$ is a fixed two-cocycle for G. Let $C(G,A)$ be the Banach space of all continuous $A$-valued functions on G. The addition and scalar multiplication in $C(G,A)$ are defined pointwise, and the norm is given by:

$$\|\phi\|_\infty = \sup_{g \in G} \|\phi(g)\| \qquad (\phi \in C(G,A)).$$

For $\phi, \psi \in C(G,A)$ one defines the element $\phi * \psi \in C(G,A)$ by

(1.4)
$$\phi * \psi(g) = \int_G \phi(gh)\psi(h^{-1})\sigma(gh,h^{-1})dm(h)$$

where m is the (normalized) Haar measure on G.

For each $\phi \in C(G,A)$ let us define $\phi^* \in C(G,A)$ by

(1.5)
$$\phi^*(g) = \sigma(g,g^{-1})^{-1}\phi(g^{-1})^*.$$

From (1.1) and (1.2) it follows easily that, under the operation (1.4) as multiplication and the operation (1.5) as involution, $C(G,A)$ becomes a Banach $*$-algebra, denoted by $A[G,\sigma]$, and referred to as the convolution algebra associated with $A$, $G$ and $\sigma$.

The algebra $A[G,\sigma_0]$, where $\sigma_0$ is the trivial two-cocycle, will be denoted simply
by $A[G]$.

If $\sigma$ and $\sigma'$ are cohomologous two-cocycles for G then the algebras $A[G,\sigma]$ and
$A[G,\sigma']$ are isomorphic.

**1.3.** A convolution algebra $A[G,\sigma]$ is unital if and only if the group G is finite. In
the general case, instead of the unit, it is possible to find a countable approximate unit.
More precisely, by a clasical construction which uses "bump functions", one obtains a
sequence $\{\epsilon_n\}$ of hermitian elements in $A[G,\sigma]$, such that

(1.6)
$$\lim_{n \to \infty} \epsilon_n * \phi = \lim_{n \to \infty} \phi * \epsilon_n = \phi \qquad (\phi \in A[G,\sigma]).$$

For a fixed g in G, let us denote by $\epsilon_n^g$ the element of $A[G,\sigma]$ defined as follows:

$$\epsilon_n^g(h) = \epsilon_n(hg) \qquad (h \in G)$$

From (1.6) we get, after a brief computation,

(1.7)
$$\lim_{n \to \infty} (\epsilon_n^g * \phi)(h) = \sigma(g^{-1},g)\sigma(g,h)^{-1}\phi(gh)$$

and

(1.8)
$$\lim_{n \to \infty} (\phi * \epsilon_n^g)(h) = \sigma(g,g^{-1})\sigma(h,g)^{-1}\phi(hg),$$

for all $h \in G$ and $\phi \in A[G,\sigma]$.

In addition, let us notice that $A[G,\sigma]$ becomes a left (resp. a right) $A$-module, if
we write

(1.9)        $(a \cdot \phi)(g) = a\phi(g)$   (resp. $(\phi \cdot a)(g) = \phi(g)a$)     $(a \in A, \phi \in A[G,\sigma], g \in G)$.

**1.4.** We introduce now the concept of projective representation. Suppose that $\sigma$ is a fixed two-cocycle for G. By a projective representation of G in $A$ with the multiplier $\sigma$, one means a continuous map $\alpha$ of G into $A$ such that

(r1)     $\alpha(gh) = \sigma(g,h)\alpha(g)\alpha(h)$     $(g,h \in G)$,

(r2)     $\alpha(e) = 1$,

(r3)     $\alpha(g)^* = \sigma(g,g^{-1})\alpha(g^{-1})$   $(g \in G)$.

In what follows we call each map $\alpha$ as above a $\sigma$-representation. The set of all $\sigma$-representations will be denoted by $R(G,\sigma,A)$, or just $R$ when no confusion may arise.

Clearly $R$ is a closed subset of $A[G,\sigma]$ and for any $\alpha \in R$ and each unitary $u \in A$ we have $u \cdot \alpha \cdot u^* \in R$.

Actually, (r3) together with (1.5) shows that $R$ is a closed subset of $A[G,\sigma]_h$, the set of all hermitian elements in $A[G,\sigma]$. Moreover, since $\sigma(g,g^{-1}) = \sigma(g^{-1},g)$ for all $g \in G$, one obtains that $\alpha(g)$ is a unitary in $A$, for any $\alpha \in R$ and each $g \in G$.

If $\sigma$ and $\sigma'$ in $Z^2(G)$ are cohomologous, the spaces $R = R(G,\sigma,A)$ and $R' = R(G,\sigma',A)$ are homeomorphic in an obvious fashion.

The space $R(G,\sigma_o,A)$ of all unitary representations of G in $A$ will be denoted by $R(G,A)$.

**1.5.** The next result gives a characterization of $\sigma$-representations as elements of the convolution algebra $A[G,\sigma]$.

**PROPOSITION.** *Let* $\alpha \in A[G,\sigma]_h$ *with* $\alpha(e) = 1$. *The following conditions are equivalent.*

   (i) $\alpha$ *is a* $\sigma$-*representation ;*

   (ii) *for all* $\phi \in A[G,\sigma]_h$ *we have*

(1.10)                             $(\phi * \alpha)(e) \cdot \alpha = \phi * \alpha$ ;

   (iii) *for all* $\phi \in A[G,\sigma]_h$ *we have*

(1.11)                             $\alpha \cdot (\alpha * \phi)(e) = \alpha * \phi$.

PROOF. The equivalence between (ii) and (iii) is a simple matter. Indeed, (1.11) is obtained by taking adjoints in (1.10).

Assume now that $\alpha$ is a $\sigma$-representation. Then, for all $\phi \in A[G,\sigma]_h$ and $g \in G$ we have

$$(\phi_* \alpha)(e)\cdot\alpha(g) = \int_G \phi(h)\alpha(h^{-1})\alpha(g)\sigma(h,h^{-1})dm(h) =$$

$$= \int_G \phi(h)\alpha(h^{-1}g)\alpha(h^{-1},g)^{-1}\sigma(h,h^{-1})dm(h).$$

But $\sigma(h,h^{-1}g)\alpha(h^{-1},g) = \sigma(h,h^{-1})$, thus

$$(\phi_* \alpha)(e)\cdot\alpha(g) = \int_G \phi(h)\alpha(h^{-1}g)\sigma(h,h^{-1}g)dm(h) = (\phi_* \alpha)(g),$$

and (1.10) is proved.

Conversely, suppose that $\alpha$ satisfies condition (ii). The only thing we have to prove is relation (r1). To this end let us first observe that equation (1.10) is complex linear in $\phi$, therefore it holds for all $\phi$ in $A[G,\sigma]$. In particular

$$(\varepsilon_n^g *\alpha)(e)\cdot \alpha = \varepsilon_n^g *\alpha \qquad (g \in G),$$

where $\{\varepsilon_n\}$ is an approximate unit in $A[G,\sigma]$. By (1.7) we find

$$\sigma(g^{-1},g)\alpha(g)\alpha(h) = \sigma(g^{-1},g)\sigma(g,h)^{-1}\alpha(gh) \qquad (h \in G),$$

and the proof is complete.

**COROLLARY.** *If* $\alpha \in R$ *then* $\alpha$ *is an idempotent of* $A[g,\sigma]_h$, *that is*

(1.12)
$$\alpha_* \alpha = \alpha.$$

PROOF. It is enough to remark that

$$\alpha_* \alpha(e) = \int_G \alpha(g)\alpha(g^{-1})\sigma(g,g^{-1})dm(g) = \int_G \alpha(e)dm(g) = 1,$$

and then to use (1.10).

**1.7.** As another application of Proposition 1.5 we indicate a simple proof of a well-known result concerning close representations.

**PROPOSITION.** *Let* $\alpha,\beta \in R$ *and assume that* $\|\alpha - \beta\|_\infty < 1$. *Then* $\alpha$ *and* $\beta$ *are unitary equivalent representations, that is,* $\alpha$ *and* $\beta$ *are related by an equality of the form* $\beta = u\cdot\alpha\cdot u^*$, *where* $u$ *is a unitary in* $A$.

PROOF. Let $a = (\beta_* \alpha)(e)$. In view of Proposition 1.5 we have

(1.13)
$$\beta\cdot a = a\cdot\alpha \quad \text{and} \quad \alpha\cdot a^* = a^*\cdot\beta.$$

By (1.12) and (r2) it follows that

$$\| 1 - a \| = \| (\beta_*(\beta - \alpha))(e) \| \leq \| \beta \|_\infty \| \beta - \alpha \|_\infty < 1,$$

hence $a$ is invertible. On the other hand, from (1.13) we deduce $\alpha \cdot (a^*a) = = a^* \cdot \beta \cdot a = (a^*a) \cdot \alpha$. Therefore, $\alpha \cdot |a| = |a| \cdot \alpha$, and also $\alpha \cdot |a|^{-1} = |a|^{-1} \cdot \alpha$, where $|a| = (a^*a)^{\frac{1}{2}}$.

Finally, let $u$ be the unitary part in the polar decomposition of $a$, that is, $u = a|a|^{-1}$. Then

$$\beta \cdot u = \beta \cdot a|a|^{-1} = a \cdot \alpha \cdot |a|^{-1} = a|a|^{-1} \cdot \alpha = u \cdot \alpha,$$

as desired.

## 2. TANGENT SPACES

In this section we treat the space $R = R(G,\sigma,A)$ as a "closed algebraic subset" of $A[G,\sigma]$, defined by the equations (r1), (r2) and (r3). Our purpose is to define and study the tangent spaces to $R$.

**2.1.** We start our discussion by introducing an appropriate definition of the tangent vectors to $R$.

In the algebraic geometry appear various descriptions of the Zariski tangent space to an algebraic set at a point (see, for instance, [7]). A concrete one relying on a trick with dual numbers is easily transferable in our setting. In order to do this, let $\mathbf{C}[\delta] = \mathbf{C} + \mathbf{C}\delta$, with $\delta^2 = 0$, be the algebra of dual numbers, and let $\tilde{A} = A \otimes_{\mathbf{C}} \mathbf{C}[\delta]$. An element $\tilde{a}$ of $\tilde{A}$ can be represented uniquely as $\tilde{a} = a_o + a_1 \delta$, where $a_o$, $a_1 \in A$. For $\tilde{a} = a_o + a_1 \delta$ and $\tilde{b} = b_o + b_1 \delta$ in $\tilde{A}$ one defines

(i) $\tilde{a} + \tilde{b} = (a_o + b_o) + (a_1 + b_1)\delta$

(ii) $\tilde{a} \cdot \tilde{b} = a_o b_o + (a_o b_1 + a_1 b_o)\delta$

(iii) $\tilde{a}^* = a_o^* + a_1^* \delta$

(iv) $\|\tilde{a}\| = \|a_o\| + \|a_1\|$

With these operations $\tilde{A}$ is a Banach $*$-algebra.

**DEFINITION.** Assume that $\alpha$ is a point in $R$. An element $\theta$ of $C(G,A)$ is said to be a tangent vector to $R$ at the point $\alpha$ if $\tilde{\alpha} = \alpha + \theta\delta$ is a projective representation with multiplier $\sigma$ of $G$ in $\tilde{A}$.

To be more specific, the definition requires $\tilde{\alpha}$ to satisfy the equations (r1), (r2), and (r3). After a short calculation we conclude that the equations of definition of a tangent vector $\theta$ at $\alpha$ are:

(t1)    $\theta(gh) = \sigma(g,h)(\theta(g)\alpha(h) + \alpha(g)\theta(h))$    $(g,h \in G)$,

(t2)    $\theta(e) = 0$,

(t3)    $\theta(g)^* = \sigma(g,g^{-1})\theta(g^{-1})$    $(g \in G)$.

Let us notice that $\theta$ satisfies (t3) if and only if it is a hermitian element of $A[G,\sigma]$.

The set $T_\alpha R$ of all tangent vectors $\theta$ as above will be called the *tangent space* to $R$ at $\alpha$. Since the conditions involved are linear, $T_\alpha R$ is a subspace of the real vector space $A[G,\sigma]_h$.

As counterparts of Proposition 1.5 and Corollary 1.6 we obtain the next results.

**2.2. PROPOSITION.** *Let* $\alpha \in R$ *and* $\theta \in A[G,\sigma]_h$ *with* $\theta(e) = 0$. *The following conditions are equivalent:*

(i) $\theta$ *is a tangent vector to $R$ at $\alpha$;*

(ii) *for all* $\phi \in A[G,\sigma]_h$ *we have*

(2.1)    $$(\phi * \theta)(e) \cdot \alpha + (\phi * \alpha)(e) \cdot \theta = \phi * \theta ;$$

(iii) *for all* $\phi \in A[G,\sigma]_h$ *we have*

(2.2)    $$\theta \cdot (\alpha * \phi)(e) + \alpha \cdot (\theta * \phi)(e) = \theta * \phi.$$

**2.3. PROPOSITION.** *If* $\alpha \in R$ *and* $\theta \in T_\alpha R$, *then*

(2.3)    $$\theta * \alpha + \alpha * \theta = \theta.$$

The proofs are left to the reader.

**2.4.** Before continuing we make a remark. Recall that by a $*$-derivation in the $C^*$-algebra $A$ one means a linear map $\partial$ of $A$ into itself, with the properties

(i)  $\partial (ab) = (\partial a)b + a(\partial b)$   $(a, b \in A)$,

(ii)  $\partial (a^*) = (\partial a)^*$   $(a \in A)$.

If $\partial$ is a $*$-derivation and $\alpha \in R$, let $\theta = \partial \alpha$ be the element of $A[G,\sigma]$ defined by

$$\theta(g) = \partial (\alpha(g))   (g \in G).$$

Then $\theta$ is a tangent vector at $\alpha$. Indeed, the relations (t1), (t2), and (t3) are obtained from (r1), (r2), and (r3) by applying $\partial$.

This remark explains how one obtains the infinitesimal equations (2.1), (2.2), and (2.3) from the corresponding equations (1.10), (1.11), and (1.12).

Actually, all tangent vectors at $\alpha$ can be obtained by using $*$-derivations of $A$ (see Corollary 2.7 below).

**2.5.** Suppose that $\alpha$ is a $\sigma$-representation and $u$ is a unitary in $A$. Let $\beta = u \cdot \alpha \cdot u^*$. Then $\beta$ is also a point in $R$ and

$$T_\beta R = u \cdot T_\alpha R \cdot u^*.$$

Combining this remark with Proposition 1.7 we find:

**COROLLARY.** *Let* $\alpha, \beta \in R$ *with* $\| \alpha - \beta \|_\infty < 1$. *Then the tangent spaces* $T_\alpha R$ *and* $T_\beta R$ *are isomorphic.*

**2.6.** We introduce now two maps which will enable us to obtain a simple description of the tangent spaces to $R$.

Let $A_{sh}$ be the set of all skew-hermitian elements of $A$. For a fixed $\sigma$-representation $\alpha$ we define the following real-linear maps:

$$(2.4) \qquad \lambda_\alpha : A_{sh} \longrightarrow A[G,\sigma]_h, \quad \lambda_\alpha(a) = a \cdot \alpha - \alpha \cdot a \quad (a \in A_{sh});$$

$$(2.5) \qquad \mu_\alpha : A[G,\sigma]_h \longrightarrow A_{sh}, \quad \mu_\alpha(\phi) = \tfrac{1}{2}(\phi_*\alpha - \alpha_*\phi)(e) \quad (\phi \in A[G,\sigma]_h).$$

Clearly $\lambda_\alpha$ and $\mu_\alpha$ are well defined. The next result points out two basic properties of the maps $\lambda_\alpha$ and $\mu_\alpha$.

**PROPOSITION.** *Let* $\alpha$ *be a* $\sigma$-*representation. Then*

(i) $\lambda_\alpha(a) \in T_\alpha R$ *for all* $a \in A_{sh}$ ;

(ii) *for any* $\phi \in A[G,\sigma]_h$ *which satisfies the conditions*

$$(2.6) \qquad \phi(e) = 0 \quad and \quad \phi_*\alpha + \alpha_*\phi = \phi$$

*we have*

$$(2.7) \qquad \lambda_\alpha(\mu_\alpha(\phi)) = \phi.$$

PROOF. The assertion (i) is a direct consequence of the fact that $\lambda_\alpha(a) = \partial \alpha$, where $\partial$ is the inner $*$-derivation corresponding to $\alpha$.

For the second assertion, we first note that

$$(\phi_*\alpha)(e) + (\alpha_*\phi)(e) = 0,$$

whence

$$\mu_\alpha(\phi) = (\phi * \alpha)(e) = -(\alpha * \phi)(e).$$

According to (1.10) and (1.11) it follows that

$$\lambda_\alpha(\mu_\alpha(\phi)) = \mu_\alpha(\phi) \cdot \alpha - \alpha \cdot \mu_\alpha(\phi) = (\phi * \alpha)(e) \cdot \alpha + \alpha \cdot (\alpha * \phi)(e) = \phi * \alpha + \alpha * \phi = \phi,$$

as asserted.

**2.7. COROLLARY.** *Let $\alpha$ be a $\sigma$-representation.*

(i) *If $\theta \in T_\alpha R$ then*

(2.8)
$$\lambda_\alpha(\mu_\alpha(\theta)) = \theta.$$

(ii) *The real linear map*

$$\pi_\alpha : A[G,\sigma]_h \to A[G,\sigma]_h, \ \pi_\alpha = \lambda_\alpha \mu_\alpha \, ,$$

*is a projection of $A[G,\sigma]_h$ onto $T_\alpha R$.*

PROOF. If $\theta \in T_\alpha R$ then, by (t2) and (2.3) we clearly find that $\theta$ satisfies condition (2.6), and (i) follows.

In order to prove (ii), in view of assertion (i) in Proposition 2.6 we first observe that $\pi_\alpha(\phi) \in T_\alpha R$ for any $\phi \in A[G,\sigma]_h$. On the other hand, if $\theta \in T_\alpha R$, then $\pi_\alpha(\theta) = \theta$, as a consequence of (2.8). By combining these two remarks we conclude that $\pi_\alpha(\pi_\alpha(\phi)) =$
$= \pi_\alpha(\phi)$ for all $\phi \in A[G,\sigma]_h$.

**2.8.** The existence of the projection $\pi_\alpha$ shows that the tangent space $T_\alpha R$ has a closed complement in $A[G,\sigma]_h$.

It is also interesting to emphasize on a rather unexpected consequence of the assertion (ii) in Proposition 2.6. Let $E = E(G,\sigma A)$ be the set of all elements $\epsilon$ of $A[G,\sigma]$ such that

(e1) $\quad \epsilon * \epsilon = \epsilon = \epsilon^*,$

(e2) $\quad \epsilon(e) = 1.$

As we already proved (see (1.12) and (r2)), $R$ is a subset of $E$ (in general a proper subset).

Since $E$ is also an "algebraic closed subset" of $A[G,\sigma]$, it is possible to define the tangent space to $E$ at a point $\epsilon$ , by an obvious modification of Definition 2.1. It follows that $T_\epsilon E$ is the space of those $\phi$ in $A[G,\sigma]$ which satisfy the equations

(t'1) $\quad \phi * \epsilon + \epsilon * \phi = \phi = \phi^*,$

(t'2)     $\phi(e) = 0$.

The unexpected fact which follows from Proposition 2.6 is that, if $\epsilon = \alpha \epsilon R$ and $\phi \epsilon T_\alpha E$, then

$$\phi = \lambda_\alpha(\mu_\alpha(\phi)) \epsilon T_\alpha R.$$

In other words, when $\epsilon = \alpha \epsilon R$ the equations (t'1) and (t'2) characterize the tangent space $T_\alpha R$, that is: $T_\alpha R = T_\alpha E$.

**2.9.** We end this section with another remark. Assume that $\alpha$ is a point in $R$ and let $\nu_\alpha = \mu_\alpha | T_\alpha R$ be the restriction of the map $\mu_\alpha$ to $T_\alpha R$.

Then (2.8) implies that $\lambda_\alpha \nu_\alpha$ is the identity map of $T_\alpha R$, whence $\chi_\alpha = \nu_\alpha \lambda_\alpha$ is a projection of $A_{sh}$ onto $(A_{sh})_\alpha = \mathbf{ran}\, \nu_\alpha$, a subspace isomorphic to the tangent space $T_\alpha R$.

The complemetary projection $\chi^\alpha = id - \chi_\alpha$ has the range $(A_{sh})^\alpha = \mathbf{ker}\, \lambda_\alpha = \{a \epsilon A_{sh} : a \cdot \alpha = \alpha \cdot a\}$. The projection $\chi^\alpha$ is given explicitly by the formula

(2.9)            $$\chi^\alpha(a) = (\alpha \cdot a * \alpha)(e) = \int_G \alpha(g) a \alpha(g)^* dm(g) \qquad (a \epsilon A_{sh}).$$

## 3. THE REAL ANALYTIC STRUCTURE AND THE CANONICAL CONNECTION

In this section we provide the space $R = R(G, \sigma, A)$ with a real analytic manifold structure. Then we introduce a linear connection on $R$ and obtain the expression for the exponential map with respect to this connection.

For terminology and a few basic results concerning (infinite dimensional) differentiable manifolds we refer in what follows to [2].

**3.1.** Let $U(A) = \{u \epsilon A : u^*u = uu^* = 1\}$ be the unitary group of the $C^*$-algebra $A$. As in the classical case, a simple proof shows that $U(A)$ is a real analytic submanifold of $A$.

Indeed, let $f : A^{-1} \to A_h$ be the function defined by $f(a) = a^*a - 1$ $(a \epsilon A^{-1})$, where $A^{-1}$ is the open set of all invertible elements of $A$ and $A_h$ denotes the (real Banach) space of all hermitian elements of $A$. We have $U(A) = f^{-1}(0)$. For any u in $U(A)$ the differential $df(u) : A \to A_h$ acts by the formula

$$df(u)(x) = x^*u + u^*x \quad (x \epsilon A).$$

For each $a \epsilon A_h$ we have $df(u)(\frac{1}{2}ua) = a$, so $df(u)$ is surjective. On the other hand, since $df(u)(x) = 0$ is equivalent with $u^*x \epsilon A_{sh}$, we obtain that $\mathbf{ker}\, df(u) = uA_{sh}$, whence $uA_h$ is a closed complement of $\mathbf{ker}\, df(u)$ in $A$. These two remarks show that the function f is a submersion at each point $u \in U(A)$. In view of a well-known theorem ([2, 5.11.7]), $U(A)$ is a real-analytic submanifold of $A$, and the tangent space to $U(A)$ at a point u in $U(A)$ is given by

$$T_u U(A) = \ker df(u) = u A_{sh}.$$

**3.2.** We return now to the space $R = R(G,\sigma,A)$. There exists a natural action

$$(3.1) \quad \rho : U(A) \times A[G,\sigma]_h \to A[G,\sigma]_h, \quad \rho(u,\phi) = u \cdot \phi \cdot u^* \quad (u \in U(A), \ \phi \in A[G,\sigma]_h).$$

Given $\alpha \in R$, let us define

$$(3.2) \quad \rho_\alpha : U(A) \to A[G,\sigma]_h, \quad \rho_\alpha(u) = \rho(u,\alpha) \quad (u \in U(A)),$$

$$(3.3) \quad R(\alpha) = \{u \cdot \alpha \cdot u^* : u \in U(A)\}.$$

By Proposition 1.7, $R(\alpha)$ is an open and closed subset of $R$. Thus, in order to introduce a (generalized) real analytic structure on $R$, it will be enough to prove that each $R(\alpha)$ as above is a real analytic submanifold of $A[G,\sigma]_h$. This can be done simply by showing that the function $\rho_\alpha$ is a subimersion at any point $u \in U(A)$ (for details see [2, Section 5.10] and [2, Section 5.12]). To be more specific, we have to prove that the subspaces $\ker d\rho_\alpha(u)$ and $\operatorname{ran} d\rho_\alpha(u)$ have closed complements, where $d\rho_\alpha(u) :$ $: T_u U(A) \to A[G,\sigma]_h$ is the differential of the function $\rho_\alpha$ at a point u in $U(A)$.

First, let us notice the relation

$$d\rho_\alpha(u)(x) = x \cdot \alpha \cdot u^* + u \cdot \alpha \cdot x^* \quad (x \in T_u U(A)).$$

Each x in $T_u U(A)$ is uniquely represented as $x = ua$, where $a \in A_{sh}$. Since

$$(3.4) \quad d\rho_\alpha(u)(ua) = u \cdot (a \cdot \alpha - \alpha \cdot a) \cdot u^* = u \cdot \lambda_\alpha(a) \cdot u^*,$$

we conclude that

$$\ker d\rho_\alpha(u) = u(\ker \lambda_\alpha)u^* = u A_{sh}^\alpha u^*,$$

whence $u(A_{sh})_\alpha u^*$ is a closed complement for $\ker d\rho_\alpha(u)$. Recall that the complemetary subspaces $A_{sh}^\alpha$ and $(A_{sh})_\alpha$ where introduced previously, at the end of Section 2.

The same formula (3.4) shows that

$$\operatorname{ran} d\rho_\alpha(u) = u \cdot (\operatorname{ran} \lambda_\alpha) \cdot u^* = u \cdot T_\alpha R \cdot u^* = T_{u \cdot \alpha \cdot u^*} R,$$

a subspace which, as we already proved (see 2.8), has a closed complemet.

In conclusion, $R(\alpha)$ is a real analytic submanifold of $A[G,\sigma]_h$, and therefore $R$ is a (generalized) real analytic manifold. Moreover, the tangent spaces to $R$ in the sense of differential geometry are exactly the tangent spaces defined in Section 2.

**3.3.** At this moment a classical object appears naturally.

**DEFINITION.** By a *smooth vector field* X over $R$ we mean a smooth map $X : R \to A[G,\sigma]_h$, such that $X(\alpha) \in T_\alpha R$, for all $\alpha \in R$.

The set $X(R)$ of all smooth vector fields over $R$ is a $C^\infty(R)$ module, in an obvious fashion.

As a simple example, let us observe that given an element a in $A_{sh}$, the map

$$X : R \to A[G,\sigma]_h, \quad X(\alpha) = \lambda_\alpha(a) = a \cdot \alpha - \alpha \cdot a \quad (\alpha \in R)$$

is a smooth vector field over $R$.

**3.4.** Recall that for any point $\alpha \in R$ these exists the projection map $\pi_\alpha$ from $A[G,\sigma]_h$ onto the tangent space $T_\alpha R$ (see 2.7). The existence of this map induces, by a standard construction, a canonical linear connection on $R$, $\nabla : X(R) \times X(R) \to X(R)$, defined as follows:

$$(3.5) \qquad \nabla(X,Y)(\alpha) = \pi_\alpha[dY(\alpha)(X(\alpha))] \quad (X,Y \in x(R), \alpha \in R),$$

where $dY(\alpha) : T_\alpha R \to A[G,\sigma]_h$ is the differential of Y at $\alpha$ in $R$.

Clearly the map $\nabla$ is well defined. It follows also easily that $\nabla$ satisfies the two axioms of a linear connection.

In some instances, the next local description of $\nabla$ is useful. Let X,Y in $X(R)$ and $\alpha$ in $R$. Consider a smooth map $\gamma : \mathbf{R} \to R$ such that $\gamma(0) = \alpha$ and $(d\gamma/dt)(0) = X(\alpha)$. Then

$$(3.6) \qquad \nabla(X,Y)(\alpha) = \pi_\alpha[(d(Y \circ \gamma)/dt)(0)].$$

**3.5.** Following the standard terminology, we say that a smooth curve $\gamma$ in $R$ is a geodesic with respect to $\nabla$ if

$$\nabla(\gamma'(t), \gamma'(t))(\gamma(t)) = 0$$

for all possible values of the real parameter t, where $\gamma' = d\gamma/dt$.

By (3.6) one obtains the equation of a geodesic in a simpler form

$$(3.7) \qquad \pi_{\gamma(t)}(\gamma''(t)) = 0.$$

A geodesic $\gamma$ is a called *total* if the parameter domain is the whole real line. As we shall see, all geodesics on $R$ are total.

**3.6.** In order to give an explicit construction of the geodesics, we define, for each point $\alpha \in R$, a real analytic map $E_\alpha : A[G,\sigma]_h \longrightarrow R$ given by

$$E_\alpha(\phi) = \exp \mu_\alpha(\phi) \cdot \alpha \cdot \exp \mu_\alpha(-\phi) \qquad (\phi \in A[G,\sigma]_h).$$

Recall that $\mu_\alpha$ is the map defined by formula (2.5). The map $E_\alpha$ turns out to be the exponential map with respect to the canonical connection $\nabla$.

**PROPOSITION.** *Given $\alpha \in R$ and $\theta \in T_\alpha R$ there exists a unique total geodesic $\gamma$ such that $\gamma(0) = \alpha$ and $\gamma'(0) = \theta$. It is given by*

$$(3.8) \qquad\qquad \gamma(t) = E_\alpha(t\theta) \qquad (t \in \mathbf{R}).$$

PROOF. We have to consider the initial value problem

$$(3.9) \qquad\qquad \pi_{\gamma(t)}(\gamma''(t)) = 0, \quad \gamma(0) = \alpha, \quad \gamma'(0) = \theta.$$

It will be enough to prove that the smooth curve $\gamma$ defined by (3.8) fulfils (3.9).

Clearly $\gamma(0) = \alpha$. We also see that

$$\gamma'(t) = \exp \mu_\alpha(t\theta) \cdot [\mu_\alpha(\theta) \cdot \alpha - \alpha \cdot \mu_\alpha(\theta)] \cdot \exp \mu_\alpha(-t\theta).$$

But $\mu_\alpha(\theta) \cdot \alpha - \alpha \cdot \mu_\alpha(\theta) = \lambda_\alpha \mu_\alpha(\theta) = \pi_\alpha(\theta) = \theta$, therefore

$$(3.10) \qquad\qquad \gamma'(t) = \exp \mu_\alpha(t\theta) \cdot \theta \cdot \exp \mu_\alpha(-t\theta),$$

hence $\gamma'(0) = \theta$. Using (3.10) we deduce

$$(3.11) \qquad\qquad \gamma''(t) = \exp \mu_\alpha(t\theta) \cdot [\mu_\alpha(\theta) \cdot \theta - \theta \cdot \mu_\alpha(\theta)] \cdot \exp \mu_\alpha(-t\theta).$$

We claim that $\gamma$ is a geodesic in $R$, that is, $\pi_{\gamma(t)}(\gamma''(t)) = 0$. Obviously, this will follow from

$$(3.12) \qquad\qquad \mu_{\gamma(t)}(\gamma''(t)) = 0.$$

By (2.5) we obtain

$$(3.13) \qquad\qquad \mu_{\gamma(t)}(\gamma''(t)) = \tfrac{1}{2} \exp \mu_\alpha(t\theta)[\phi * \alpha - \alpha * \phi](e)\exp \mu_\alpha(-t\theta),$$

where $\phi = \exp \mu_\alpha(-t\theta) \cdot \gamma''(t) \cdot \exp \mu_\alpha(t\theta)$. From (3.11) we have

$$\phi = \mu_\alpha(\theta) \cdot \theta - \theta \cdot \mu_\alpha(\theta),$$

so that

$$\phi * \alpha - \alpha * \phi = \mu_\alpha(\theta) \cdot \theta * \alpha - \theta \cdot \mu_\alpha(\theta) * \alpha - \alpha * \mu_\alpha(\theta) \cdot \theta + \alpha * \theta \cdot \mu_\alpha(\phi).$$

Using again the equality $\mu_\alpha(\theta) \cdot \alpha - \alpha \cdot \mu_\alpha(\theta) = \theta$ we find

$$\phi * \alpha - \alpha * \phi = \mu_\alpha(\theta) \cdot \theta * \alpha - \theta * (\theta + \alpha \cdot \mu_\alpha(\theta)) - (\mu_\alpha(\theta)\alpha - \theta) * \theta + \alpha * \theta \cdot \mu_\alpha(\theta) =$$

$$= \mu_\alpha(\theta) \cdot (\theta * \alpha - \alpha * \phi) - (\theta * \alpha - \alpha * \theta) \cdot \mu_\alpha(\theta),$$

hence

(3.14)     $$(\phi * \alpha - \alpha * \phi)(e) = 2\mu_\alpha(\theta)^2 - 2\mu_\alpha(\theta)^2 = 0.$$

Comparing (3.13) and (3.14) we conclude that condition (3.12) is fulfilled, as asserted.

**3.7.** The explicit form of the exponential map $E_\alpha : T_\alpha R \to R$ enables us to verify simply that $E_\alpha$ is a real analytic diffeomorphism from a neighbourhood of 0 in $T_\alpha R$ onto a neighbourhood of $\alpha$ in $R$. Indeed, $E_\alpha(0) = \alpha$, and

$$dE_\alpha(0)(\theta) = \mu_\alpha(\theta) \cdot \alpha - \alpha \cdot \mu_\alpha(\theta) = \lambda_\alpha \mu_\alpha(\theta) = \pi_\alpha(\theta) = \theta,$$

for all $\theta \in T_\alpha R$. This remark shows that, if the $\sigma$-representations $\alpha$ and $\beta$ are near each other, then there exists a unitary $u = \exp \mu_\alpha(\theta)$, with $\theta \in T_\alpha R$, such that $\beta = u \cdot \alpha \cdot u^*$. In the general case this unitary differs from that produced in the proof of Proposition 1.7.

However, assuming that $R$ is the Grassmann manifold (i.e. $G = \mathbf{Z}/2$ is the cyclic group of order 2 and $\sigma = \sigma_0$ is the trivial two-cocycle), and $\| \beta - \alpha \|_\infty < 1$ $(\alpha, \beta \in R)$, then the element $a = (\beta * a)(e)$ turns out to be normal, and, in this case it can be proved that the unitary $u = a|a|^{-1}$ is of the form $\exp \mu_\alpha(\theta)$ with $\theta$ in $T_\alpha R$ (for details see [9]). Since $\theta$ can be computed simply in terms of $\alpha$ and $\beta$, namely

$$\theta = \lambda_\alpha(\log a|a^{-1}|),$$

(3.8) yields an explicit equation of a total geodesic $\gamma$ such that $\gamma(0) = \alpha$ and $\gamma(1) = \beta$. This enables one to obtain further informations about the geodesics in the Grassmann manifold (see [9] and [8]).

## 4. AN EXAMPLE

This section is concerned with the case when $G$ is a finite cyclic group and $\sigma$ is the trivial cocycle. Our aim will be to present an alternative description of the space $R(G,A)$. As a matter of fact all assertions below are direct consequences of the previous general results, therefore we omit the proofs.

**4.1.** In this section we take $G = \mathbb{Z}/(n + 1)$, where

$$\mathbb{Z}/(n + 1) = \{0, 1, \ldots n\}$$

is the cyclic group of order $n + 1$ with addition "**mod**$(n + 1)$".

The convolution algebra $A[\mathbb{Z}/(n + 1)]$ has a simple realization. Namely, let $A^{n+1}$ be the direct product of $n + 1$ copies of $A$, with its standard structure of a $C^*$-algebra, and let $\zeta = \exp(2\pi i/(n + 1))$ be a primitive $(n + 1)$-th root of unity. For each $\mathbf{a} = (a_0, a_1, \ldots, a_n)$ in $A^{n+1}$ let us define the element $\phi$ of $A[\mathbb{Z}/(n + 1)]$ by

$$(4.1) \qquad \phi(r) = \sum_{s=0}^{n} \zeta^{rs} a_s \qquad (r \in \mathbb{Z}/(n + 1)).$$

By straightforward computations one obtains that the correspondence $\mathbf{a} \mapsto \phi$ is an isomorphism of involutive Banach algebras (and also an isomorphism of left or right $A$-modules).

The inverse map is given by

$$(4.2) \qquad a_s = [1/(n + 1)] \sum_{r=0}^{n} \lambda^{-sr} \phi(r) \qquad (0 \le s \le n).$$

**4.2.** Let $\alpha : \mathbb{Z}/(n + 1) \to A$ be a representation of $\mathbb{Z}/(n + 1)$ in $A$. Then there exists a unitary $u \in U(A)$, with $u^{n+1} = 1$, such that

$$(4.3) \qquad \alpha(r) = u^r \qquad (r \in \mathbb{Z}/(n + 1)).$$

The spectrum of $u$ is contained in the set $\{\zeta^s : 0 \le s \le n\}$.

By the isomorphism described above, the representation $\alpha$ is related to an element $\mathbf{e} = (e_0, e_1, \ldots, e_n)$ of $A^{n+1}$. A brief verification shows that $e_s$ is exactly the spectral projection of the unitary $u$ corresponding to the point $\zeta^s$ in its spectrum, so $e_s$ $(0 \le s \le n)$ are an orthogonal family of hermitian idempotents in $A$, with $\sum_{s=0}^{n} e_s = 1$ ($e_s = 0$ is allowed for some $s$).

Conversely, each point $\mathbf{e} = (e_0, e_1, \ldots, e_n) \in A^{n+1}$ such that:

$$(4.4) \qquad e_s^2 = e_s = e_s^* \qquad (0 \le s \le n),$$

$$(4.5) \qquad e_s e_t = 0 \qquad (0 \le s, t \le n, s \ne t),$$

$$(4.6) \qquad \sum_{s=0}^{n} e_s = 1,$$

defines a unique representation $\alpha \in R(\mathbb{Z}/(n + 1), A)$, given by (4.3), where $u = \sum_{s=0}^{n} \zeta^s e_s$.

In conclusion we obtain a homeomorphism from $R(\mathbb{Z}/(n + 1), A)$ to $E_n = E_n(A)$, the closed set of all points $e \in A^{n+1}$ which satisfy the equations (4.4), (4.5), and (4.6). By using this homeomorphism we translate the real analytic structure of $R(\mathbb{Z}(n + 1), A)$ to $E_n$, and $E_n$ becomes a real analytic submanifold of $A^{n+1}$.

**4.3.** In the same circle of ideas, we continue by considering the tangent spaces to the manifold $E_n$.

Given $e = (e_0, e_1, \ldots, e_n)$ in $E_n$, the tangent space $T_e E_n$ to $E_n$ at $e$ is the space of all elements $x = (x_0, x_1, \ldots, x_n)$ of $A^{n+1}$, such that

$$(4.7) \qquad x_s e_s + e_s x_s = x_s = x_s^* \qquad (0 \le s \le n),$$

$$(4.8) \qquad x_s e_t + e_s x_t = 0 \qquad (0 \le s, t \le n, s \neq t),$$

$$(4.9) \qquad \sum_{s=0}^{n} x_s = 0.$$

In Section 2 we associated to any representation $\alpha$ the maps $\lambda_\alpha$ and $\mu_\alpha$. In our present situation these maps have a peculiar form. More precisely, to each $e = (e_0, e_1, \ldots, e_n)$ in $E_n$ we associate the maps $\lambda_e : A_{sh} \to (A^{n+1})_h$ and $\mu_e : (A^{n+1})_h \to A_{sh}$ given by

$$(4.10) \qquad \lambda_e(a) = ([a, e_0], [a, e_1], \ldots, [a, e_n]),$$

$$(4.11) \qquad \mu_e(a_0, a_1, \ldots, a_n) = \frac{1}{2} \sum_{s=0}^{n} [a_s, e_s],$$

for all $a \in A_{sh}$ and $(a_0, a_1, \ldots, a_n) \in (A^{n+1})_h$, where $[a, b] = ab - ba$ denotes the commutator of the elements $a, b \in A$.

As a consequence of Proposition 2.6 we obtain

(i) $\lambda_e(a) \in T_e E_n$ for all $a \in A_{sh}$,

(ii) $\lambda_e \mu_e(x) = x$ for any $x \in T_e E_n$, and

(iii) the map $\pi_e = \lambda_e \mu_e$ is a projection of $(A^{n+1})_h$ onto $T_e E_n$.

By appropriate changes in (3.6) and (3.7) we can find easily the expressions for the canonical connection and the exponential map of $E_n$.

In a different context, the maps defined by (4.10) and (4.11) where used in [4] and [5] for the study of smooth manifolds with almost product structures.

**4.4.** Actually, each point $e = (e_0, e_1, \ldots, e_n)$ in $E_n$ is uniquely defined by its projection $e' = (e_1, \ldots, e_n)$ in $A^n$. This allows us to consider the manifold $E_n$ as a submanifold of $A^n$. For example, the Grassmannn manifold of $A$, $E_1$ in our notation, becomes a submanifold of $A$.

In [9] the Grassmann manifold $E_1$ was used to set up an abstract framework for a natural generalization of the class of operators introduced by Cowen and Douglas in [3]. The manifold $E_n$ ($n \geq 2$) seems to be also available for extensions of the Cowen - - Douglas theory. This aspect, and other questions related to $E_n$, will be discussed elsewhere.

## REFERENCES


1. **Beyl, F.R. ; Tappe, J. :** *Group extensions, representations, and the Schur multiplicator*, Lecture Notes in Mathematics, vol. **958**, Springer Verlag, Berlin, Heidelberg, New-York, 1982.

2. **Bourbaki, N. :** *Variété différentielles et analytiques, Fascicule de resultats*, Hermann, Paris, 1967.

3. **Cowen, M.J. ; Douglas, R.G. :** Complex geometry and operator theory, *Acta Math.* **141**(1979), 187-261.

4. **Martin, M. :** Almost product structures and derivations, *Bull. Math. Soc. Sci. Math. R.S. Roumanie* **23**(1979), 171-176.

5. **Martin, M. :** Geometric structures on vector bundles (in Romanian), *Stud. Cerc. Mat.* **36**(1984): I, (2), 113-142; II, (3), 171-192.

6. **Martin, M. :** Projective group representations and geometric structures, Preprint INCREST **53**(1986).

7. **Mumford, D. :** *Algebraic geometry. I: Complex projective varieties*, Springer Verlag, Berlin, Heidelberg, New-York, 1976.

8. **Recht, L. ; Potra, H. :** On the minimality of geodesics in Grassmann manifolds, *Proc. Amer. Math. Soc.*

9. **Salinas, N. :** The Grassmann manifold of a $C^*$-algebra and hermitian holomorphic bundles, in *Special classes of linear operators and other topics*, Birkhäuser-Verlag, 1988, pp. 267-289.

10. **Varadarajan, V.S. :** *Geometry of quantum theory*, Springer Verlag, Berlin, Heidelberg, New-York, 1985.



**Mircea Martin**

Department of Mathematics, INCREST,
Bdul Păcii 220, 79622 Bucharest,
Romania.

Operator Theory:
Advances and Applications, Vol. 43
© 1990 Birkhäuser Verlag Basel



# ON THE GENERAL STABLE RANK FOR PAIRS

**Gabriel Nagy**

Recent works of M.A. Rieffel ([4], [5]) on the so-called "cancellation" property for projective modules over $C^*$-algebras give sufficient motivation to consider his theory of stable ranks as a fruitful instrument in the non-stable $K$-theory of $C^*$-algebras.

For example, in order to give a sufficient condition for a stably free module to be actually free, Rieffel used the general stable rank (**gsr**) of a $C^*$-algebra.

But if one wants to study the behaviour of this invaraint in extensions, one finds that the inequality "**gsr**$(A) \leq$ **max(gsr**$(J)$, **gsr**$(A/J))$" does not hold in the general case (see the example given in this paper). The situation is the same if "**gsr**" is replaced by "**sr**" (stable rank), but in this case Rieffel proved the inequality

$$sr(A) \leq max(sr(J), sr(A/J), csr(A/J)).$$

In [3] we introduced the *general stable rank of a pair* (**gsr**$_J(A)$) and proved the following inequalities

$$(1) \qquad\qquad gsr(A) \leq max(gsr(J), gsr_J(A))$$

$$(2) \qquad\qquad sr(A) \leq max(sr(J), sr(A/J), gsr_J(A)).$$

In this paper we give further properties for our invariant which show that this has a "sub-homological" behaviour.

We being by recalling some notations and definitions.

For a unital $C^*$-algebra $A$ and an integer $n$ one takes

$$Lg_n(A) = \{(a_1, \ldots, a_n) \in A^n \mid \exists\, b_1, \ldots, b_n \in A \text{ such that } b_1 a_1 + \ldots + b_n a_n = 1\}.$$

Considering $Lg_n(A)$ as a set of column vectors one has an action of the group $GL_n(A)$ (invertible matrices of $M_n(A)$) by left multiplication. Rieffel's definition for the *general stable rank* of $A$ is

$$gsr(A) = min\{m \in \mathbf{N} \mid GL_n(A) \text{ acts transtively on } Lg_n(A) \text{ for every } n \geq m\}.$$

If the above set if empty one takes $\mathbf{gsr}(A) = \infty$, and in the non-unital case one works in the algebra $\tilde{A}$ obtained by adjoining a unit.

To show that we cannot expect to have "$\mathbf{gsr}(A) \leq \max(\mathbf{gsr}(J),\mathbf{gsr}(A/J))$" we give the following

**EXAMPLE.** Consider $T$ the Toeplitz $C^*$-algebra and the ideal $K \subseteq T$ of compact operators. One has $T/K \simeq C(\mathbb{T})$ ([2]) and $\mathbf{gsr}(K) = \mathbf{gsr}(C(\mathbb{T})) = 1$ but $\mathbf{gsr}(T) = 2$ ($\mathbf{gsr}(T)$ cannot be 1 since $T$ contains a non-unitary isometry).

We now recall our definition for the "$\mathbf{gsr}$" *of a pair* (with a little change of notation).

**DEFINITION.** Suppose $J$ is an ideal (by this we mean closed and two-sided ideal) of the unit $C^*$-algebra $A$. Then $\mathbf{gsr}(A,J)$ is the least integer m such that for every $n \geq m$ the set $\{x \in \mathbf{GL}_n(A/J) \mid \partial[x] = 0\}$ acts transitively on $\mathbf{Lg}_n(A/J)$. Again if no such m exists one takes $\mathbf{gsr}(A,J) = \infty$ and in the non-unital case one proceeds as before. $\partial : \mathbf{K}_1(A/J) \to \mathbf{K}_0(J)$ is the "index" homomorphism in $\mathbf{K}$-theory ([1], [6]). In fact our invariant gives some information about $\partial$, namely we have

**LEMMA.** *Suppose* $\mathbf{gsr}(A,J) \leq n < \infty$. *Then the map*

$$\mathbf{GL}_{n-1}(A/J) \ni x \longmapsto \partial[x] \in \mathbf{Im}\ \partial$$

*is onto.*

PROOF. Take $m > n - 1$ and $y \in \mathbf{GL}_m(A/J)$. Take $y^m$ the last column of y. Of course $y^m \in \mathbf{Lg}_m(A/J)$ and since $m \geq \mathbf{gsr}(A,J)$, there exists $T \in \mathbf{GL}_m(A/J)$ with $\partial[T] = 0$ such that $Ty^m = (0, \ldots, 1)^t$ ("t" stands for transpose). Ty will have the form $Ty = \begin{bmatrix} x & 0 \\ a & 1 \end{bmatrix}$ with $a \in M_{1\times(m-1)}(A/J)$ and $x \in \mathbf{GL}_{m-1}(A/J)$. It is clear that there is a continuous path in $\mathbf{GL}_m(A/J)$ joining Ty with $\begin{bmatrix} x & 0 \\ 0 & 1 \end{bmatrix}$ and so $[Ty]_{K_1(A/J)} = [x]_{K_1(A/J)}$. Of course $\partial[x] = \partial[Ty] = \partial[T] + \partial[y] = \partial[y]$. Consequently we found $x \in \mathbf{GL}_{m-1}(A/J)$ such that $\partial[y] = \partial[x]$. If we still have $m - 1 > n - 1$ we continue this procedure. After $m - n + 1$ steps we obtain an element of $\mathbf{GL}_{n-1}(A/J)$ with same index as y.

**REMARKS.** Since $K_0(\{0\}) = \{0\}$ we have $\mathbf{gsr}(A) = \mathbf{gsr}(A,\{0\})$. Also one easily gets

$$(3) \qquad \qquad \mathbf{gsr}(A/J) \leq \mathbf{gsr}(A,J).$$

To prove one of the main results we have to recall one fact from [3].

**THEOREM 1.** *Suppose* $x \in \mathbf{GL}_n(A/J)$ *and* $n \geq \mathbf{gsr}(J) - 1$. *Then the following conditions are equivalent:*

   (i) *there exists a lifting* $X \in \mathbf{GL}_n(A)$ *for* $x$;

   (ii) $\partial [x] = 0$.

Now we can prove that inequalities (1) and (3) hold in a more general setting, namely:

**THEOREM 2.** *Suppose* $K$ *and* $J$ *are ideal of* $A$ *such that* $K \subset J$. *Then*

   (i) $\mathbf{gsr}(A/K, J/K) \leq \mathbf{gsr}(A,J)$

   (ii) $\mathbf{gsr}(A,K) \leq \mathbf{max}(\mathbf{gsr}(A,J), \mathbf{gsr}(J,K))$.

PROOF. There is no difficulty so show that we can suppose $A$ unital. In the following diagram of $C^*$-algebras and $*$-homomorphisms

$$
\begin{array}{ccccccccc}
0 & \longrightarrow & K & \overset{i_1}{\hookrightarrow} & J & \overset{\pi_1}{\longrightarrow} & J/K & \longrightarrow & 0 \\
 & & \| & & \downarrow{i_3} \ \ \pi_2 & & \downarrow{i_4} & & \\
0 & \longrightarrow & K & \overset{i_2}{\hookrightarrow} & A & \overset{\pi_2}{\longrightarrow} & A/K & \longrightarrow & 0 \\
 & & \downarrow{i_1} \ \ i_3 & & \| & & \downarrow{\pi_4} & & \\
0 & \longrightarrow & J & \overset{i_3}{\hookrightarrow} & A & \overset{\pi_3}{\longrightarrow} & A/J & \longrightarrow & 0 \\
 & & \downarrow{\pi_1} \ \ i_4 & & \downarrow{\pi_2} \ \ \pi_4 & & \| & & \\
0 & \longrightarrow & J/K & \overset{i_4}{\longrightarrow} & A/K & \overset{\pi_4}{\longrightarrow} & A/J & \longrightarrow & 0
\end{array}
$$

the four rows are exact and all the squares are commutative. If we denote by $\partial_1$, $\partial_2$, $\partial_3$, $\partial_4$ the "index" homomorphisms corresponding to each row, by naturality of the exact sequence of K-theory ([1], [6]), we obtain the following diagram of groups and homomorphisms in which each square is commutative:

$$
\begin{array}{ccccccc}
K(J/K) & \overset{i_{4*}}{\longrightarrow} & K(A/K) & \overset{\pi_{4*}}{\longrightarrow} & K(A/J) & = & K(A/J) \\
\downarrow{\partial_1} & & \downarrow{\partial_2} & & \downarrow{\partial_3} & & \downarrow{\partial_4} \\
K_0(K) & = & K_0(K) & \overset{i_{1*}}{\longrightarrow} & K_0(J) & \overset{\pi_{1*}}{\longrightarrow} & K_0(J/K) .
\end{array}
$$

To prove (i) take $n \geq \mathbf{gsr}(A,J)$ and $(a_1, \ldots, a_n) \in \mathbf{Lg}_n((A/K)/(J/K)) = \mathbf{Lg}_n(A/J)$. From the

the definition there exists $S \in \mathbf{GL}_n(A/J)$ such that $\partial_3[S] = 0$ and $S(a_1, \ldots, a_n)^t =$
$= (1, 0, \ldots, 0)^t$. But $\partial_4[S] = \pi_{1*} \circ \partial_3[S] = 0$ which shows that $\{S \in \mathbf{GL}_n(A/J) \mid \partial_4[S] = 0\}$
acts trasitively on $\mathbf{Lg}_n(A/J)$ and so we obtain (i).

To prove (ii) let $n \geq \mathbf{max}(\mathbf{gsr}(A,J), \ \mathbf{gsr}(J,K))$ and $(a_1, \ldots, a_n) \in \mathbf{Lg}_n(A/K)$.
Consider $b = (\pi_4(a_1), \ldots, \pi_4(a_n))^t$. Clearly $b \in \mathbf{Lg}_n(A/J)$ and, since $n \geq \mathbf{gsr}(A,J)$, there
exists $T \in \mathbf{GL}_n(A/J)$ such that $\partial_3[T] = 0$ and $Tb = (1, 0, \ldots, 0)^t$. Again we have $\partial[T] = 0$
and, since $n \geq \mathbf{gsr}(J,K) \geq \mathbf{gsr}(J/K)$, by Theorem 1 there exists $S \in \mathbf{GL}_n(A/K)$ such that
$\pi_4(S) = T$ (for a $*$-homomorphism we keep the same notation for its extension to
matrices). But this means that $\pi_4(S(a_1, \ldots, a_n)^t) = (1, 0, \ldots, 0)^t$, and we obtain
$S(a_1, \ldots, a_n)^t \in \mathbf{Lg}_n(J/K)$ (of course $J \subset A$ and so $J/K \subset A/K$). On the other hand
$i_{1*} \circ \partial_2[S] = \partial_3 \circ \pi_{4*}[S] = \partial_3[T] = 0$. Exactness of the sequence of $\mathbf{K}$-theory for the
first row gives $\partial_2[S] \in \mathbf{Ker} \ i_{1*} = \mathbf{Im} \ \partial_1$ $(\partial_1 : \mathbf{K}_1(J/K) \longrightarrow \mathbf{K}_0(K))$. Since $n \geq \mathbf{gsr}(J,K)$, by
Lemma 1, there exists $R \in \mathbf{GL}_n(J/K)$ such that $\partial_1[R] = \partial_2[S]$. Take $Y = R^{-1}S$. Since
$R^{-1} \in \mathbf{GL}_n(J/K)$ and $S(a_1, \ldots, a_n)^t \in \mathbf{Lg}_n(J/K)$, we conclude that $Y(A_1, \ldots$
$\ldots, a_n)^t \in \mathbf{Lg}_n(J/K)$. From $n \geq \mathbf{gsr}(J,K)$ we get the existence of $Z \in \mathbf{GL}_n(J/K)$ such that
$\partial_1[Z] = 0$ and $ZY(a_1, \ldots, a_n)^t = (1, 0, \ldots, )^t$. We have $\partial_2[ZY] = \partial_2[ZR^{-1}S] =$
$= \partial_2 \circ i_{4*}[Z] - \partial_2 \circ i_{4*}[R] + \partial_2[S] = \partial_1[Z] - \partial_1[R] + \partial_2[S] = \partial_2[Z] = 0$. This shows that
the set $\{X \in \mathbf{GL}_n(A/K) \mid \partial_2[X] = 0\}$ acts transitively on $\mathbf{Lg}_n(A/K)$, which proves (ii).

For the following result we introduce the next

**DEFINITION.** Suppose J is an ideal in A and K is an ideal in B. By a $*$-*homo-*
*morphism of pairs* $\phi : (A,J) \longrightarrow (B,K)$ we mean a $*$-homomorphism $\phi : A \longrightarrow B$ such that
$\phi(J) \subset K$. Given two $*$-homomorphisms of pairs $\phi, \psi : (A,J) \longrightarrow (B,K)$, we say that they are
*homotopic* if there exists $\Phi : [0,1] \times A \longrightarrow B$ such that $\Phi(0, \cdot) = \phi$, $\Phi(1, \cdot) = \psi$ and

(i) for every $t \in [0,1]$, $\Phi(t, \cdot) : (A,J) \longrightarrow (B,K)$ is a $*$-homomorphism of pairs;

(ii) for every $a \in A$ the map $\Phi(\cdot, a) : [0,1] \longrightarrow B$ is continuous.

The pairs $(A,J)$ and $(B,K)$ are *homotopically equivalent* if there exist two
$*$-homomorphisms of pairs $\phi : (A,J) \longrightarrow (B,K)$ and $\psi : (B,K) \longrightarrow (A,J)$ such that $\psi \circ \phi$ and
$\mathbf{Id}_A$ are homotopic and also $\phi \circ \psi$ and $\mathbf{Id}_B$ are homotopic.

The following result shows the homotopy invariance for "gsr".

**THEOREM 3.** *If the pair $(A,J)$ and $(B,K)$ are homotopically equivalent then*
$\mathbf{gsr}(A,J) = \mathbf{gsr}(B,K)$.

PROOF. Obviously we can suppose A and B unital and also that the $*$-homomorphisms $\phi$ and $\psi$ in the definition are unit-preserving. By simmetry it suffices to prove only the inequality $\mathbf{gsr}(A,J) \leq \mathbf{gsr}(B,K)$. Let us denote by $\hat{\phi}: A/J \to B/K$ and $\hat{\psi}: B/K \to A/J$ the $*$-homomorphisms induced by $\phi$ and $\psi$. Take $n \geq \mathbf{gsr}(B,K)$ and $(a_1, \ldots, a_n) \in \mathbf{Lg}_n(A/J)$. Consider $(\hat{\phi}(a_1), \ldots, \hat{\phi}(a_n)) \in \mathbf{Lg}_n(B/K)$. Since $n \geq \mathbf{gsr}(B,K)$ there exists $T \in \mathbf{GL}_n(B/K)$ such that $\partial_2[T] = 0$ and $T(\hat{\phi}(a_1), \ldots, \hat{\phi}(a_n))^t = (1, 0, \ldots, 0)^t$. By naturality we have a commutative diagram of groups and homomorphisms

$$
\begin{array}{ccc}
K_1(B/K) & \xrightarrow{\hat{\psi}_*} & K_1(A/J) \\
\Big\downarrow{\partial_2} & & \Big\downarrow{\partial_1} \\
K_0(K) & \xrightarrow{\psi_*} & K_0(J)
\end{array} \quad .
$$

Let $S = \hat{\psi}(T) \in \mathbf{GL}_n(A/J)$. Since $\hat{\psi}$ induces an isomorphism on $\mathbf{K}$-theory we get $\partial_1[S] = 0$. On the other hand $(\hat{\psi} \circ \hat{\phi}(a_1), \ldots, \hat{\psi} \circ \hat{\phi}(a_n))$ can be obviously be joined by a continuous path in $\mathbf{Lg}_n(A/J)$ with $(a_1, \ldots, a_n)$ and, consequently (Corollary 8.5 of [4]) there exists $R \in \mathbf{GL}_n^0(A/J)$ such that $(\hat{\psi} \circ \hat{\phi}(a_1), \ldots, \hat{\psi} \circ \hat{\phi}(a_n))^t = R(a_1, \ldots, a_n)^t$. $(\mathbf{GL}_n^0(\cdot)$ stands for the connected component of the identity in $\mathbf{GL}_n(\cdot))$. In particular we have $[R]_{K_1(A/J)} = 0$. So we get $\partial_1[SR] = \partial_1[S] + \partial_1[R] = 0$ and $SR(a_1, \ldots, a_n)^t = (1, 0, \ldots 0)^t$ which means that $\{X \in \mathbf{GL}_n(A/J) \mid \partial_1[X] = 0\}$ acts transitively on $\mathbf{Lg}_n(A/J)$.

For matrix algebras we have:

**PROPOSITION 1.** $\mathbf{gsr}(M_n(A), M_n(J)) \leq \{(\mathbf{gsr}(A,J) - 1)/n\} + 1$

$(\{t\}$ stands for the least integer greater than t).

PROOF. Take $k \geq \{(\mathbf{gsr}(A,J) - 1)/n\} + 1$, that is $kn - n + 1 \geq \mathbf{gsr}(A,J)$. Let $b \in \mathbf{Lg}_k(M_n(A)/M_n(J)) = \mathbf{Lg}_k(M_n(A/J))$ considered as a left-invertible $kn \times n$ matrix. Of course $(b_{11}, b_{21}, \ldots, b_{kn,1})^t \in \mathbf{Lg}_{kn}(A/J)$ (it is the first column of b). Since $kn \geq \mathbf{gsr}(A,J) + n - 1 \geq \mathbf{gsr}(A,J)$ there exists $T_1 \in \mathbf{GL}_{kn}(A/J)$ such that $\partial[T_1] = 0$ and $T_1(b_{11}, b_{21}, \ldots, b_{kn,1})^t = (1, 0, \ldots, 0)^t$. Let $a = T_1 b \in \mathbf{Lg}_k(M_n(A/J))$, and $c \in M_{n \times kn}(A/J)$ a left inverse for a, that is $ca = I_n$. Put $S_1 = \begin{bmatrix} c_{11} & c_{12} & \cdots & c_{1,kn} \\ 0 & & I_{kn-1} & \end{bmatrix}$ $((c_{11}, \ldots, c_{1,kn})$ is the first row of c). Since the first column of a is $(1, 0, \ldots, 0)^t$ we have $c_{11} = 1$ and so $S_1 \in \mathbf{GL}_{kn}(A/J)$. Of course $[S_1]_{K_1(A/J)} = 0$. On the other hand since $c_{11}a_{1p} + \ldots \ldots + c_{1,kn}a_{kn,p} = 0$ for every $p \neq 1$, if we let $X_1 = S_1 T_1$, we have $\partial[X_1] = 0$ and

$X_1 b = \begin{bmatrix} 1 & 0 \\ 0 & b^1 \end{bmatrix}$ with $b^1$ a left-invertible $(kn - 1) \times (n - 1)$ matrix. If we still have $kn - 1 \geq \mathbf{gsr}(A,J)$ we continue this procedure and find $X_2 \in \mathbf{GL}_{kn-1}(A/J)$ with $\partial [X_2] = 0$ such that $X_2 b^1 = \begin{bmatrix} 1 & 0 \\ 0 & b^2 \end{bmatrix}$ with $b^2$ a left-invertible $(kn - 2) \times (n - 2)$ matrix, and so on. After $n$ steps if we take $X = X_1 \cdot \begin{bmatrix} 1 & 0 \\ 0 & X_2 \end{bmatrix} \cdot \ldots \cdot \begin{bmatrix} I_{n-1} & 0 \\ 0 & X_n \end{bmatrix}$ we obtain $\partial [X] = 0$ and $Xb = \begin{bmatrix} I_n \\ 0 \end{bmatrix}$ which ends the proof since $\overline{\mathbf{GL}}_k(M_n(A/J)) = \mathbf{GL}_{kn}(A/J)$ and the "index" homomorphism for the pair $(M_n(A), M_n(J))$ is exactly $\partial$.

For inductive limits we have

**THEOREM 4.** *Suppose* $A = \varinjlim A_n$ *and* $J = \varinjlim J_n$ *where* $A_n \subset A_{n+1}$, $J_n = J \cap A_n$ *and* $J_n$ *is an ideal of* $A_n$ *for every* n. *Then* $\mathbf{gsr}(A,J) \leq \lim \inf \mathbf{gsr}(A_n, J_n)$.

PROOF. We may suppose $1_A \in A_n$ for every n. Take $k \geq \lim \inf \mathbf{gsr}(A_n, J_n)$. Choosing a subsequence we restrict to the situation $k \geq \mathbf{gsr}(A_n, J_n)$ for every n. Let us denote by $\pi_n : A_n/J_n \longrightarrow A/J$ the $*$-homomorphisms induced by the embeddings $(A_n, J_n) \longrightarrow (A,J)$ and by $i_n : J_n \longrightarrow J$. Take $(a_1, \ldots, a_k) \in \mathbf{Lg}_k(A/J)$. Since $\bigcup_{n \in \mathbf{N}} \pi_n(A_n/J_n)$ is dense in $A/J$ there exists some $n$ and $(b_1, \ldots, b_k) \in \mathbf{Lg}_k(\pi_n(A_n/J_n))$ close enough to $(a_1, \ldots, a_k)$ such that there exists $R \in \mathbf{GL}_k^0(A/J)$ for which $(a_1, \ldots, a_k)^t = R(b_1, \ldots, b_k)^t$ (Corollary 8.5 of [4]). Under the assumptions about $J, \pi_n$ will be injective and so there exists $(x_1, \ldots, x_k) \in \mathbf{Lg}_k(A_n/J_n)$ such that $(b_1, \ldots, b_k) = (\pi_n(x_1), \ldots, \pi_n(x_k))$. Inequality $k \geq \mathbf{gsr}(A_n, J_n)$ gives the existence of an invertible $X \in \mathbf{GL}_k(A_n/J_n)$ with $\partial_n[X] = 0$ such that $X(x_1, \ldots, x_k)^t = (1, 0, \ldots, 0)^t$. Take $T = R\pi_n(X)^{-1}$. The commutativity of the following diagram

$$
\begin{array}{ccc}
K_1(A_n/J_n) & \xrightarrow{\ \pi_{n*}\ } & K_1(A/J) \\
\Big\downarrow{\partial_n} & & \Big\downarrow{\partial} \\
K_0(J_n) & \xrightarrow{\ i_{n*}\ } & K_0(J)
\end{array}
$$

shows that $\partial [T] = 0$. Since $(a_1, \ldots, a_k)^t = T(1, 0, \ldots, 0)^t$ we conclude that $\{T \in \mathbf{GL}_k(A/J) / \partial [T] = 0\}$ acts transitively on $\mathbf{Lg}_k(A/J)$ which ends the proof.

**REFERENCES**

1.      **Blackadar, B.** : *K-theory for operator algebras*, Springer -Verlag, 1985.

2.      **Douglas, R.G.** : *Banach algebra techniques in the theory of Toeplitz operators*,

CBMS Regional Conference Series  in Math., nr. **15**, Amer. Math. Soc., 1972.

3.    **Nagy, G.** : Some   remarks   on   lifting   invertible   elements   from   quotient $C^*$-algebras, *J. Operator Theory* (to appear).

4.    **Rieffel, M.A.** : Dimension and stable rank in the **K**-theory of $C^*$-algebras, *Proc. London Math. Soc.* **46**(1983), 301-333.

5.    **Rieffel, M.A.** : The cancellation theorem for projective modules over irrational rotation $C^*$-algebras, *Proc. London Math. Soc.* **47**(1983), 285-302.

6.    **Taylor, J.L.** : Banach algebras and topology, *Algebras in Analysis,*  Academic Press, 1975, pp. 118-186.

**Gabriel Nagy**

Department of Mathematics, INCREST,
Bdul Păcii 220, 79622 Bucharest
Romania.

Operator Theory:
Advances and Applications, Vol. 43
© 1990 Birkhäuser Verlag Basel



# WIENER-HOPF OPERATORS ON THE POSITIVE SEMIGROUP OF A HEISENBERG GROUP

Alexandru Nica

## 1. INTRODUCTION

The classical Wiener-Hopf operators are obtained by compressing the left-
-convolution operators on $L^2(\mathbf{R})$ to the space $L^2([0,\infty))$. One can make such a compression in the general context of a locally compact group, with $[0,\infty)$ replaced by a semigroup which is the closure of its interior. The most often considered examples of generalized Wiener-Hopf operators obtained in this manner are the Euclidean ones, where the group is $\mathbf{R}^n$ and the compression is made to a closed convex cone with non-void interior.

An interesting non-Euclidean example is given by the Heisenberg group $H_n \subseteq \mathbf{Mat}_n(\mathbf{R})$ of upper-triangular matrices having 1 on the diagonal and its "positive semigroup" $P_n$, obtained by intersection with the set of matrices with non-negative entries. This is the example studied in the present paper in the cases $n = 3$ and $n = 4$ (we note that $n = 2$ gives the classical Wiener-Hopf operators).

The suggestion of considering Wiener-Hopf operators on the Heisenberg groups was given to us in 1985 by Dan Voiculescu; we express him our most profound gratitude.

The instrument we use in our study is the groupoid theory. The observation that Wiener-Hopf operators can be derived from groupoids was made by P. Muhly and J. Renault in [1]; we shall use here the groupoid construction made in [3], which is briefly recalled in Section 2 of the paper.

It is known that the $C^*$-algebra generated by the classical Wiener-Hopf operators contains the compact operators on $L^2([0,\infty))$ and it is natural to ask under what conditions is this assertion true in more general situations. P. Muhly and J. Renault show in [1] that this is the case when the semigroup is pointed and the set of units of the groupoid involved is a regular compactification of the semigroup (see Section 3.1). In Proposition 3.2.1 we present two "nice" conditions on the order relation induced by the semigroup which imply together "regular compactification" and which are satisfied by the positive semigroup of any $H_n$. It is noteworthy that these conditions are also satisfied in any pointed Euclidean case. Hence, in all these cases, the $C^*$-algebra of the

Wiener-Hopf operators contains the compact operators.

Section 4 is devoted to $H_3$. Using groupoid techniques, we find without difficulty a composition series of the $C^*$-algebra of the Wiener-Hopf operators on $P_3$, which has easily tractable quotients between consecutive ideals. In particular, this $C^*$-algebra is found to be of type I.

Finally, in Section 5, we make the same discussion for $H_4$. The corresponding $C^*$-algebra is also of type I, and this result is also obtained by exhibiting a composition series. The general idea is the same as in Section 4; but some new complications occur, which indicate that the generalization to the positive semigroup of an arbitrary $H_n$ is not immediate.

## 2. THE GROUPOID CONSTRUCTION

In this section we recall the construction made in [3] of a groupoid whose associated $C^*$-algebra is isomorphic to the one generated by the Wiener-Hopf operators.

Our setting is as follows: let G be a locally compact second countable unimodular group and let $\mu$ be a fixed Haar measure on G. We shall call a subset A of G "solid" if $A \neq \emptyset$ and $A = \mathbf{clos}\,\overset{\circ}{A}$; it is easy to see that if A is solid, then $\mathbf{supp}\,\mu\,|\,A = A$ (so it is worth considering $L^2(\mu\,|\,A)$). Let P be a solid semigroup of G. For any f in $C_c(G)$ we define the Wiener-Hopf operator with symbol f on P to be: $W_p(f) = pL(f)j \in B(L^2(\mu\,|\,P))$, where $L(f) \in B(L^2(\mu))$ is the left-convolution operator with f, $p : L^2(\mu) \longrightarrow L^2(\mu\,|\,P)$ is projection and $j = p^* : L^2(\mu\,|\,P) \longrightarrow L^2(\mu)$ is inclusion. The $C^*$-subalgebra of $B(L^2(\mu\,|\,P))$ generated by $\{W_p(f) \mid f \in C_c(G)\}$ is called the $C^*$-algebra of Wiener-Hopf operators on P and is denoted by $W(P)$.

We say that P satisfies Condition (**M**) if every element of $w^*\text{-}\mathbf{clos}\{\chi_{tP^{-1}} \mid t \in P\} \subseteq L^\infty(\mu)$ is of the form $\chi_A$ with A a solid subset of G. (Remark: A is uniquely determined by $\chi_A$, i.e. A, B solid and $\chi_A = \chi_B$ $\mu$-a.e. imply A = B – see Observation 2.3.3 of [3].) If Condition (**M**) is satisfied, we can construct a groupoid $G$ having $C^*(G) \simeq W(P)$ in the following manner (for details, see Section 2 of [3]):

a) the set of units of $G$ is $U = w^*\text{-}\mathbf{clos}\{\chi_{tP^{-1}} \mid t \in P\} \subset L^\infty(\mu)$;

b) the set of arrows of $G$ is given by left translations with elements of G; that is, whenever $\chi_A \in U$ and $t \in G$ are such that $\chi_{tA} \in U$ (this is shown to be equivalent to $t \in A^{-1}$), we have an arrow $x = (t,A) \in G$ with $d(x) = \chi_A$ and $r(x) = \chi_{tA}$;

c) the multiplication on $G$ is defined by $(s,tA)(t,A) = (st,A)$; the identity at $\chi_A$ is $(e,A)$, with e the unit of G, and the inverse of $(t,A)$ is $(t^{-1},tA)$;

d) the topology on $G$: since $G \subseteq G \times L^\infty(\mu)$, we can take the product between the topology of G and the $w^*$-topology on $L^\infty(\mu)$ and reduce it to $G$. We remark that the groupoid topology induced on U coincides with the $w^*$-topology; it is compact, as we see from the Alaoglu theorem. U is also metrizable, because we assumed G second countable;

e) the Haar system: for any $\chi_A \in U$, the set of arrows leaving $\chi_A$ is $\{(t,A) \mid t \in A^{-1}\}$, canonically isomorphic to $A^{-1}$; we take on it the measure obtained from $\mu \mid A^{-1}$. In this way we get a right Haar system on $G$.

As it is shown in Section 3 of [3], any closed convex cone with non-void interior in $\mathbf{R}^n$ satisfies Condition (**M**), so the groupoid construction is available in the Euclidean case.

We consider now the positive semigroup $P_n$ of $H_n$. We shall identify $H_n$ with $\mathbf{R}^{(n-1)n/2}$. It is easy to see that the Lebesgue measure on $\mathbf{R}^{(n-1)n/2}$ is both left and right invariant with respect to the multiplication on $H_n$; this is the Haar measure we are going to work with. By our identification $P_n$ becomes $[0,\infty)^{(n-1)n/2}$. It is clear that $H_n$ and $P_n$ are situated in the above considered setting. We prove that, in addition, the Condition (**M**) is satisfied, so that the groupoid construction can be used to describe $W(P_n)$.

**PROPOSITION 2.1.** *For any* $n \geq 2$, $P_n$ *satisfies Condition* (**M**).

PROOF. By Corollary 3.4.5 of [3], the set $T = \{\chi_A \mid [0,\infty)^{(n-1)n/2} \subseteq A \subseteq \mathbf{R}^{(n-1)n/2}$, A closed and convex$\}$ is $w^*$-compact. But for any t in $P_n$, $P_n t^{-1}$ is closed and convex and contains $[0,\infty)^{(n-1)n/2}$; so $w^*$-$\mathbf{clos}\{\chi_{P_n t^{-1}} \mid t \in P_n\} \subseteq T$. Using unimodularity we obtain that any element of $w^*$-$\mathbf{clos}\{\chi_{t P_n^{-1}} \mid t \in P_n\}$ is of the form $\chi_B$ with $B^{-1}$ closed and convex with non-void interior (this clearly implies B solid).

## 3. SUFFICIENT CONDITIONS FOR $W(P) \supseteq K(L^2(\mu \mid P))$

**3.1. The condition of "regular compactification".** We shall assume that, in the setting of Section 2, the semigroup P is pointed, i.e. $P \cap P^{-1} = \{e\}$. Then the map h : $P \longrightarrow U$ defined by $h(t) = \chi_{t P^{-1}}$ is one-to-one, because $t P^{-1} = s P^{-1} \Longrightarrow t^{-1} s \in P \cap P^{-1} \Longrightarrow t = s$. The map h is easily seen to be continuous, and has dense range by the very definition of U. But U is compact, so (h,U) is a compactification of P.

Let us denote by V the range of h. We recall that the compactification (h,U) is called regular if V is open in U and if h : $P \longrightarrow V$ is a homeomorphism. It was remarked by P. Muhly and J. Renault in Corollary 3.7.2 of [1] that "(h,U) regular" implies that $W(P)$ contains the compact operators. As a matter of fact, they use a groupoid construction different from ours, but in order to make their proof to work it is sufficient to show that:

a) V is an invariant set of units; indeed, for any t in P, the arrows leaving $X_{tP^{-1}}$ are of the form $(a, tP^{-1})$ with a in $Pt^{-1}$, hence $a = st^{-1}$ for some s in P, and the range of $(st^{-1}, tP^{-1})$ is $X_{sP^{-1}} \in V$.

b) the reduced groupoid $G|V$ is transitive and principal, i.e. for any s, t in P there exists a unique arrow from $X_{sP^{-1}}$ to $X_{tP^{-1}}$; this arrow is $(ts^{-1}, sP^{-1})$.

c) $(t,s) \to (ts^{-1}, sP^{-1})$ is a groupoid isomorphism between the groupoid $E$ of the trivial equivalence relation on P and $G|V$. If $h : P \to V$ is a homeomorphism, then this isomorphism is topological. In addition, one can easily check that it transforms the Haar system induced by $\mu|P$ on $E$ into the Haar system inherited from $G$ on $G|V$.

We mention that in the proof of Corollary 3.7.2 of [1], $K$ is found as the ideal of $W(P)$ corresponding by the canonical isomorphism $C^*(G) \simeq W(P)$ to the ideal of $C^*(G)$ produced by the open invariant subset V of U; this fact will be used in Propositions 4.3 and 5.2.

We also note that as a corollary to Theorem 3.1 of [2], the $C^*$-algebra of a transitive and principal groupoid is always isomorphic to $K$; so even if $h : P \to V$ is not a homeomorphism, the fact alone that V is open in U implies the existence of an ideal of $W(P)$ which is isomorphic to $K$.

**3.2. Conditions on the order relation.** The relation induced by P on G is defined by $x \leq y \Leftrightarrow x^{-1}y \in P$. We still assume that P is pointed, and this implies that $\leq$ is antisymmetric. The semigroup properties of P imply that $\leq$ is reflexive and transitive, so that it is an order relation on G. The "strict order relation associated to $\leq$" is defined by $x < y \Leftrightarrow x^{-1}y \in \overset{\circ}{P}$. Let us record some simple properties of $\leq$ and $<$ which will be used in the sequel:

$1^\circ$ if $x \leq y < z$ or if $x < y \leq z$, then $x < z$; this happens because $P\overset{\circ}{P} \subseteq \overset{\circ}{P}$ and $\overset{\circ}{P}P \subseteq \overset{\circ}{P}$, being open subsets of P;

$2^\circ$ for any x in G we have $x \nleq x$, because $P \cap P^{-1} = \{e\}$ implies $e \notin \overset{\circ}{P}$;

$3^\circ$ for any x in G the sets $\{y \in G | y \geq x\}$ and $\{y \in G | y \leq x\}$ are closed, being in fact $xP$ and $xP^{-1}$ respectively. Similarly, $\{y \in G | y > x\} = x\overset{\circ}{P}$ and $\{y \in G | y < x\} = x\overset{\circ}{P}{}^{-1}$ are non-void open sets;

$4^\circ$ if for some s, $t \in G$ it is true that "$x < s \Rightarrow x < t$", then $s \leq t$. Indeed, we have

$x < s \Leftrightarrow x \in s\overset{o}{P}{}^{-1}$, $x \leq t \Leftrightarrow x \in tP^{-1}$, so the hypothesis becomes $s\overset{o}{P}{}^{-1} \subseteq tP^{-1}$ and implies in turn: $sP^{-1} = \mathbf{clos}\, s\overset{o}{P}{}^{-1} \subseteq tP^{-1} \Longrightarrow Ps^{-1}t \subseteq P \Rightarrow s^{-1}t \in P \Rightarrow s \leq t$.

Our result is the following:

**PROPOSITION 3.2.1.** *Let us assume that (besides the conditions imposed above) we have that for any* $x$ *in* P:

(i) *the set* $\{y \in P \mid y \leq x\}$ *is compact, and*

(ii) *there exists a continuous path* $\lambda : [0,1] \to P$ *such that* $\lambda(0) = e$, $\lambda(1) = x$ *and* $0 \leq a \leq b \leq 1$ *implies* $\lambda(a) \leq \lambda(b)$.

*Then* $(h,U)$ *is a regular compactification of* P, *and hence* $W(P)$ *contains* $K(L^2(\mu \mid P))$.

PROOF. In order to prove $(h,U)$ regular it suffices to show that whenever $t$ and $(t_n)_{n=1}^{\infty}$ of $P$ are such that $h(t_n) \underset{n \to \infty}{\longrightarrow} h(t)$, there exists a compact subset of $P$ which contains every $t_n$. This is a general fact from topology, and we leave its proof to the reader. (N.B.: The proof makes use of the metrizability of $U$.)

So, for the rest of the proof, we fix $t$ and $(t_n)_{n=1}^{\infty}$ of $P$ such that $X_{t_nP^{-1}} \underset{n \to \infty}{\longrightarrow} X_{tP^{-1}}$. We want to exhibit a compact subset of $P$ which contains every $t_n$. In order to do this, we also fix a $t'$ in $t\overset{o}{P}$ ($\subseteq \overset{o}{P}$), i.e. such that $t < t'$. We shall prove that $t_n < t'$ for sufficiently large $n$; this fact, together with the compactness of $\{s \in P \mid s \leq t'\}$, ensured by the hypothesis (i), will clearly finish the proof.

We first prove a related statement:

**LEMMA 1.** *Let* $s \in G$ *be such that* $s \npreceq t$. *Then* $s \npreceq t_n$ *for sufficiently large* $n$.

PROOF OF LEMMA 1. We define $D = \{x \in G \mid x < s,\ x \npreceq t\}$; $D$ is obviously open and it is non-void because otherwise the implication "$x < s \Leftrightarrow x < t$" would hold, leading to $s \leq t$ by Observation $4^{\circ}$ above. We have $D \cap tP^{-1} = D \cap \{x \in G \mid x \leq t\} = \emptyset$. On the other hand, for any $n$ satisfying $s \leq t_n$ we clearly have $D \subseteq \{x \in G \mid x \leq t_n\} = t_nP^{-1}$. So if we consider an open relatively compact non-void subset $D_o$ of $D$ we see that $\int_{tP^{-1}} X_{D_o}\, d\mu = 0$, while $\int_{t_nP^{-1}} X_{D_o}\, d\mu = \mu(D_o) > 0$ for any $n$ satisfying $s \leq t_n$. But we know that $\int_{t_nP^{-1}} X_{D_o}\, d\mu \underset{n \to \infty}{\longrightarrow} \int_{tP^{-1}} X_{D_o}\, d\mu$, and this makes the statement of the lemma clear.

Now we consider the compact subset $K = \{x \in P \mid x \leq t',\ x \npreceq t'\}$ of $P$ and prove the following lemma concerning it:

**LEMMA 2.** *For any* $y$ *in* $G$ *such that* $y \nleq t'$ *there exists an* $x$ *in* $K$ *such that* $x \leq y$.

PROOF OF LEMMA 2. If $y \leq t'$ we may take $x = y$, so we shall assume that $y \nleq t'$. We define $D = \{z \in P \mid z < t'\} = P \cap t'\hat{P}^{-1}$. $D$ is open in $P$ and we have $\mathbf{clos}_P D = \mathbf{clos}_G D \subseteq P \cap t'P^{-1} = \{z \in P \mid z \leq t'\}$, hence the boundary of $D$ relative to $P$ is contained in $K$. Let $\lambda : [0,1] \to P$ be a continuous path connecting $e$ and $y$, which is increasing relatively to $\leq$ (hypothesis (ii) of the theorem). We have $\lambda(0) \in D$, $\lambda(1) \notin \mathbf{clos}_P D$, so by the connectedness of $[0,1]$ there must exist an $a$ in $(0,1)$ such that $x = \lambda(a) \in \partial_P D \subseteq K$. But $\lambda(a) \leq \lambda(1)$ means exactly that $x \leq y$, and this ends the proof of the lemma.

Observe now that $\underset{s \in G, s \nleq t}{\cup} \{x \in G \mid x > s\}$ is an open cover of $K$; indeed, if $x \in G$ does not belong to this union, then it is true that "$s \nleq t \Longrightarrow s \nleq x$" which is equivalent to "$s < x \Rightarrow s \leq t$" and hence leads to $x \leq t$, by Observation $4^\circ$ above; but $x \leq t < t'$ implies $x < t'$, so $x$ cannot be in $K$.

Let us take a finite subcover of this open cover of $K$; that is, we pick $s_1, \ldots, s_m$ of $G$ such that $s_j \nleq t$ for any $1 \leq j \leq m$ and such that $K \subseteq \underset{j=1}{\overset{m}{\cup}} \{x \in G \mid x > s_j\}$. Using Lemma 2 it is clear that $\{x \in G \mid x \nleq t'\} \subseteq \underset{j=1}{\overset{m}{\cup}} \{x \in G \mid x > s_j\}$, hence that $s_1 \nleq x, \ldots, s_m \nleq x$ imply together $x < t'$. Using Lemma 1 we find $n_o$ with the property that $s_j \nleq t_n$ for any $n \geq n_o$ and $1 \leq j \leq m$. Then $n \geq n_o$ implies $t_n < t'$ and the proof is over.

**PROPOSITION 3.2.2.** *The hypothesis of Proposition 3.2.1 are satisfied:*

a) *by any closed convex pointed cone with non-void interior in an Euclidean space.*

b) *by the positive semigroup* $P_n$ *of any* $H_n$.

PROOF. a) Let $P \subseteq \mathbf{R}^n$ be a closed convex pointed cone with non-void interior and let us fix $x \in P$. Defining $\lambda(a) = ax$ on $[0,1]$ we see that hypothesis (ii) is satisfied $(0 \leq a \leq b \leq 1 \Longrightarrow -ax + bx = (b - a)x \in P)$. In order to verify (i) we consider the dual of $P$, $\hat{P} = \{\xi \in \mathbf{R}^n \mid \langle y, \xi \rangle \geq 0, \forall y \in P\}$, which is also a closed convex pointed cone with non-void interior. We have $\mathbf{sp}\,\hat{P} = \mathbf{R}^n$, hence we can choose $n$ linearly independent vectors $\xi_1, \xi_2, \ldots, \xi_n$ of $\hat{P}$. It is known that $\hat{\hat{P}} = P$, that is $y \in P \Longleftrightarrow \langle y, \xi \rangle \geq 0, \forall \xi \in \hat{P}$; as a consequence we see that $y \leq x \Longleftrightarrow -y + x \in P \Longleftrightarrow \langle -y + x, \xi \rangle \geq 0, \forall \xi \in \hat{P} \Rightarrow \langle y, \xi_j \rangle \leq \langle x, \xi_j \rangle, \forall 1 \leq j \leq n$. Denoting $\alpha_j = \langle x, \xi_j \rangle$ $(1 \leq j \leq n)$, we obtain that $\{y \in P \mid y \leq x\} \subseteq \{y \in \mathbf{R}^n \mid |\langle y, \xi_j \rangle| \leq \alpha_j, \forall 1 \leq j \leq n\}$; the last set is easily seen to be bounded.

b) If $x = (x_{i,j})_{i,j}$ and $y = (y_{i,j})_{i,j}$ of $P_n$ are such that $y \leq x$, then $x_{i,j} \leq y_{i,j}$ for any $1 \leq i < j \leq n$; indeed, denoting $z = (z_{i,j})_{i,j} = y^{-1}x \in P_n$, we have $x_{i,j} = \sum_{k=i}^{j} y_{i,k} z_{k,j} \geq y_{i,j} z_{j,j} = y_{i,j}$. This makes clear the hypothesis (i) of Proposition 3.2.1.

We pass to (ii). We shall prove that for $s = (s_{i,j})_{i,j}$ and $t = (t_{i,j})_{i,j}$ of $P_n$ placed in any of the following two situations it is true that $s \leq t$ and there exists an increasing continuous path connecting s with t:

$\alpha$) s is obtained from t by replacing a component of the first line with 0;

$\beta$) there exist $2 \leq p < q \leq n$ such that $t_{i,p} = 0$ for any $1 \leq i \leq p - 1$ and such that s is obtained from t by replacing $t_{p,q}$ with 0.

Once we have done this, it is easy to see how an arbitrary $x \in P_n$ can be connected with $e \in P_n$ by a continuous decreasing path made of $(n-1)n/2$ pieces.

Proof for the situation $\alpha$: we can write $t = s + ce_{1,q}$ for some $d \in [0,\infty)$ and $2 \leq q \leq n$, where $e_{1,q} \in \mathbf{Mat}_n(R)$ has the 1,q-entry equal to 1 and the others equal to 0. $\lambda(a) = s + ace_{1,q}$ is then a continuous increasing path connecting a with t, because $0 \leq a \leq b \leq 1 \Rightarrow \lambda(a)^{-1}\lambda(b) = e + (b-a)ce_{1,q} \in P_n$, (here e is the unit of $H_n$).

The proof for the situation $\beta$ is similar to the one for $\alpha$.

## 4. THE $C^*$-ALGEBRA $W(P_3)$

In this section we deal with the Wiener-Hopf operators on the positive semi-group $P_3$ of $H_3$. We shall explicitly describe in this particular case the unit space of the groupoid construction of Section 2, and we shall use it to obtain a composition series for $W(P_3)$.

We make the identification of $H_3$ with $\mathbf{R}^3$ by writing (a,b,c) instead of $\begin{bmatrix} 1 & a & c \\ 0 & 1 & b \\ 0 & 0 & 1 \end{bmatrix}$; so we are in fact working with $\mathbf{R}^3$ endowed with the multiplication $(a,b,c)(a',b',c') = (a + a', b + b', c + c' + ab')$. As we remarked at the end of Section 2, the Haar measure coincides with the Lebesgue measure and $P_3$ becomes $[0,\infty)^3$.

For any t in $\mathbf{R}^3$, $tP_3^{-1}$ is easily seen to be $\{x \in \mathbf{R}^3 \mid x_1 \leq t_1, x_2 \leq t_2, x_3 + (t_2 - x_2)x_1 \leq t_3\}$, where $x_1, x_2, x_3$ and $t_1, t_2, t_3$ are the components of x and t respectively; it is convenient to denote $tP_3^{-1}$ by $S_{t_1,t_2,t_3}$. So the set of units of the groupoid construction of Section 2 is $U = w^*\text{-}\mathbf{clos}\{x_{S_{t_1,t_2,t_3}} \mid t_1, t_2, t_3 \in [0,\infty)\}$. We also make the following notations ($t_1, t_2, t_3 \in \mathbf{R}$ are arbitrary):

$$S_{t_1,t_2,\cdot} = \{x \in \mathbf{R}^3 \mid x_1 \leq t_1, x_2 \leq t_2\};$$

$$S_{\cdot,t_2,t_3} = \{x \in \mathbf{R}^3 \mid x_2 \le t_2,\ x_3 + (t_2 - x_2)x_1 \le t_3\};$$
$$S_{t_1,\cdot,\cdot} = \{x \in \mathbf{R}^3 \mid x_1 \le t_1\};$$
$$S_{\cdot,t_2,\cdot} = \{x \in \mathbf{R}^3 \mid x_2 \le t_2\}.$$

**PROPOSITION 4.1.** U *is made of six orbits, which are:*

$$U_{1,2,3} = \{X_{S_{t_1,t_2,t_3}} \mid t_1, t_2, t_3 \in [0,\infty)\};$$
$$U_{1,2} = \{X_{S_{t_1,t_2,\cdot}} \mid t_1, t_2 \in [0,\infty)\};$$
$$U_{2,3} = \{X_{S_{\cdot,t_2,t_3}} \mid t_2, t_3 \in [0,\infty)\};$$
$$U_1 = \{X_{S_{t_1,\cdot,\cdot}} \mid t_1 \in [0,\infty)\};$$
$$U_2 = \{X_{S_{\cdot,t_2,\cdot}} \mid t_2 \in [0,\infty)\};$$
$$U_0 = \{X_{\mathbf{R}^3}\} = \{1\}.$$

*Moreover, if we place these six orbits on four levels as in the Table 1, then the closure of each one consists of itself and the orbits situated (strictly) below it.*

| Level 0 | $U_{1,2,3}$ |
|---|---|
| Level 1 | $U_{1,2};\ U_{2,3}$ |
| Level 2 | $U_1;\ U_2$ |
| Level 3 | $U_0$ |

Table 1.

PROOF. It is easier to compute $U' = w^*\text{-}\mathbf{clos}\{X_{P_3 t^{-1}} \mid t \in P_3\}$; $U$ and $U'$ are related by the equality $U = \{X_{A^{-1}} \mid X_A \in U^1\}$. For any $t \in \mathbf{R}^3$, $P_3 t^{-1} = \{x \in \mathbf{R}^3 \mid x_1 \ge -t_1,\ x_2 \ge -t_2,\ x_3 + x_1 t_2 \ge -t_3\}$ is a closed convex set containing $P_3$.

Let $(t^{(k)})_{k=1}^{\infty}$ be a sequence of $P_3$ such that $X_{P_3(t^{(k)})^{-1}} \xrightarrow[k\to\infty]{w^*} X_A$ for a solid set A.

(In fact A must be convex and must contain $P_3$, as we saw in the proof of Proposition 2.1). We write, for any k, $t^{(k)} = (t_1^{(k)}, t_2^{(k)}, t_3^{(k)})$. Passing to a subsequence, we shall suppose that $(t_j^{(k)})_{k=1}^{\infty}$ converges to $c_j \in [0,\infty]$ for any $j \in \{1, 2, 3\}$. If $c_2 = \infty$, we shall assume, in addition, that $t_2^{(k)} \ne 0$ for any k and that there exists $c = \lim_{k\to\infty} t_3^{(k)}/t_2^{(k)} \in [0,\infty]$.

There are eight possible cases, given by the following tree:

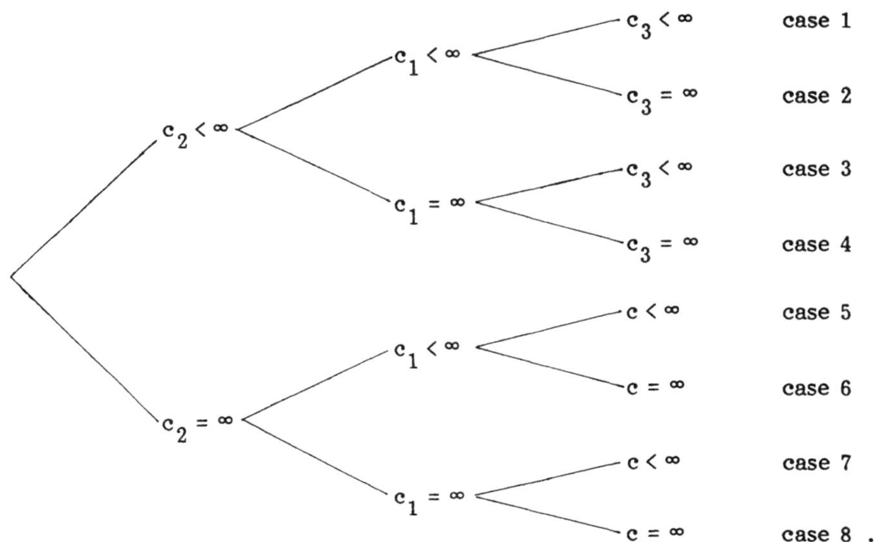


In each case, the limit set A can be written explicitly. Let us take for instance the case 6. We claim that we have $A = \{x \in \mathbf{R}^3 \,|\, x_1 \geq -c_1\}$. Indeed, let us denote the last set by A'. If $x \in \overset{\circ}{A'}$, that is if $x_1 > -c_1$, then $x \in P_3(t^{(k)})^{-1}$ for all sufficiently large k because: $-t_1^{(k)} \xrightarrow[k]{} -c_1 < x_1$, $-t_2^{(k)} \xrightarrow[k]{} -\infty < x_2$ and $x_3 + x_1 t_2^{(k)} \geq -t_3^{(k)} \Longleftrightarrow x_3/t_2^{(k)} + x_1 \geq -t_3^{(k)}/t_2^{(k)}$ is valid for sufficiently large k, since the left part of the inequality tends to $x_1$, while its right part tends to $-\infty$. If $x \notin A'$, then $x \notin P_3(t^{(k)})^{-1}$ for all sufficiently large k, because $-t_1^{(k)} \xrightarrow[k]{} -c_1 > x_1$. Taking into account that $\partial A'$ has null Lebesgue measure, we can apply the dominated convergence theorem to see that $\chi_{P_3(t^{(k)})^{-1}} \xrightarrow[k \to \infty]{w^*} \chi_{A'}$.

The rest of the proof is a mechanical computation. The reader may convince himself that if one will write the form of A in the other seven cases and operate the inversion $A \to A^{-1}$, then he will obtain the results stated in the proposition.

At this moment we have at our disposal a general machinery used by P. Muhly and J. Renault in [1] (Theorems 4.7 and 6.6) for some special Euclidean cases, which provides a composition series for a groupoid $C^*$-algebra. This machinery starts with a locally compact groupoid with Haar system, $G$, and with a partition of its set U of units into invariant subsets. The partition is written with double-index, $U = \bigcup_{\ell=0}^{n} \left( \bigcup_{j=1}^{m_\ell} U_{\ell,j} \right)$, because its members are placed on $n + 1$ levels (the index $\ell$ comes from "level"), and the following condition concerning closures is satisfied: for any $\ell$ and j we have

$$\mathbf{clos}\, U_{\ell,j} \subseteq U_{\ell,j} \cup \left( \bigcup_{\ell'=\ell+1}^{n} \left( \bigcup_{j'=1}^{m_{\ell'}} U_{\ell',j'} \right) \right).$$ This condition is obviously weaker than the one observed for $P_3$ in Proposition 4.1.

Under these hypothesis, we see that each $U_{\ell,j}$ is locally closed, being the difference of the two closed sets $U_{\ell,j} \cup \left( \bigcup_{\ell'=\ell+1}^{n} \left( \bigcup_{j'=1}^{m_{\ell'}} U_{\ell',j'} \right) \right)$ and $\bigcup_{\ell'=\ell+1}^{n} \left( \bigcup_{j'=1}^{m_{\ell'}} U_{\ell',j'} \right)$; hence $G|U_{\ell,j}$ is a locally compact groupoid, endowed with an inherited Haar system (because $U_{\ell,j}$ is invariant). The result used by P. Muhly and J. Renault is the following:

**PROPOSITION 4.2.** *One can find the closed two-sided ideals* $I_0 \subseteq I_1 \subseteq \cdots \subseteq I_n = C^*(G)$ *of* $C^*(G)$ *such that* $I_0$ *is isomorphic to* $\displaystyle \mathop{+}_{j=1}^{m_0} C^*(G|U_{0,j})$, *and for any* $1 \leq \ell \leq n$, $I_\ell/I_{\ell-1}$ *is isomorphic to* $\displaystyle \mathop{+}_{j=1}^{m_\ell} C^*(G|U_{\ell,j})$.

The proof of Proposition 4.2 is based on the fact that open invariant sets of units give rise to ideals of $C^*(G)$ and other related results (see for instance the proof of Theorem 4.7 in [1]).

As a consequence of Propositions 4.1 and 4.2, we see that $W(P_3)$ has a sequence of ideals $I_0 \subseteq I_1 \subseteq I_2 \subseteq I_3 = W(P_3)$, such that:

$$I_0 \simeq C^*(G|U_{1,2,3});$$
$$I_1/I_0 \simeq C^*(G|U_{1,2}) \oplus C^*(G|U_{2,3});$$
$$I_2/I_1 \simeq C^*(G|U_1) \oplus C^*(G|U_2);$$
$$I_3/I_2 \simeq C^*(G|U_0).$$

What remains to be done is the description of the $C^*$-algebras of the six reduced groupoids. Reviewing the discussion of 3.1 we see that $C^*(G|U_{1,2,3}) \simeq K$, and that in fact $I_0 = K(L^2(\mu_3|P_3))$ with $\mu_3$ the Lebesgue measure. At the other extreme, we have $C^*(G|U_0) \simeq C^*(H_3)$, because $U_0$ has a single element, with isotropy group $H_3$, hence $G|U_0$ is in fact $H_3$. In what concernes the other four groupoids, each of them can be seen to be isomorphic (as a locally compact groupoid with Haar system) to the product of a group and a trivial equivalence relation on a certain set. For example, if $E$ denotes the trivial equivalence relation on $[0,\infty)^2$ considered with the Lebesgue measure $\mu_2$, then

$$(a, (t_2,t_3), (s_2,s_3)) \longrightarrow ((a, t_2 - s_2, t_3 - s_3 - as_2), s_{.,s_2,s_3})$$

establishes an isomorphism between $R \times E$ and $G|U_{2,3}$, which implies $C^*(G|U_{2,3}) \simeq C_0(R) \otimes K(L^2(\mu_2|[0,\infty)^2))$. In a similar way we find that $C^*(G|U_{1,2})$ is isomorphic to

$C_0(R)\otimes K$, too, and that $C^*(G\,|\,U_1)$ and $C^*(G\,|\,U_2)$ are isomorphic to $C_0(R^2)\otimes K$. Let us make the remark that if we use Theorem 3.1 of [2], then we are exempt from establishing the groupoid isomorphisms, and we only need to compute the isotropy groups of the groupoids involved.

Finally, we have come to the following result:

**PROPOSITION 4.3.** $W(P_3)$ *is of type* I, *and has a composition series of length* 3, *such that:*

a) *the first ideal is* $K$;

b) *the last quotient is isomorphic to* $C^*(H_3)$;

c) *the intermediate quotients are direct sums of terms of the form* $C^*(M)\otimes K$, *with* M *a subgroup of* $H_3$ ($M = \mathbf{R}$ *or* $M = \mathbf{R}^2$).

## 5. THE $C^*$-ALGEBRA $W(P_4)$

Computations similar to those of the preceding section can be made in the case of $P_4 \subseteq H_4$. The main difference is that, unlike the result of Proposition 4.1, U has now an infinite set of orbits. However, we can gather the orbits in a natural way into a finite number of invariant sets, such that the hypothesis of Proposition 4.2 are fulfilled. After doing this, we still have the problem of describing the $C^*$-algebras of those reduced groupoids which are not transitive. Fortunately, all of them are seen to be isomorphic to the product of a group and a principal groupoid of the type described in the next proposition.

**PROPOSITION 5.1.** *Let* B *and* Y *be second countable locally compact spaces. We consider the equivalence relation defined by* $(b,x,y) \simeq (b',x',y') \Longleftrightarrow b = b'$ *on* $B \times Y \times Y$, *which gives a locally compact groupoid* H. *If* $\mu$ *is a positive Radon measure on* Y *having* **supp** $\mu = Y$, *and* $\gamma : B \longrightarrow (0,\infty)$ *is a continuous function, then the family of Radon measures* $\Lambda = (\lambda^{(b,y)})_{(b,y) \in B \times Y}$ *defined by:*

$$\int_{H^{(b,y)}} f(b,y,x)d\lambda^{(b,y)}(b,y,x) = \gamma(b)\int_Y f(b,y,x)d\mu(x)$$

*for any* f *in* $C_c(H^{(b,y)})$, *is a left Haar system on* H, *and the reduced* $C^*$-*algebra associated to* H *and* $\Lambda$ *is isomorphic to* $C_0(B)\otimes K(L^2(\mu))$.

**REMARK.** We do not need to study the amenability of $(H,\Lambda)$, because, using the proposition, we shall obtain the reduced $C^*$-algebras of some groupoids which are known to be amenable (by Propositions 3.7 and 3.9 of Chapter II of [4]).

PROOF OF PROPOSITION 5.1. We recall that the set of units of $H$ is $B \times Y$; its set of arrows is $B \times Y \times Y$, the domain and range of $(b,y,x)$ being $(b,x)$ and $(b,y)$ respectively. The multiplication on $H$ is given by $(b,z,y)(b,y,x) = (b,z,x)$.

We consider and fix a positive Radon measure $\beta$ on B, such that **supp** $\beta = B$, and an element $x_o \in X$. We have **supp** $\beta \times \delta_{x_o} = B \times \{x_o\}$, hence the invariant support of $\beta \times \delta_{x_o}$ is $B \times Y$. This implies, by Proposition 2.17 of [1], that the induced representation **Ind** $\beta \times \delta_{x_o}$ is isometric on $C^*_{red}(H, \Lambda)$. The space of this representation is $L^2(\nu^{-1})$, where the Radon measure $\nu^{-1}$ on $B \times Y \times Y$ is given by:

$$\int_{B \times Y \times Y} f d\nu^{-1} = \int_{B \times Y} \left( \int_{H_{(b,x)}} f(b,y,x) d\lambda_{(b,x)}(b,y,x) \right) d(\beta \times \delta_{x_o})(b,x) =$$

$$= \int_{B \times Y} \gamma(b) f(b,y,x_o) d(\beta \times \mu)(b,y), \quad \forall\, f \in C_c(B \times Y \times Y).$$

It is easy to see that the mapping $f \to \gamma^{\frac{1}{2}}(\cdot) f(\cdot, \cdot, x_o)$ from $C_c(B \times Y \times Y)$ onto $C_c(B \times Y)$ extends to a unitary $T : L^2(\nu^{-1}) \to L^2(\beta \times \mu)$. We conjugate **Ind** $\beta \times \delta_{x_o}$ with T and obtain an isometric representation $\Pi : C^*_{red}(H, \Lambda) \to B(L^2(\beta \times \mu))$, which is found to act by the formula:

$$[(\Pi f)\xi](b,y) = \gamma(b) \int_Y f(b,y,z)\xi(b,z) d\mu(z),$$

for any $f \in C_c(B \times Y \times Y)$, $\xi \in C_c(B \times Y)$, $b \in B$, $y \in Y$. If we make the identification $L^2(\beta \times \mu) \simeq L^2(\beta) \overline{\otimes} L^2(\mu)$, the above formula shows that when $f = g \otimes h_1 \otimes \overline{h_2}$ with $g \in C_c(B)$, $h_1, h_2 \in C_c(Y)$, $\Pi f$ is $M_{\gamma g} \otimes (\langle \cdot\, | h_2 \rangle h_1)$. ($M_{\gamma g}$ is the multiplication operator with $\gamma g$ on $L^2(\beta)$ and $\langle \cdot\, | h_2 \rangle h_1$ is the corresponding rank one operator on $L^2(\mu)$.) The elements of the form $g \otimes h_1 \otimes \overline{h_2}$ generate $C^*(H, \Lambda)$ as a $C^*$-algebra, while the $C^*$-algebra of $B(L^2(\beta) \overline{\otimes} L^2(\mu))$ generated by the operators $M_{\gamma g} \otimes (\langle \cdot\, | h_2 \rangle h_1)$ is $\{M_g \,| g \in C_o(B)\} \otimes K(L^2(\mu))$, isomorphic to $C_o(B) \otimes K(L^2(\mu))$.

For the sake of completeness we shall present a table comprising the invariant sets of units which appear and the manner in which they are arranged on levels. As in the case of $H_3$, it is easier to compute $U' = w^*\text{-}\mathbf{clos}\{X_{P_4 t^{-1}} \,| t \in P_4\}$ instead of $U = w^*\text{-}\mathbf{clos}\{X_{t P_4^{-1}} \,| t \in P_4\}$. But this time, because the inversion operation is more arduous, we prefer to replace the groupoid $G$ of Section 2 with the groupoid $G'$ whose units are U' and whose arrows are given by right translation with elements of G; that is, if $x_B \in$ U' and s $\in$ G are such that $x_{Bs^{-1}}$ is still in U' (which is equivalent to s $\in$ B), then we have an arrow (s,B) from $x_B$ to $x_{Bs^{-1}}$. Multiplication, topology and Haar system are defined on $G'$ by symmetry with the case of $G$. It is obvious that $G$ and $G'$ are isomorphic, so that $C^*(G') \simeq W(P_4)$.

The identification of $H_4$ with $\mathbf{R}^6$ is made by writing $(x_1, x_2, \ldots, x_6)$ instead of

$$\begin{bmatrix} 1 & x_1 & x_4 & x_6 \\ 0 & 1 & x_2 & x_5 \\ 0 & 0 & 1 & x_3 \\ 0 & 0 & 0 & 1 \end{bmatrix}$$ . If we write the formula of the multiplication which is obtained on

$\mathbf{R}^6$ and take into account that $x \in P_4 t^{-1} \Longleftrightarrow xt \in P_4$, we find that, for any t in $\mathbf{R}^6$, $P_4 t^{-1}$ can be expressed as $R_{t_1} \cap S_{t_2, t_4} \cap T_{t_3, t_5, t_6}$, with $t_1, t_2, \ldots, t_6$ the components of t, and where we use the following notations (a, b, c $\in \mathbf{R}$ are arbitrary):

$$R_a = \{x \in \mathbf{R}^6 \mid x_1 \geq -a\};$$

$$S_a = \{x \in \mathbf{R}^6 \mid x_2 \geq -a\};$$

$$T_a = \{x \in \mathbf{R}^6 \mid x_3 \geq -a\};$$

$$S_{a,b} = \{x \in \mathbf{R}^6 \mid x_2 \geq -a, x_4 + ax_1 \geq -b\};$$

$$\Sigma_{a,b} = \{x \in \mathbf{R}^6 \mid x_4 + ax_1 \geq -b\};$$

$$T_{a,b} = \{x \in \mathbf{R}^6 \mid x_3 \geq -a, x_5 + ax_2 \geq -b\};$$

$$T_{a,b,c} = \{x \in \mathbf{R}^6 \mid x_3 \geq -a, x_5 + ax_2 \geq -b, x_6 + bx_1 + ax_4 \geq -c\}.$$

With these notations, the family of sets whose characteristic functions appear in U' is listed in Table 2. U' is divided into 28 invariant sets, placed on seven levels; unless otherwise stated, the coefficients which appear are arbitrary in $[0, \infty)$.

### Table 2

| Level 0 | 1. $R_{c_1} \cap S_{c_2, c_4} \cap T_{c_3, c_5, c_6}$ |
|---|---|
| Level 1 | 2. $R_{c_1} \cap S_{c_2, c_4} \cap T_{c_3, c_5}$ |
| | 3. $R_{c_1} \cap S_{c_2} \cap T_{c_3, c_5, c_6}$ |
| | 4. $S_{c_2, c_4} \cap T_{c_3, c_5, c_6}$ |
| | 5. $R_{c_1} \cap S_{c_2, c_4} \cap \Sigma_{d_2, d_4}$, with $d_2 > c_2$ and $d_4 < c_4 + c_1(d_2 - c_2)$ |

| | |
|---|---|
| Level 2 | 6. $R_{c_1} \cap S_{c_2} \cap T_{c_3,c_5}$ |
| | 7. $R_{c_1} \cap S_{c_2,c_4} \cap T_{c_3}$ |
| | 8. $R_{c_1} \cap T_{c_3,c_5,c_6}$ |
| | 9. $S_{c_2,c_4} \cap T_{c_3,c_5}$ |
| | 10. $S_{c_2} \cap T_{c_3,c_5,c_6}$ |
| | 11. $R_{c_1} \cap S_{c_2} \cap \Sigma_{d_2,d_4}$, with $d_2 > c_2$ |
| | 12. $S_{c_2,c_4} \cap \Sigma_{d_2,d_4}$, with $d_2 > c_2$ |
| Level 3 | 13. $R_{c_1} \cap S_{c_2} \cap T_{c_3}$ |
| | 14. $R_{c_1} \cap S_{c_2,c_4}$ |
| | 15. $R_{c_1} \cap T_{c_3,c_5}$ |
| | 16. $S_{c_2} \cap T_{c_3,c_5}$ |
| | 17. $S_{c_2,c_4} \cap T_{c_3}$ |
| | 18. $T_{c_3,c_5,c_6}$ |
| | 19. $S_{c_2} \cap \Sigma d_2,d_4$, with $d_2 > c_2$ |
| Level 4 | 20. $R_{c_1} \cap S_{c_2}$ |
| | 21. $R_{c_1} \cap T_{c_3}$ |
| | 22. $S_{c_2} \cap T_{c_3}$ |
| | 23. $S_{c_2,c_4}$ |
| | 24. $T_{c_3,c_5}$ |
| Level 5 | 25. $R_{c_1}$ |
| | 26. $S_{c_2}$ |
| | 27. $T_{c_3}$ |
| Level 6 | 28. $\mathbf{R}^6$ |

The reduced groupoids corresponding to the lines 5, 11, 12 and 19 of Table 2 are not transitive; they are handled with the aid of Proposition 5.1. For example, in the case of line 5, we put $B=(0,\infty)^2$, $Y=\{x \in [0,\infty)^3 \mid x_1 + x_3 \geq 1\}$, $\mu=$Lebesgue measure on $Y$, $\gamma(b_1,b_2) = b_2^2/b_1$ on $B$; we construct $H$ and $\Lambda$ as in Proposition 5.1, and we make the product $R^3 \times H$. A topological isomorphism, which preserves the canonical Haar systems, from the reduced groupoid corresponding to the line 5 onto $R^3 \times H$ can be defined by the formula $x = (t, R_{c_1} \cap S_{c_2,c_4} \cap \Sigma_{d_2,d_4}) \to ((a_1(x),a_2(x),a_3(x)) \in R^3$, $b(x) \in B$, $y_1(x) \in Y$, $y_2(x) \in Y)$, where:

$$a_1(x) = t_3;$$
$$a_2(x) = t_5 - (c_2 + t_2)t_3;$$
$$a_3(x) = t_6 - t_5(t_1 + c_1) - t_3(t_4 + c_4) + t_3(c_1 c_2 + c_1 t_2 + t_1 t_2);$$
$$b(x) = (d_2 - c_2, c_4 - d_4 + c_1(d_2 - c_2));$$
$$y_1(x) = ((d_2 - c_2)(t_1 + c_1)/(c_4 - d_4 + c_1(d_2 - c_2)), t_2 + c_2, (t_4 + c_4 + t_1 c_2)/(c_4 -$$
$$- d_4 + c_1(d_2 - c_2));$$
$$y_2(x) = ((d_2 - c_2)c_1/(c_4 - d_4 + c_1(d_2 - c_2)), c_2, c_4/(c_4 - d_4 + c_1(d_2 - c_2))).$$

The reduced groupoids corresponding to the rest of the lines are transitive, hence, by Theorem 3.1 of [2], their $C^*$-algebras are determined by their isotropy groups.

We finally obtain:

**PROPOSITION 5.2.** $W(P_4)$ *is of type* I, *and has a composition series* $K = I_0 \subset$ $\subset I_1 \subset \ldots \subset I_6 = W(P_4)$ *such that:*

$$I_1/I_0 \simeq (\bigoplus_{j=1}^{3} C_0(\mathbf{R})\otimes K)\oplus(C_0(\mathbf{R}^3)\otimes C_0((0,\infty)^2)\otimes K);$$

$$I_2/I_1 \simeq (\bigoplus_{j=1}^{5} C_0(\mathbf{R}^2)\otimes K)\oplus(\bigoplus_{j=1}^{2} (C_0(\mathbf{R}^3)\otimes C_0((0,\infty))\otimes K);$$

$$I_3/I_2 \simeq (\bigoplus_{j=1}^{4} C_0(\mathbf{R}^3)\otimes K)\oplus(\bigoplus_{j=1}^{2} C^*(H_3)\otimes K)\oplus(C^*(\mathbf{R}\otimes H_3)\otimes C_0((0,\infty))\otimes K);$$

$$I_4/I_3 \simeq (\bigoplus_{j=1}^{4} C^*(\mathbf{R} \times H_3)\otimes K)\oplus(C_0(\mathbf{R}^4)\otimes K);$$

$$I_5/I_4 \simeq \bigoplus_{j=1}^{3} C^*(M_j)\otimes K,$$

*where for any* $1 \leq j \leq 3$, $M_j$ *is the subgroup of* $H_4$ *obtained by forcing the j-th component to be zero;*

$$I_6/I_5 \simeq C^*(H_4).$$

## REFERENCES


1.  **Muhly, P. ; Renault J.** : $C^*$-algebras of multivariable Wiener-Hopf operators, *Trans. Amer. Math. Soc.* **274**(1982), 1-44.

2.  **Muhly, P. ; Renault, J. ; Williams, D.** : Equivalence and isomorphism for groupoid $C^*$-algebras, *J. Operator Theory* **17**(1987), 3-22.

3.  **Nica, A.** : Some remarks on the groupoid approach to the Wiener-Hopf operators, *J. Operator Theory* **18**(1987), 163-198.

4.  **Renault, J.** : *A groupoid approach to $C^*$-algebras*, Lecture Notes in Mathematics, Vol. **793**, Springer Verlag, New York, 1980.



**Alexandru Nica**

Department of Mathematics, INCREST
Bdul Păcii 220, 79622 Bucharest,
Romania.

Operator Theory:
Advances and Applications, Vol. 43
© 1990 Birkhäuser Verlag Basel



# DERIVATIONS OF CERTAIN NEST-SUBALGEBRAS OF
# VON NEUMANN ALGERBAS

Florin Pop

## INTRODUCTION

In their celebrated papers [7] and [5] S. Sakai and R.V. Kadison proved that every derivation of a von Neumann algebra is inner.

In [1] E. Christensen proved that nest algebras have the same property.

Intersections between von Neumann algebras and nest algebras (called nest-subalgebras of von Neumann algebras) were introduced by F. Gilfeather and D.R. Larson [3].

These algebras turned out to be very interesting and quite different both from von Neumann algebras and from nest algebras. We refer the reader to [3], [4] and [6] for more information about these problems.

In this paper we investigate derivations of nest-subalgebras of von Neumann algebras.

We prove that every derivation of an atomic nest-subalgebra of a type $\text{II}_1$ factor is inner (Theorem III 2) and that ultraweakly continuous derivations of atomic nest-subalgebras of hyperfinite factors are inner (Theorem III 4).

Positive results are also obtained for nest-subalgebras of type $\text{II}_\infty$ factors associated to increasing sequences of finite projections (Theorem III 5) as well as for "diagonal" derivations (Theorem III 3).

I am grateful to Professor Şerban Strătilă for many fruitful conversations and constant encouragement.

Throughout this paper all Hilbert spaces H are assumed to be separable. Let B(H) denote the algebra of bounded operators on H, $M \subseteq B(H)$ a von Neumann algebra and M' the commutant of M.

Let L be a totally ordered strongly closed family of (selfadjoint) projections in M (i.e. L is a nest of projections). Denote by R the (abelian) von Neumann algebra generated by L. L is said to be atomic if R is a purely atomic abelian algebra. Define

$$\text{Alg } L = \{x \in B(H); (I - p)xp = 0 \quad (\forall) p \in L\}$$

Dually, if $B \subseteq B(H)$,

$$\textbf{Lat } B = \{p = p^2 = p^* \in B(H); (I - p)xp = 0 \quad (\forall) x \in B\}.$$

Define $\textbf{Alg}_M L = M \cap \textbf{Alg } L$ the nest-subalgebra of $M$ with respect to $L$ and $\textbf{Lat}_M B = M \cap$ $\cap \textbf{Lat } B$. If $A = \textbf{Alg}_M L$ then $\textbf{Alg Lat } A = A$ (A is reflexive), $\textbf{Alg}_M \textbf{Lat}_M A = A$ and if $M$ is a factor then $\textbf{Lat}_M A = L$. A is said to be atomic if $L$ is an atomic nest. If $B \subseteq B(H)$ is an arbitrary subalgebra, a derivation of $B$ is a linear mapping $\delta : B \rightarrow B$ satisfying

$$\delta (ab) = a\delta (b) + \delta (a)b \quad (\forall) a,b \text{ in } B.$$

$\delta$ is said to be inner if there exists an operator $a$ in $B$ such that $\delta (x) = [a,x] = ax -$ $- xa \ (\forall) x$ in $B$.

### I. THE RELATIVE PROPERTY P

We assume that $M$ is a factor ($M \cap M' = \textbf{C}\cdot I$). Since $A' = M'$ ([4], 2.5) it follows that $A' \cap M = \textbf{C}\cdot I$. Define $N = R' \cap M$ and assume that $L$ is atomic. Then $N$ can be written as a countable direct sum $N = \oplus \{eMe; e \text{ minimal projection in } R\}$.

**PROPOSITION I 1:** *For every $x$ in $M$ the ultraweakly closed convex hull of the set*

$$\{u^* xu \ ; \ u \in N \text{ unitary operator} \}$$

*has non-void intersection with $N' \cap M = R$.*

**REMARKS.** If two von Neumann algebras $M_1 \subseteq M_2$ satisfy the above property then $M_1$ is said to have the relative property P with respect to $M_2$. The arguments in ([8]) show that:

i) If $(M_\alpha)$ is a totally ordered family of subalgebras of $M$ such that each $M_\alpha$ has the relative property P, then the von Neumann algebra generated by all $(M_\alpha)$ has the relative property P (with respect to $M$).

ii) The same is true if $(M_\alpha)$ is a mutually commuting family of subalgebras.

By the above remarks, Proposition I 1 is a consequence of the following result:

**LEMMA I 1.** *Let $M$ be a factor and $0 \neq p$ be a projection in $M$. Then the subalgebra $M_o = pMp \oplus \textbf{C}(I - p)$ has the relative property P with respect to $M$.*

PROOF. We shall prove a stronger result, namely that the norm-closed convex hull of the set $\{u^* xu, u \in M_o \text{ unitary}\}$ has non-void intersection with $M_o' \cap M$ for every $x$

in M. Let then x in M and $\epsilon > 0$ be given. By ([8], 2.1.16) choose $u_1, \ldots, u_n$ unitary operators in pMp, $\lambda \in \mathbf{C}$ and $\lambda_1, \ldots, \lambda_n \geq 0$, $\sum_{i=1}^{n} \lambda_i = 1$ such that

$$\| \sum_{i=1}^{n} \lambda_i u_i^*(pxp)u_i - \lambda_p \| \leq \epsilon.$$

Consider the unitary operators in $M_o$, $v_i = u_i \oplus (I - p)$ $(i = 1, \ldots n)$ and $w = p \oplus$ $\oplus(p - I)$. Since for every operator y in M $\frac{1}{2}(y + w^*yw) = pyp \oplus (I - p)y(I - p)$ it follows that for x in M, $z = \sum_{i=1}^{n} \lambda_i u_i^*(pxp)u_i$, $y = \sum_{i=1}^{n} \lambda_i v_i^* x v_i$

$$\| (\tfrac{1}{2}y + \tfrac{1}{2}w^*yw) - (\lambda p + (I - p)x(I - p)) \| = \| z - \lambda p \| \leq \epsilon.$$

Since $\lambda p \oplus (I - p)x(I - p) \in M_o'$, the lemma is proved.

**PROPOSITION I 2.** *For every* x *in* M

$$\inf\{ \| x - \lambda I \| ; \lambda \in \mathbf{C} \} \leq 2 \sup \{ \| xa - ax \| ; a \in A \ \| a \| \leq 1 \}.$$

**REMARK.** Not that the assumption that M is a factor is essential, since xa = ax for every a in A implies $x \in A' \cap M = \mathbf{C} \cdot I$.

PROOF. Let $y \in R$ be given by Proposition I 1. If **ad**(x) denotes the derivation implemented by x, then $\| [uxu^*, a] \| = \| [x, u^*au] \|$ for every $u \in N$ unitary and $a \in A$. Since $u^*au$ belongs to A, it follows that

$$\| \mathbf{ad}(y)|A \| \leq \| \mathbf{ad}(x)|A \| .$$

We show that for every operator y in R

$$\inf\{ \| y - \lambda I \| ; \lambda \in \mathbf{C} \} \leq \| \mathbf{ad}(y)|A \| .$$

But $y = \oplus \lambda_n e_n$, $\lambda_n \in \mathbf{C}$ and $e_n$ are atoms of L. For every $m \neq n$, either $e_m M e_n$ or $e_n M e_m$ are included in A. Suppose that $e_m M e_n \subset A$. Since M is a factor, $e_m M e_n \neq \{0\}$ so choose $a \in e_m M e_n$ with $\| a \| = 1$.

Then $\| ya - ay \| = |\lambda_m - \lambda_n|$. It follows that $\sup\{|\lambda_m - \lambda_n|\} \leq \| \mathbf{ad}(y)|A \| $. This means that the spectrum of the normal operator y is included in the ball of center $\lambda_1$ and radius $\| \mathbf{ad}(y)|A \| $. Consequently

$$\inf\{ \| y - \lambda I \| , \lambda \in \mathbf{C} \} \leq \| y - \lambda_1 I \| \leq \| \mathbf{ad}(y)|A \|$$

and

$$\inf\{\| x - \lambda I\| , \lambda \in \mathbf{C}\} \le 2\| \mathbf{ad}(y)| A\| .$$                        **Q.E.D.**

## II. SOME RESULTS ON HOMOMORPHISMS AND DERIVATIONS

**LEMMA II 1.** *Let* $M_1 \subset M_2$ *be two semifinite von Neumann algebras and* $\phi$ : $M_1 \to M_2$ *be a (not necessarily involutive) homomorphism such that* $\| id - \phi\| < 1$. *Suppose that* $M_1' \cap M_2$ *contains a family of finite projections in* $M_2$ *with supremum equal to the identity. Then the ultraweakly closed convex hull of the set*

$$\{\phi(u)u^* ; u \in M_1 \text{ unitary operator}\}$$

*contains a point a satisfying*

$$\phi(x)a = ax \ (\forall) \ x \in M_1 \quad and \quad \| I - a\| \le \| id - \phi\| .$$

PROOF. Let $\tau$ denote a faithful semifinite normal trace on $M_2$. For every unitary operator u in $M_1$ define

$$T_u : M_2 \to M_2 \quad T_u(x) = \phi(u)xu^*$$

$$T_u T_v(x) = \phi(u) \phi(v)xv^* u^* = \phi(uv)x(uv)^* = T_{uv}(x)$$

hence the set $K_o = \{T_u(I), u \in M_1 \text{ unitary}\} = \{\phi(u)u^*, u \in M_1 \text{ unitary}\}$ is invariant under all operators $T_u$.

If we regard $M_2$ as the locally convex space endowed with the strong-star topology, then, if K denotes the ultraweak convex closure of $K_o$, K is weakly compact.

For every projection e' in $M_1' \cap M_2$, e' finite in $M_2$, the mapping $x \to \tau(x^* xe')^{1/2}$ $x \in M_2$ defines a strongly-star seminorm, since $e'xe' \to \tau(e'xe')$ is strongly continuous. Suppose that $x_1$ and $x_2$ are in K, $x = x_1 - x_2 \ne 0$ and

$$\inf\{\tau(ux^* \phi(u)^* \phi(u)xu^* e') ; u \in M_1 \text{ unitary}\} = 0.$$

Let $\varepsilon > 0$ be given and choose $u \in M_1$ unitary operator such that $\tau(ux^* \phi(u)^* \phi(u)xu^* e') =$ $= \tau(x^* \phi(u)^* \phi(u)xe') = \| \phi(u)xe'\|_2^2 \le \varepsilon^2$, so $\| \phi(u)xe'\|_2 \le \varepsilon$. Now if $\| id - \phi\| = \alpha < 1$ then

$$\| xe'\|_2 = \| uxe'\|_2 \le \| (\phi(u) - u)xe'\|_2 + \| \phi(u)xe'\|_2$$

hence $(1 - \alpha)\| xe'\|_2 \le \varepsilon$. Since $\varepsilon$ was arbitrary, $\| xe'\|_2 = 0$, that is $\tau(x^* xe') = 0$ hence $\tau(x^* x) = 0$ by normality and $x = 0$ by faithfulness.

Consequently the semigroup $(T_u)$ is noncontractive so, by the Ryll-Nardzewski fixed point theorem ([9], Appendix 3) there is an operator a in K such that $T_u(a) =$

= a $(\forall)$ u $\in M_1$ unitary, hence $\phi(x)a = ax$ $(\forall) x \in M_1$ by linearity. Finally,

$$\left\| I - \sum_{i=1}^{n} \lambda_i \phi(u_i)u_i^* \right\| = \left\| \sum_{i=1}^{n} \lambda_i u_i u_i^* - \sum_{i=1}^{n} \lambda_i \phi(u_i)u_i^* \right\| =$$

$$= \left\| \sum_{i=1}^{n} \lambda_i (u_i - \phi(u_i)) \right\| \leq \| \mathbf{id} - \phi \|$$

if $\lambda_i \geq 0$ and $\sum_{i=1}^{n} \lambda_i = 1$. It follows that $\| I - a \| \leq \| \mathbf{id} - \phi \|$ which conclude the proof.

**LEMMA II 2.** *Let* $M_1 \subset M_2$ *be von Neumann algebras,* $M_1$ *hyperfinite, and* $\phi : M_1 \longrightarrow M_2$ *a (non necessarily involutive) bounded ultraweakly continuous homomorphism. Then the ultraweakly closed convex hull of the set* $\{\phi(u)u^* ; u \in M_1$ *unitary operator* $\}$ *contains a point* a *such that* $\phi(x)a = ax$ $(\forall) x \in M_1$ *and* $\| I - a \| \leq \| \mathbf{id} - \phi \|$.

PROOF. Since $M_1$ is hyperfinite, there is an amenable subgroup U of the unitary group of $M_1$ such that $U'' = M_1$. Let $\mu$ denote an invariant mean on U. Define $a = \int_U \phi(u)u^* d\mu(u)$. It follows that $\phi(u)au^* = a$ $(\forall)$ u $\in M_1$ unitary and, as in the proof of Lemma II 1, $\| I - a \| \leq \| \mathbf{id} - d \|$.

**PROPOSITION II 1.** *Every derivation of a nest-subalgebra of a von Neumann algebra is bounded.*

We shall omit the proof, since it is almost identical with the one in the case B(H). We refer the reader to E. Christensen's paper ([1], Paragraph 2).

With the notations in the introduction, let $\delta : A \longrightarrow A$ be a derivation and define

$$e^\delta : A \longrightarrow A \qquad e^\delta(x) = x + \delta(x)/1! + \delta(\delta(x))/2! + \dots .$$

Then $e^\delta$ is a homomorphism and $\| e^\delta \| \leq e^{\| \delta \|}$.

**LEMMA II 3.** *Let* M *be a semifinite von Neumann algebra,* $A \subset M$ *be a nest-subalgebra,* $N = R' \cap M$ *and* $\delta : A \longrightarrow A$ *be a strongly-star continuous derivation. Assume that* $\phi = e^\delta$ *satisfies* $\| \mathbf{id} - \phi \| < 1$. *Then there is an operator* a *in* M *satisfying* $\phi(x)a = ax$ $(\forall) x \in N$ *and* $\| I - a \| \leq \| \mathbf{id} - \phi \|$.

PROOF. Choose $(f_n)$ to be an increasing sequence of finite projections in M, $f_n \in N$ and $\bigvee f_n = I$. Define $\delta_n : f_n A f_n \longrightarrow f_n A f_n$ by $\delta_n(f_n a f_n) = f_n \delta(f_n a f_n)f_n$ and $\phi_n = e^{\delta_n}$.

By Lemma II 1 there are operators $f_n y_n f_n$ in $f_n M f_n$ such that

(1) $$\phi_n(f_n x f_n) f_n y_n f_n = f_n y_n f_n x f_n \qquad (\forall)\, x \in N$$

and

(2) $$\| f_n - f_n y_n f_n \| \leq \| \mathbf{id} - \phi \|.$$

Let $a \in M$ be a ultraweakly adherent point of the sequence $(f_n y_n f_n)$. By (2) $\| I - a \| \leq \| \mathbf{id} - \phi \|$ and we may assume that $a = \text{wo} - \lim f_n y_n f_n$, hence the right side in (1) converges in the wo-topology to $ax$. Since $\delta$ is strongly-star continuous, it follows that $\phi_n(f_n x f_n)$ tends strongly-star to $\phi(x)$. Now (1) shows that $\phi(x)a = ax$ $(\forall)\, x \in N$.    **Q.E.D.**

**COROLLARY.** *The above lemma is true for every derivation* $\delta : A \longrightarrow A$ *such that* $\delta(N) \subset N$.

PROOF. By ([2]), $\delta \mid N$ is strongly-star continuous.

**LEMMA II 4.** *Let* $M_o$ *be a factor with unit* $I_o$, $H_2$ *be a two-dimensional Hilbert space with orthonormal basis* $\{e_1, e_2\}$, $M = M_o \otimes B(H_2)$ *and* $p = I_o \otimes e_1$. *If* $\phi$ *is an automorphism of* $A = M \cap \mathbf{Alg}\{0, p, I\}$ *such that* $\phi(x) = x$ $(\forall)\, x \in pMp \oplus (I - p)M(I - p)$ *then* $\phi$ *is implemented by an invertible operator* $a = \lambda p \oplus \mu(I - p)$.

PROOF. Let $\alpha : M \longrightarrow M$ be the linear mapping such that

$$\phi \begin{bmatrix} x & y \\ 0 & z \end{bmatrix} = \begin{bmatrix} x & \alpha(y) \\ 0 & z \end{bmatrix}.$$

It follows that for every $x, y$ and $z$ in $M$, $\alpha(xy) = x\alpha(y)$ and $\alpha(yz) = \alpha(y)z$ hence $\alpha(x) = x\alpha(I) = \alpha(I)x$ for every $x$ in $M$. Consequently there is $\theta \in \mathbf{C}$ such that $\alpha(x) = \theta x$, so $\phi$ is implemented by any invertible operator of the from $a = \lambda p \oplus \mu(I - p)$ with the condition $\lambda^{-1}\mu = \theta$.

**LEMMA II 5.** *Let* $M$ *be a factor,* $p \in M$ *be a projection and* $A = M \cap \mathbf{Alg}\{0, p, I\}$. *If* $\phi$ *is an automorphism of* $A$ *such that* $\phi(x) = x$ $(\forall)\, x$ *in* $pMp \oplus (I - p)M(I - p)$ *then* $\phi$ *is implemented by an invertible operator of the form* $a = \lambda p \oplus \mu(I - p)$.

PROOF. If both $p$ and $(I - p)$ are infinite projections, then $M$ and $A$ are spatially isomorphic with the algebras $M$ and $A$ in the preceding lemma, so the conclusion follows. If one and only one of the projections $p$ and $I - p$ is finite (suppose that $p$ is) then $M$ and $A$ are spatially isomorphic with $M_o \otimes B(K)$ ($K$ is a Hilbert space with orthonormal basis $\{e_1, e_2, \ldots\}$ and $M_o$ is a factor with unit $I_o$) and with $M_o \otimes B(K) \cap \mathbf{Alg}\{0, I_o \otimes e_1, I\}$, respectively. Let $\phi_i : M_o \longrightarrow M_o$ $(i \geq 2)$ be the linear mappings such that

$$\phi \begin{pmatrix} x_{11} & x_{12} & x_{13} & \cdots \\ 0 & x_{22} & x_{23} & \cdots \\ 0 & x_{32} & x_{33} & \cdots \\ \vdots & \vdots & \vdots & \end{pmatrix} = \begin{pmatrix} x_{11} & \phi_2(x_{12}) & \phi_3(x_{13}) & \cdots \\ 0 & x_{22} & x_{23} & \cdots \\ 0 & x_{32} & x_{33} & \cdots \\ \vdots & \vdots & \vdots & \end{pmatrix} .$$

By Lemma II 4, $\phi_i(x_{1i}) = \lambda_i x_{1i}$, $\lambda_i \in \mathbf{C}$, $i \geq 2$. But since $\phi$ is a homomorphism, it follows that actually all the numbers $\lambda_i$ are equal, which concludes the proof. Suppose that $p$ and $(I - p)$ are both finite projections. Suppose that $p \precsim I - p$. If $I - p$ is a multiple of $p$, the proof follows the same lines as above. In other case, let $p_1, \ldots, p_n$ be mutually orthogonal projections in $M$, all equivalent with $p$ and let $q \in M$ be a projection equivalent with $p$ such that

$$q + \sum_{i=1}^{n} p_i = I - p$$

and

$$(I - p) - \sum_{i=1}^{n} p_i \leq q.$$

As before, there are $\lambda, \mu \in \mathbf{C}$ such that

$$\phi(pxp_i) = \lambda pxp_i \quad (\forall) 1 \leq i \leq n \quad (\forall) x \in M$$

and

$$\phi(pxq) = \mu pxq \quad (\forall) x \in M.$$

Since there is a projection $p_i$ such that $qp_i \neq 0$, it follows that $\lambda = \mu$, which concludes the proof.

**COROLLARY.** *Let* $M$ *be a factor and* $L = \{0 = p_o \leq p_1 \ldots \leq p_n = I\}$ *be a finite nest of projections in* $M$. *If* $\phi$ *is an automorphism of* $M \cap \mathbf{Alg}\, L$ *such that* $\phi(x) = x$ $(\forall)\, x \in$ $\in \bigoplus_{i=0}^{n-1} (p_{i+1} - p_i)M(p_{i+1} - p_i)$ *then* $\phi$ *is implemented by an invertible operator* $a = \bigoplus_{i=0}^{n-1} \lambda_{i+1} \cdot$ $\cdot (p_{i+1} - p_i).$

PROOF. For every $i < j$

$$\phi((p_{i+1} - p_i)x(p_{i+1} - p_i)) = \lambda_{ij}(p_{i+1} - p_i)x(p_{i+1} - p_i).$$

Since $\phi$ is a homomorphism, a routine computation shows that for $j - i \geq 2$

$$\lambda_{ij} = \prod_{k=i}^{j-1} \lambda_{k,k+1}.$$

Hence, if we define $\lambda_1 = 1$ and $\lambda_i = \lambda_{i-1,i}$ $(2 \leq i \leq n)$, $\phi$ is seen to be implemented by the invertible operator $a = \overset{n-1}{\underset{i=0}{+}} \lambda_{i+1}(p_{i+1} - p_i)$.

**PROPOSITION II 2.** *Let $M$ be a factor, $A \subseteq M$ be an atomic nest-subalgebra and $\phi$ be an automorphism of $A$ such that $\phi(x) = x$ $(\forall)$ $x \in N$ and $\|\mathbf{id} - \phi\| < 1/2$. Then $\phi$ is implemented by an invertible operator $a \in R$, $\|I - a\| \leq 2\|\mathbf{id} - \phi\|$.*

PROOF. The proof of the preceding corollary shows that $\phi$ is implemented by an invertible operator $a \in R$   $a = \sum \lambda_n e_n$, the sum being taken over all the atoms $e_n$ of $L$.

We may assume that one of the numbers $\lambda_n$ is equal to 1, hence for every n one has either $|1 - \lambda_n| \leq \|\mathbf{id} - \phi\| < 1/2$ or $|1 - \lambda_n^{-1}| \leq \|\mathbf{id} - \phi\| < 1/2$. In both cases $|1 - \lambda_n| \leq 2\|\mathbf{id} - \phi\|$ and the conclusion follows.

### III. THE MAIN RESULTS

**THEOREM III 1.** *Let $M$ be a finite factor and $A \subseteq M$ be an atomic nest--subalgebra. For every automorphism $\phi$ of $A$ such that $\|\mathbf{id} - \phi\| < 1/24$ there exists an invertible operator $x$ in $A$ such that $x$ implements $\phi$ and $\|I - x\| < 14\|\mathbf{id} - \phi\|$.*

PROOF. Let $y \in M$ be given by Lemma II 1 $(M_1 = N, M_2 = M)$, $\|I - y\| \leq \|\mathbf{id} - \phi\|$, $\phi(x)y = yx$ $(\forall)$ $x \in N$. Since $\|\mathbf{id} - \phi\| < 1/24$, $y$ is invertible and $\psi : A \longrightarrow A$ $\psi(x) = y^{-1}\phi(x)y$ is an automorphism of $A$ with the property that $\phi(x) = x$ for every operator $x$ in $N$. Indeed, for every $x$ in $N$ $\psi(x) = y^{-1}\phi(x)y = y^{-1}yx = x$ and for every $a$ in $A$ and $p$ in $L$

$$\psi(a)p = \psi(a)\psi(p) = \psi(ap) = \psi(pap) = \psi(p)\psi(ap) = p\psi(ap)$$

hence $\psi(A) \subseteq A$. Moreover if $t = \|\mathbf{id} - \phi\|$ then $\|I - y\| \leq t \Rightarrow \|y^{-1}\| \leq (1 - t)^{-1}$ and $\|y\| \leq 1 + t$, $\|I - y^{-1}\| = \|y^{-1}(I - y)\| \leq t(1 - t)^{-1}$ and for every $a$ in $A$

$$\|\psi(a) - a\| \leq \|y^{-1}(\phi(a) - a)y\| + \|y^{-1}ay - ay\| +$$

$$+ \|ay - a\| \leq (3t + t^2)(1 - t)^{-1}\|a\|$$

and since $t < 1/24$ we obtain $\|\mathbf{id} - \psi\| < 4\|\mathbf{id} - \phi\|$.

Since $\psi$ leaves $N$ elementwise fixed, there is an invertible operator $u \in R$ such that $\| u \| = 1$ and $\psi(a) = uau^{-1}$ for every a in A.

Now, for a in A $\| ua - au \| \leq \| u \| \| a - uau^{-1} \|$ hence $\| \mathbf{ad}(u) | A \| \leq \| \mathbf{id} - \psi \| < 4 \| \mathbf{id} - \phi \|$. Choose, by Proposition I 2, $\lambda \in \mathbf{C}$ such that $\| u - \lambda I \| \leq 8t$ hence $|\lambda| > 1 - 8t > 2/3$ and $\| \lambda^{-1} u - I \| = |\lambda^{-1}| \| u - \lambda \| \leq 12t < 1/2$. Finally, define $x = \lambda^{-1} yu$, $x \in A$

$$\| I - x \| \leq \| y(I - \lambda^{-1} u) \| + \| y - I \| \leq$$

$$\leq (1 + t)12t + t = (13 + 12t)t \leq 14 \| \mathbf{id} - \phi \| .$$

Moreover $xax^{-1} = \phi(a)$ $(\forall)$ $a \in A$, which concludes the proof.

**THEOREM III 2.** *Every derivation of an atomic nest-subalgebra of a finite factor is inner.*

PROOF. It follows from Theorem III 1 and Theorem 3.2 in ([1]).

**THEOREM III 3.** *Every derivation $\delta$ of an atomic nest-suabalgera of a type $II_\infty$ factor M which satisfy $\delta(N)^\subset N$ is inner.*

PROOF. If $\delta$ is as above, then by Lemma II 3, for every $t > 0$, $\phi_t = e^{t\delta}$ has the property in Theorem III 1. One uses again Theorem 3.2 in ([1]).

**THEOREM III 4.** *Every ultraweakly continuous derivation of an atomic nest-subalgebra of a hyperfinite factor is inner.*

PROOF. It follows from Lemma II 2, the proof of Theorem III 1 and Theorem 3.2 in ([1]).

**THEOREM III 5.** *Let M be a type $II_\infty$ factor and L be an increasing sequence of finite projections in M converging strongly to the identity. Then every derivation of the algebra $A = M \cap \mathbf{Alg}\ L$ is inner.*

PROOF. It follows from Lemma II 1, the proof of Theorem III 1 and Theorem 3.2 in ([1]).

**REFERENCES**

1.    **Christensen, E.** : Derivations of nest algebras, *Math. Ann.* **229**(1977), 155-161.

2.    **Christensen, E.** : Extensions of derivations. II, *Math. Scand.* **50**(1982), 111-122.

3.    **Gilfeather, F. ; Larson, D.R.** : Nest-subalgebras of von Neumann algebras, *Adv. in Math.* **46**(1982), 171-199.

4.    **Gilfeather, F. ; Larson, D.R.** : Nest-subalgebras of von Neumann algebras: Com-

mutants modulo compacts and distance estimates, *J. Operator Theory* **7**(1982), 279-302.

5.    **Kadison, R.V.** : Derivations of operator algebras, *Ann. of Math.* **83**(1966), 280-
      -293.

6.    **Pop, F.** : Perturbations of nest-subalgebras of von Neumann algebras, *J. Operator Theory* **21**(1989), 139-144.

7.    **Sakai, S.** : Derivations of $W^*$-algebras, *Ann. of Math.* **83**(1966), 273-279.

8.    **Sakai, S.** : $C^*$-*algebras and* $W^*$-*algebras*, Springer Verlag, 1971.

9.    **Strătilă, Ş. ; Zsidó, L.** : *Lectures on von Neumann algebras*, Editura Academiei and Abacus Press, 1979.

**Florin Pop**

Department of Mathematics, INCREST
Bdul Păcii 220, 79622 Bucharest
Romania.

Operator Theory:
Advances and Applications, Vol. 43
© 1990 Birkhäuser Verlag Basel



# BOOLEAN ALGEBRAS OF PROJECTIONS AND SPECTRAL MEASURES IN DUAL SPACES

**Werner J. Ricker**

The work of W. G. Badé [1], [2], [3] on Boolean algebras (briefly, **B.a.**) of projections in Banach spaces is well known; a comprehensive treatment of these results can be found in [8]. In recent years there has been a revival of interest in this topic and most of Badé's program (together with the reflexivity result of T. A. Gillepsie [10]) has been extended to the setting of **B.a.**'s of projections in locally convex spaces (briefly, **lcs**); see [4], [5], [6], [7], [13], [17], [19]. In this setting phenomena arise which cannot be overcome by replacing norms with seminorms and then using Banach space arguments. These difficulties were overcome by developing and extending the theory of integration with respect to spectral measures and its connections with the theory of order and **lc**-Riesz spaces. An (almost) universal assumption for all of the results in the **lc**-setting is that the **B.a.** should be equicontinuous and that the space of continuous linear operators on the underlying **lcs** should be sequentially complete for the strong operator topology. Although these hypotheses are mild enough to include most spaces and **B.a.**'s of interest they are, nevertheless, too stringent to admit the consideration of at least one natural class of spaces, namely dual spaces equipped with their weak-star (briefly, $w^*$) topology. Indeed, the continuous linear operators on such spaces need not be sequentially complete for the strong operator topology and **B.a.**'s of projections are almost never equicontinuous. The aim of this paper is to investigate Badé's classical results for such $w^*$-dual spaces. Despite the difficulties indicated most of the expected analogues are valid. We will not attempt the utmost generality but restrict our attention to the duals of Fréchet spaces.

Let Y be a **lcs**, always assumed to be Hausdorff. The continuous dual space of Y is denoted by Y'. By $Y'_\sigma$ we mean Y' equipped with its $w^*$-topology. Let L(Y) denote the space of all continuous linear operators of Y into itself. Given $T \in L(Y)$, denote by $T' \in L(Y'_\sigma)$ the dual operator to T. If Y has its Mackey topology, then for every $S \in L(Y'_\sigma)$, the unique element of L(Y) whose dual is S will be denoted by $^*S$. For any **lcs** Y, let $L_s(Y)$ and $L_b(Y)$ denote L(Y) equipped with the topology $\rho_s$ of pointwise convergence in Y and the topology $\rho_b$ of uniform convergence on bounded sets in Y, respectively.

The notion of a **B.a.** of projections in Y, say $M$, is standard; see [2], [19]. It is assumed that the unit element of $M$ is the identity operator I. Then $M$ is complete ($\sigma$-complete) if it is complete ($\sigma$-complete) as an abstract **B.a.** with respect to the lattice operations $\wedge$ and $\vee$ induced by range inclusion and for any set (sequence) $\{A_\alpha\}$ in $M$ we have

$$(\wedge_\alpha A_\alpha)(Y) = \cap_\alpha A_\alpha(Y) \quad \text{and} \quad (\vee_\alpha A_\alpha)(Y) = \overline{\mathrm{sp}}(\cup_\alpha A_\alpha(Y)),$$

the closed subspace of Y generated by $\cup_\alpha A_\alpha(Y)$; see [2], [19]. The identity

$$(1) \qquad\qquad \wedge_\alpha A_\alpha = I - \vee_\alpha (I - A_\alpha)$$

is valid for any set (sequence) of elements $\{A_\alpha\}$ in a **B.a.** which is complete ($\sigma$-complete) as an abstract **B.a.** If M (resp. N) is a subspace of Y (resp. Y'), then the annihilator $M^\perp$ (resp. $^\perp N$) is defined by

$$M^\perp = \cap_{y \in M} \mathbf{ker}(\langle y, \cdot \rangle) \quad \text{and} \quad {}^\perp N = \cap_{\xi \in N} \mathbf{ker}(\langle \cdot, \xi \rangle),$$

respectively, where $\mathbf{ker}(\cdot)$ refers to the null space of the linear functionals $\langle \cdot, \xi \rangle$ on Y and $\langle y, \cdot \rangle$ on Y'. Then $M^\perp$ is a closed subspace of $Y'_\sigma$ and $^\perp N$ is a closed subspace of Y. Furthermore, $^\perp(M^\perp)$ is the closed subspace of Y generated by M and $(^\perp N)^\perp$ is the closed subspace of $Y'_\sigma$ generated by N. If Y has its Mackey topology and $P \in L(Y'_\sigma)$ and $Q \in L(Y)$ are projections, then $^\perp(P(Y')) = (I - {}^*P)(Y)$ and $(Q(Y))^\perp = (I - Q')(Y')$. In addition, if $P_1 \leq P_2$ are commuting projections from $L(Y'_\sigma)$ and $Q_1 \leq Q_2$ are commuting projections from L(Y), then $^*P_1 \leq {}^*P_2$ and $Q'_1 \leq Q'_2$. These remarks together with (1) can be used to establish the following result.

**LEMMA 1.** *Let* Y *be a* **lsc.** *If* M *is a complete ($\sigma$-complete)* **B.a.** *in* L(Y), *then* $M' = \{A' ; A \in M\}$ *is a complete ($\sigma$-complete)* **B.a.** *in* $L(Y'_\sigma)$. *Similarly, if* N *is a complete ($\sigma$-complete)* **B.a.** *in* $L(Y'_\sigma)$ *and* Y *has its Mackey topology, then* $^*N = \{^*B ; B \in N\}$ *is a complete ($\sigma$-complete)* **B.a.** *in* L(Y).

In the Banach space setting a complete or $\sigma$-complete **B.a.** , say $M$, can always be realized as the range of a spectral measure (i.e. a $\sigma$-additive map $P : \Sigma \rightarrow L_s(Y)$, defined on a $\sigma$-algebra of sets $\Sigma$ of some set $\Omega$, satisfying $P(\Omega) = I$ and $P(E \cap F) = P(E)P(F)$, for every E, F $\in \Sigma$). This is possible because the completeness ($\sigma$-complete-of $M$ is equivalent to the existence of $A = \lim_\alpha A_\alpha$ in $L_s(Y)$, for every monotonic net (sequence) $\{A_\alpha\} \subseteq M$, where A equals $\vee_\alpha A_\alpha$ (resp. $\wedge_\alpha A_\alpha$) if $\{A_\alpha\}$ is increasing (resp.

decreasing); see [8; XVII, Lemma 3.4]. It is this equivalent formulation of completeness and σ-completeness which permits the construction of a spectral measure whose range is $M$ (even for non—normable spaces Y [19; p. 299]). A crucial fact is that the **B.a.** is uniformly bounded (i.e. equicontinuous), a property which follows from the σ-completeness of $M$ merely as an abstract **B.a.** [2; Theorem 2.2]. This is also the case if Y is a Fréchet **lcs** [19; Proposition 1.2], but is not true in general. Accordingly, the equicontinuity of $M$ is usually assumed to ensure that $M$ can be realized as the range of a spectral measure; see Proposition 1.3 of [19] and the remarks following it. There are examples of σ-complete and complete **B.a.**'s which are range of spectral measures but are not equicontinuous. For instance, if $N$ is any complete (σ-complete) **B.a.** in a Fréchet **lcs** Y, then $N$ is equi-continuous and hence is the range of some spectral measure $P : \Sigma \to L_s(Y)$. Then $M = N'$ is the range of the (dual) spectral measure $P' : \Sigma \to L_s(Y'_\sigma)$ defined by $E \to P(E)'$, $E \epsilon \Sigma$, and $M$ is complete (σ-complete) by Lemma 1. However, $M$ will rarely be equicontinuous, even if Y is a Banach space or Hilbert space.

An examination of the proof of [19; Corollary 4.7] shows that in a separable Fréchet **lcs** every σ-complete **B.a.** is complete. The range of a spectral measure P is always a σ-complete **B.a.**; if P is a closed measure [5; §1], then its range is actually a complete **B.a.** (this follows from [5; Proposition 1.1]). If the underlying **lcs** is barrelled, then the range of any $L_s(Y)$-valued measure is equicontinuous.

**PROBLEM.** Let Y a barrelled **lcs** which is quasicomplete and let $M \subseteq L(Y)$ be a σ-complete **B.a.** Is $M$ equicontinuous? Maybe every **B.a.** which is σ-complete as an abstract **B.a.** is already equicontinuous?

So, the role of equicontinuity in the theory of **B.a.**'s of projections in **lc**-spaces is, in the first instance, to guarantee that the theory of integration with respect to spectral measures can be invoked. However, as will be indicated at various stages in the sequel, equicontinuity is also associated with other important aspects of the theory.

The sequential completeness of $L_s(Y)$ guarantees the existence of various limits and that bounded measurable functions are integrable for any $L_s(Y)$-valued measure. However, for **B.a.**'s in $w^*$-dual spaces even this property is not generally available. For example, if Y is a Banach space which is not weakly sequentially complete, then $L_s(Y'_\sigma)$ is not sequentially complete.

The organization of this paper will be to examine those Banach space results which are known to hold for equicontinuous, complete **B.a.**'s in **lc**-spaces and see whether they are still valid in the setting of $w^*$-dual spaces.

### I. IMBEDDING A σ-COMPLETE B.a. IN A COMPLETE B.a.

A σ-complete **B.a.**, say $M$, in a Banach space can be imbedded in a complete **B.a.**, namely the $\rho_s$-closure of $M$, [2; Theorem 2.7]. The same statement is true for any σ-complete **B.a.** in a **lcs** provided that it is equicontinuous, [19; Proposition 3.17]. Without the equicontinuity hypothesis this is false; the $\rho_s$-closure need not even be a **B.a.** Indeed, let X be a Hilbert space and $M \subseteq L(X)$ be a complete **B.a.** which is non-atomic and consists of selfadjoint projections. Let Y denote X' equipped with its (dual) norm topology. Then $L_s(Y'_\sigma)$ is just L(X) equipped with the weak operator topology. It follows from Lemma 1 that $M$ is a complete **B.a.** in $L(Y'_\sigma)$. However, $M$ is not a closed set in $L_s(Y'_\sigma)$; it follows from a result of H. Dye [9; Lemma 2.3] that the $\rho_s$-closure of $M$ in $L(Y'_\sigma)$ (i.e. the weak operator closure of $M$ in L(X)) is not even a B.a. Similar examples can be constructed in reflexive Banach spaces [2; p.354]. We will show that in $w^*$-dual spaces every σ-complete **B.a.** can be imbedded in a complete **B.a.** in a natural way, not by taking its $\rho_s$-closure but by taking its closure with respect to a stronger topology; see Proposition 1.

Let Y be a **lcs.** The seminorms generating the topology of $L_s(Y'_\sigma)$ are of the form

$$T \rightarrow |\langle y, T\xi \rangle|, \qquad T \in L_s(Y'_\sigma),$$

for some $y \in Y$ and $\xi \in Y'$. The seminorms generating the topology of $L_b(Y'_\sigma)$ are of the form

$$(2) \qquad T \rightarrow \sup\{|\langle y, T\xi \rangle| \; ; \xi \in B\}, \qquad T \in L_b(Y'_\sigma),$$

for some $y \in Y$ and bounded set $B \subseteq Y'_\sigma$. If Y is barrelled, then bounded subsets of $Y'_\sigma$ are equicontinuous and the topology of Y is generated by the seminorms

$$P_U : y \rightarrow \sup\{|\langle y, \xi \rangle| \; ; \xi \in U\}, \quad y \in Y,$$

as U varies through the equicontinuous subsets of Y'. Since Y has its Mackey topology a linear operator on Y is continuous iff it is continuous for the weak topology in Y. These observations can be combined to establish the following result.

**LEMMA 2.** *Let Y be a barrelled* **lcs**, *in which case* $Y'_\sigma$ *is quasicomplete. The mapping* $S \rightarrow S'$ *is a bicontinuous isomorphism of* $L_s(Y)$ *onto* $L_b(Y'_\sigma)$ *satisfying* $(RS)' = = S'R'$. *Its inverse is the mapping* $T \rightarrow {}^*T$, *for each* $T \in L(Y'_\sigma)$. *If Y is sequentially (resp. quasi) complete, then* $L_s(Y)$ *and hence, also* $L_b(Y'_\sigma)$, *is sequentially (resp. quasi) complete. The* **lcs** $L_s(Y'_\sigma)$ *is just* $L_b(Y'_\sigma)$ *equipped with its weak topology as a* **lcs**.

Let Y be a **lcs** and $P : \Sigma \rightarrow L_s(Y)$ be a spectral measure. Then the notion of a closed vector measure [11] applies to P also. We remark that P is a closed measure in $L_s(Y)$ iff it is a closed measure considered in $(L(Y))_\tau$ where $\tau$ is any **lc**-topology in $L(Y)$ consistent with the duality between $L_s(Y)$ and $(L_s(Y))'$, [14; Proposition 2]. If P is an equicontinuous measure (i.e. its range $R(P) = \{P(E) \; ; \; E \in \Sigma\}$ is an equicontinuous part of $L(Y)$) and Y is quasicomplete, then P is a closed measure iff $R(P)$ is a closed subset of $L_s(Y)$, [13; Proposition 3]. Combining these remarks with Lemma 2 it is possible to establish the following result.

**LEMMA 3.** *Let Y a barrelled* **lcs** *and* $P : \Sigma \rightarrow L_s(Y)$ *be a spectral measure. Let* $P' : \Sigma \rightarrow L_s(Y'_\sigma)$ *be the dual spectral measure. If* $P'_b$ *denotes* $P'$ *considered as taking its values in* $L_b(Y'_\sigma)$, *then* $P'_b$ *is also* $\sigma$-additive. *Furthermore,* $P'$ *is a closed measure in* $L_s(Y'_\sigma)$ *iff* $P'_b$ *is a closed measure in* $L_b(Y'_\sigma)$ *iff* $P$ *is a closed measure in* $L_s(Y)$. *If Y is quasicomplete, then these criteria are also equivalent to* $R(P)$ *being a complete* **B.a.** *in* $L(Y)$, *to* $R(P)$ *being a closed subset of* $L_s(Y)$ *and to* $R(P')$ *being a closed subset of* $L_b(Y'_\sigma)$.

In Fréchet spaces every $\sigma$-complete **B.a.** is the range of a spectral measure. Combining this fact with Lemmas 1 – 3 yields the result alluded to earlier.

**PROPOSITION 1.** *Let Y be a Fréchet* **lcs** *and* $M \subseteq L(Y'_\sigma)$ *be a* $\sigma$-complete **B.a.** *Then M can be imbedded into the complete* **B.a.** $\tilde{M}$ *consisting of the closure of M in* $L_b(Y'_\sigma)$. *Equivalently,* $\tilde{M} = \{A' \; ; \; A \in N\}$ *where N is the closure of* $^*M = \{^*B \; ; \; B \in M\}$ *in* $L_s(Y)$. *In general,* $\tilde{M}$ *is not a closed subset of* $L_s(Y'_\sigma)$.

## II. UNIFORMLY CLOSED ALGEBRAS

One of Badé's results in Banach spaces states that the closed algebra generated by a complete B.a. with respect to the uniform operator topology coincides with the closed algebra that it generates with respect to the strong (or, equivalently, weak) operator topology [2; Theorem 4.5]. If Y is a **lcs** and $M \subseteq L(Y)$ is an equicontinuous, complete **B.a.**, then the closed algebra generated by $M$ in $L_b(Y)$ coincides with the closed algebra that it generates in $L_s(Y)$; see [17]. An examination of the proof of Theorem 1 in [17] shows that the equicontinuity of $M$ is an essential ingredient in the argument; see Lemmas 1 and 2 in [17]. In dual spaces with their $w^*$-topology, the same result holds. It is a consequence of Lemma 2 and the fact that the closure of a convex set in a **lcs** is the same for the weak topology and the initial topology.

**PROPOSITION 2.** *Let Y be a barrelled* **lcs** *and* $M \subseteq L(Y'_\sigma)$ *be a complete* **B.a.**

*Then the closed algebra generated by M in $L_b(Y'_\sigma)$ coincides with the closed algebra that it generates in $L_s(Y'_\sigma)$.*

The space Y in Proposition 2 need not be a Fréchet **lcs** since the argument does not require $M$ to be the range of a spectral measure. Actually, $M$ can be any **B.a.** in $L(Y'_\sigma)$; not even $\sigma$-completeness is needed!

## III. INTEGRATION AND SPECTRAL MEASURES

Let Y be a **lcs** and $P : \Sigma \to L_s(Y)$ be a spectral measure. Then P is $\sigma$-additive iff the **C**-valued set function

$$\langle Py, \xi \rangle : E \to \langle P(E)y, \xi \rangle, \qquad E \in \Sigma,$$

is $\sigma$-additive, for every $y \in Y$ and $\xi \in Y'$. A **C**-valued, $\Sigma$-measurable function f on $\Omega$ is said to be P-integrable if it is $\langle Py, \xi \rangle$-integrable, for each $y \in Y$ and $\xi \in Y'$, and there exists an element P(f) in L(Y), also denoted by $\int_\Omega f dP$, satisfying

$$\langle P(f)y, \xi \rangle = \int_\Omega f d\langle Py, \xi \rangle, \qquad y \in Y, \xi \in Y'.$$

The indefinite integral of f with respect to P is the $L_s(Y)$-valued measure given by

$$E \to \int_E f dP = P(f)P(E) = P(E)P(f), \qquad E \in \Sigma.$$

A P-integrable function f is called P-null iff P(f) = 0. If f and g are P-integrable then so is their pointwise product fg and

$$(3) \qquad \int_E fg dP = P(f)P(g)P(E) = P(g)P(f)P(E), \qquad E \in \Sigma.$$

The space of all P-integrable functions is denoted by L(P). A typical seminorm generating the topology of $L_s(Y)$ is of the form

$$S \to p(Sy), \quad S \in L_s(Y),$$

for some $y \in Y$ and continuous seminorm p in Y. We generate a topology $\tau(P)$ in L(P) by specifying the seminorms

$$q(P) : f \to \sup\{q(\textstyle\int_E f dP) ; E \in \Sigma\}, \qquad f \in L(P),$$

for each $\rho_s$-continuous seminorm q in $L_s(Y)$. The resulting **lcs** is not necessarily Hausdorff. The quotient space of L(P) modulo the P-null functions is denoted by $L^1(P)$.

The resulting Hausdorff topology in $L^1(P)$ is again denoted by $\tau(P)$.

Let $P : \Sigma \to L_s(Y)$ be an equicontinuous spectral measure. Then all bounded, $\Sigma$-measurable functions are P-integrable. Furthermore, $L^1(P)$ is a commutative, locally convex (briefly, **lc-**) algebra with unit. If, in addition, Y is quasicomplete and $L_s(Y)$ is sequentially complete, then P is a closed measure iff R(P) is a closed subset of $L_s(Y)$ iff $L^1(P)$ is complete with respect to $\tau(P)$. In this case,

$$\Phi_P : f \to P(f) = \int_\Omega f \, dP, \qquad f \in L^1(P),$$

is a bicontinuous isomorphism of the (complete) **lc**-algebra $L^1(P)$ onto the closed algebra $\langle R(P) \rangle_s$ generated by R(P) in $L_s(Y)$. The sequential completeness of $L_s(Y)$ ensures that the closed measure P generates a complete space $(L^1(P), \tau(P))$. All of the above statements about spectral measures can be found in [5].

The aim of this section is to investigate the nature of $L^1(P)$ and its associated integration map $\Phi_P$ in dual spaces with their $w^*$-topology. So, let Y be a **lcs**. If $Q : \Sigma \to L_s(Y)$ is a spectral measure, then the (dual) set function $P = Q' : \Sigma \to L_s(Y'_\sigma)$ is also a spectral measure and satisfies

(4) $$\langle Q(E)y, \xi \rangle = \langle y, P(E)\xi \rangle, \qquad E \in \Sigma,$$

for every $y \in Y$ and $\xi \in Y'$. The next result is a consequence of (4), Lemma 2 and the definition of integrability with respect to a spectral measure.

**LEMMA 4.** *Let Y be a* **lcs** *and* $Q : \Sigma \to L_s(Y)$ *be a spectral measure. If* $Q' : \Sigma \to L_s(Y'_\sigma)$ *is the dual spectral measure, then every Q-integrable function f is Q'-integrable and*

$$\int_E f \, dQ' = Q(E)'Q(f)' = Q(f)'Q(E)', \qquad E \in \Sigma.$$

*If Y is also a Mackey space, then every Q'-integrable function g is Q-integrable and*

$$\int_E g \, dQ = {}^*(Q'(g))Q(E) = Q(E).{}^*(Q'(g)), \qquad E \in \Sigma.$$

*In particular, $L(Q) = L(Q')$ as vector spaces and they have the same null functions. Similar statements apply to the set function ${}^*P : \Sigma \to L_s(Y)$ whenever Y is a* **lcs** *with its Mackey topology and $P : \Sigma \to L_s(Y'_\sigma)$ is a given spectral measure.*

Let Y be a **lcs** and $P : \Sigma \to L_s(Y'_\sigma)$ be a spectral measure. By $P_b$ is meant the set function P considered as taking its values in $L_b(Y')$. If Y is barrelled, then it follows from Lemma 2 and the Orlicz-Pettis lemma that $P_b$ is also $\sigma$-additive. Clearly P and $P_b$ have the same null sets in $\Sigma$. Since the topology $\rho_b$ is compatible with the duality between $L_s(Y'_\sigma)$ and $(L_s(Y'_\sigma))'$ it is clear that a $\Sigma$-measurable function is P-integrable iff it is $P_b$-integrable (with the same indefinite integral). It follows from (3) applied to P in $L_s(Y'_\sigma)$ that the pointwise product of two $P_b$-integrable functions, say f and g, is also $P_b$-integrable and

$$\int_E fg\,dP_b = P(f)P(g)P(E), \qquad E \in \Sigma.$$

Typical seminorms generating the topology of $L_b(Y'_\sigma)$ are given by (2). A topology $\tau(P_b)$ is determined in $L^1(P_b)$ by specifying the seminorms

$$q(P_b) : f \to \sup_E\{q(\textstyle\int_E f\,dP_b)\ ;\ E \in \Sigma\}, \qquad f \in L^1(P_b),$$

for each $\rho_b$-continuous seminorm q in $L_b(Y'_\sigma)$. To show that $L^1(P_b)$ is an **lc**-algebra it must be shown that the mapping $(f,g) \to fg$ is separately continuous for $\tau(P_b)$. This follows as in the proofs of Lemma 4.2 and Corollary 4.1.2 in [16] once it is established that the set $A = 4\{P(E)\xi\ ;\ E \in \Sigma,\ \xi \in B\}$ is bounded in $Y'_\sigma$ whenever $B \subseteq Y'_\sigma$ is bounded. But, Y is barrelled and so $R(^*P) = \{^*P(E)\ ;\ E \in \Sigma\}$ is an equicontinuous part of L(Y). It follows that R(P) is bounded in $L_s(Y'_\sigma)$ and hence, also in $L_b(Y'_\sigma)$. Since B is bounded in $Y'_\sigma$ it follows that $A = 4R(P)(B)$ is bounded in $Y'_\sigma$. This establishes the following result.

**LEMMA 5.** *Let* Y *be a barrelled* **lcs** *and* $P : \Sigma \to L_s(Y'_\sigma)$ *be a spectral measure. Then* $L^1(P_b)$ *is a commutative* **lc**-*algebra with unit. If* Y *is sequentially complete and* $P_b$ *is a closed measure, then* $L^1(P_b)$ *is* $\tau(P_b)$-*complete and the integration map*

$$\Phi_{P_b} : f \to P_b(f) = \int_\Omega f\,dP_b = P(f), \qquad f \in L^1(P_b),$$

*is a bicontinuous isomorphism of* $L^1(P_b)$ *onto the closed algebra* $\langle R(P)\rangle_b$ *generated by* R(P) *in* $L_b(Y'_\sigma)$.

The isomorphism of Lemma 2 can be used, together with Lemma 4 and the definitions of the seminorms in $L^1(^*P)$ and $L^1(P_b)$, to show that these spaces are isomorphic. Since, by Lemma 3, $P_b$ is a closed measure in $L_b(Y'_\sigma)$ iff $^*P$ is a closed measure in $L_s(Y)$, it is also possible to deduce Lemma 5 from the fact the $\Phi_{^*P}$ is a bicontinuous isomorphism of $L^1(^*P)$ onto $\langle R(^*P)\rangle_s$. So, if $P_b$ is a closed measure, then we have the following diagram where all maps are bicontinuous isomorphisms and **Id** is the identity map.

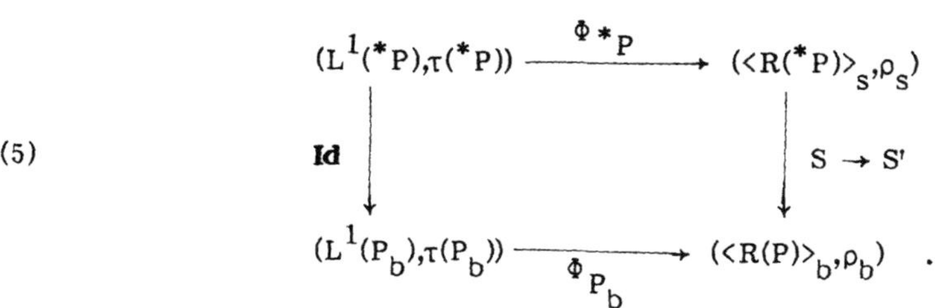

$$\begin{array}{ccc}
(L^1(^*P),\tau(^*P)) & \xrightarrow{\ \Phi_{^*P}\ } & (\langle R(^*P)\rangle_s,\rho_s) \\
\ \Big\downarrow \mathbf{Id} & & \ \Big\downarrow S \to S' \\
(L^1(P_b),\tau(P_b)) & \xrightarrow[\ \Phi_{P_b}\ ]{} & (\langle R(P)\rangle_b,\rho_b)
\end{array} \quad .$$

**PROPOSITION 3.** *Let* $Y$ *be a sequentially complete, barrelled* **lcs** *and* $P : \Sigma \to L_s(Y'_\sigma)$ *be a closed spectral measure. Then* $L^1(P)$ *is a commutative* **lc**-*algebra with identity and the integration map*

$$\Phi_P : f \to \int_\Omega f\,dP, \qquad f \in L^1(P),$$

*is a vector space, multiplicative isomorphism of* $L^1(P)$ *onto* $\langle R(P)\rangle_s$. *In addition,* $\Phi_P$ *is continuous but, in general,* $(\Phi_P)^{-1}$ *is not continuous.*

PROOF. It can be calculated that multiplication is separately continuous and so $L^1(P)$ is a **lc**-algebra. A direct computation, using the specific form of the seminorms generating the topologies in $L^1(P)$ and $L_s(Y'_\sigma)$, verifies that $\Phi_P$ is continuous. That $\Phi_P$ is a vector space isomorphism onto $\langle R(P)\rangle_s$ follows from the fact that the identity map between $L^1(P)$ and $L^1(P_b)$ is a vector space isomorphism (cf. Lemma 4), the identity map between $\langle R(P)\rangle_s$ and $\langle R(P)\rangle_b$ is a vector space isomorphism (cf. Proposition 2 and the comments preceding it), and the integration map $h \to \int_\Omega h\,dP_b$ is a bicontinuous isomorphism of $L^1(P_b)$ onto $\langle R(P)\rangle_b$; see Lemma 5. $\qquad\square$

**REMARKS.** (1) There is the question of whether or not $L^1(P)$ is complete? If $L_s(Y'_\sigma)$ is sequentially complete (e.g. $Y$ is reflexive), then the completeness of $L^1(P)$ follows from the closedness of $P$, [5; p.139]. Of course, $L_s(Y'_\sigma)$ may not be sequentially complete. Nevertheless, in the case when $Y$ is a Banach space it is known that $L^1(P)$ is always complete for any closed spectral measure $P : \Sigma \to L_s(Y'_\sigma)$ and, in addition, the **lc**-algebra $\langle R(P)\rangle_s$ is quasicomplete for the topology induced by $\rho_s$; this follows from the results of [18] and is based on Theorem 2 of [12].

**PROBLEM.** Let $Y$ be a Banach space and $P : \Sigma \to L_s(Y'_\sigma)$ be a closed spectral measure. Is $\langle R(P)\rangle_s$ actually $\rho_s$-complete? If $Y$ is a quasicomplete, barrelled **lcs** is $L^1(P)$ complete?

(2) The set $\{\chi_E ; E \in \Sigma\}$ is always a $\tau(P)$-closed subset of $L^1(P)$. If the map

$(\Phi_P)^{-1} : \langle R(P) \rangle_s \longrightarrow L^1(P)$ is continuous, then $R(P) = \{(\Phi_P)^{-1}(\chi_E) ; E \in \Sigma\}$ is a closed subset of $L_s(Y'_\sigma)$. But, we exhibited examples in (I) where this is not the case. Accordingly, $(\Phi_P)^{-1}$ need not be continuous. This is in contrast to the situation for equicontinuous measures, [5; Proposition 1.5].

(3) Since $\langle R(^*P) \rangle_s$ is an inverse-closed subalgebra of $L(Y)$ (by [15; Lemma 3] and the fact that $\langle R(^*P) \rangle_s$ is isomorphic to $L^1(^*P)$) it follows from (5) that $\langle R(P) \rangle_b$ is inverse-closed in $L(Y'_\sigma)$. Then Proposition 2 implies that $\langle R(P) \rangle_s$ is also inverse-closed in $L(Y'_\sigma)$. So, if Y is a Fréchet **lcs** and $M \subseteq L(Y'_\sigma)$ is a complete **B.a.**, then $\langle M \rangle_s$ (which equals $\langle M \rangle_b$) is an inverse-closed subalgebra of $L(Y'_\sigma)$.

So, in $w^*$-dual spaces, $L^1(P)$ and its associated integration map $\Phi_P$ do not exhibit the strong properties as for the case of equicontinuous spectral measures in spaces Y for which $L_s(Y)$ is sequentially complete.

## IV. **THE REFLEXIVITY THEOREM**

One version of the Badé reflexivity theorem in Banach spaces [8; XVII Corollary 3.24] states that if $M$ is a $\sigma$-complete **B.a.**, then a bounded operator belongs to $\langle M \rangle_s$ iff it leaves invariant each closed $M$-invariant subspace (the $\sigma$-completeness requirement can be weakened). This result is also valid in the **lcs** setting if Y is quasicomplete, $L_s(Y)$ is sequentially complete and $M$ is equicontinuous, [5; Theorem 3.1]. This result, together with Propositions 1 and 2, the isomorphisms in (5) and the fact for a Fréchet **lcs** every $\sigma$-complete **B.a.** $M \subseteq L(Y'_\sigma)$ is the range of a spectral measure can be used to establish the following result. The properties of annihilators enter in showing that $^*T$ leaves invariant every closed $^*M$-invariant subspace of Y whenever $T \in L(Y'_\sigma)$ leaves invariant each closed $M$-invariant subspace of $Y'_\sigma$.

**PROPOSITION 4.** *Let* Y *be a Fréchet* **lcs** *and* $M \subseteq L(Y'_\sigma)$ *be a $\sigma$-complete* **B.a.** *Then an element of* $L(Y'_\sigma)$ *belongs to* $\langle M \rangle_s$ *iff it leaves invariant each closed M-invariant subspace of* $Y'_\sigma$. *The same statement is true for* $\langle M \rangle_b$ *in place of* $\langle M \rangle_s$.

If Y is a **lcs**, then a closed subalgebra N of $L_s(Y)$ is called reflexive if N consists of those elements in $L(Y)$ which leave invariant all closed N-invariant subspaces of Y.

Let Y be a Fréchet **lcs** and $M \subseteq L(Y'_\sigma)$ be a $\sigma$-complete **B.a.** The map $\Lambda : T \longrightarrow {}^*T$ is a bicontinuous isomorphism of $L_s(Y'_\sigma)$ onto $L_w(Y)$, where w is the weak operator topology in $L(Y)$. Then $L_w(Y)$ is just $L_s(Y)$ equipped with its weak topology as a **lcs**. The map $\Lambda$ is also a bicontinuous isomorphism of $L_b(Y'_\sigma)$ onto $L_s(Y)$. Now $^*M$ is a $\sigma$-complete

**B.a.** in $L(Y)$ and $\Lambda$ induces (by restriction) a bicontinuous isomorphism of $\langle M \rangle_b$ onto $\langle {}^* M \rangle_s$. Using these remarks and Proposition 3.2 of [6] applied to ${}^* M$ in $L(Y)$ it can be shown that every element in the dual space of $\langle M \rangle_s$ is of the form

$$T \longrightarrow \langle y, T\xi \rangle, \qquad T \in \langle M \rangle_s,$$

for some $y \in Y$ and $\xi \in Y'$. This fact, together with Proposition 4, makes it possible to argue as on p. 371 and p. 374 of [6] to establish the following reflexivity result (cf. [6; Theorem 3.8] for the case of equicontinuous measures).

**PROPOSITION 5.** *Let* $Y$ *be a Fréchet* **lcs** *and* $M \subseteq L(Y'_\sigma)$ *be a $\sigma$-complete* **B.a.** *Then each closed unital subalgebra of* $\langle M \rangle_s$ *is a reflexive subalgebra of* $L(Y'_\sigma)$.

## V. SCALAR-TYPE SPECTRAL OPERATORS

An element $S \in L(Y)$ is called a scalar-type spectral operator (briefly, a scalar operator) if there exists an $L_s(Y)$-valued spectral measure, say $P$, and $f \in L(P)$ such that $S = P(f)$. If $Y$ is a Banach space and $N \subseteq L(Y)$ is a $\sigma$-complete **B.a.**, then every element of $\langle N \rangle_s$ is a scalar operator, [8; XVII Corollary 3.25]. This is also true in a quasicomplete **lcs** $Y$ for which $L_s(Y)$ is sequentially complete provided that $N$ is an equicontinuous, $\sigma$-complete **B.a.**, [13].

**PROPOSITION 6.** *Let* $Y$ *be a Fréchet* **lcs** *and* $M \subseteq L(Y'_\sigma)$ *be a $\sigma$-complete* **B.a.** *Then every element of* $\langle M \rangle_s$ *is a scalar-type spectral operator.*

PROOF. Imbed $M$ into a complete **B.a.** $\tilde{M}$ as in Proposition 1 and then apply Proposition 3 to any spectral measure $P : \Sigma \longrightarrow L_s(Y'_\sigma)$ for which $R(P) = \tilde{M}$. $\qquad \square$

An operator $S \in L(Y)$ is called reflexive if the closed subalgebra of $L_s(Y)$ generated by $I$ and $S$ is a reflexive algebra. Using Propositions 4 and 5 it is possible to argue as on p. 374 of [6] to establish the following result; see [6; Corollary 3.9] if $Y$ is a quasicomplete and $L_s(Y)$ is sequentially complete.

**PROPOSITION 7.** *Let* $Y$ *be a Fréchet* **lcs**. *Then every scalar-type spectral operator in* $L_s(Y'_\sigma)$ *is a reflexive operator.*

**ACKNOWLEDGEMENT.** The support of an Alexander von Humboldt Fellowship is gratefully acknowledged.

## REFERENCES


1.  **Badé, W.G.** : Unbounded spectral operators, *Pacific J. Math.* **4**(1954), 373-392.

2.  **Badé, W.G.** : On Boolean algebras of projections and algebras of operators, *Trans. Amer. Math. Soc.* **80**(1955), 345-359.

3.  **Badé, W.G.** : A multiplicity theory for Boolean algebras of projections in Banach spaces, *Trans. Amer. Math. Soc.* **92**(1959), 508-530.

4.  **Dodds, P.G. ; de Pagter, B.** : Orthomorphisms and Boolean algebras of projections, *Math. Z.* **187**(1984), 361-381.

5.  **Dodds, P.G. ; Ricker, W.J.** : Spectral measures and the Badé reflexivity theorem, *J. Funct. Anal.* **61**(1985), 136-163.

6.  **Dodds, P.G. ; de Pagter, B. ; Ricker, W.J.** : Reflexivity and order properties of scalar-type spectral operators in locally convex spaces, *Trans. Amer. Math. Soc.* **293**(1986), 355-380.

7.  **Dodds, P.G. ; de Pagter, B.** : Algebras of unbounded scalar-type spectral operators, *Pacific J. Math.* **130**(1987), 41-74.

8.  **Dunford, N. ; Schwartz, J.T.** : *Linear operators. Part III: Spectral operators*, Wiley-Interscience, New York, 1971.

9.  **Dye, H.** : The unitary structure in finite rings of operators, *Duke Math. J.* **20** (1953), 55-70.

10. **Gillespie, T.A.** : Boolean algebras of projections and reflexive algebras of operators, *Proc. London Math. Soc.* **37**(1978), 56-74.

11. **Kluvánek, I.** : The range of a vector-valued measure, *Math. Systems Theory*, **7**(1973), 44-54.

12. **Orhon, M.** : Boolean algebras of commuting projections, *Math. Z.* **183**(1983), 531-537.

13. **Ricker, W.J.** : On Boolean algebras of projections and scalar-type spectral operators, *Proc. Amer. Math. Soc.* **87**(1983), 73-77.

14. **Ricker, W.J.** : Criteria for closedness of vector measures, *Proc. Amer. Math. Soc.* **91**(1984), 75-80.

15. **Ricker, W.J.** : A spectral mapping theorem for scalar-type spectral operators in locally convex spaces, *Integral Equations Operator Theory*, **8**(1985), 276-288.

16. **Ricker, W.J.** : Spectral measures, boundedly $\sigma$-complete Boolean algebras and applications to operator theory, *Trans. Amer. Math. Soc.* **304**(1987), 819-838.

17. **Ricker, W.J. ; Schaefer, H.H.** : The uniformly closed algebras generated by a complete Boolean algebra of projections, *Math. Z.*, to appear.

18. **Ricker, W.J.** : Completeness of the $L^1$-space of closed vector measures, *Proc. Edinburgh Math. Soc.*, to appear.

19. **Walsh, B.** : Structure of spectral measures on locally convex spaces, *Trans. Amer. Math. Soc.* **120**(1965), 295-326.



**Werner J. Ricker**

School of Mathematics
University of N.S.W.
Kensington, N.S.W., 2033
Australia.

Operator Theory:
Advances and Applications, Vol. 43
© 1990 Birkhäuser Verlag Basel



# A VECTOR EXTENSION OF LOEWNER'S DIFFERENTIAL EQUATION

**James Rovnyak** [*]

Loewner's differential equation [7]

$$(1) \qquad t\frac{\partial}{\partial t}f(t,z) = \phi(t,z)z\frac{\partial}{\partial z}f(t,z)$$

arises in the estimation theory of Riemann mapping functions. We consider (1) in a
vector setting with operator valued Herglotz functions $\phi(t,z)$. Notions of geometric
function theory are lost in this generality, but the intuitive idea of an expansive flow
remains. Mathematically, it is more convenient to think of the flow as contractive in
reverse time. In the scalar case, a classical subordination theorem [9] implies that the
evolution transformations for (1), $T(a,b) : f(b,z) \rightarrow f(a,z)$, are contractive in the
Dirichlet space if $a \leq b$. For different reasons, a similar result holds in the vector
extension. The proof of the Bieberbach conjecture [2] and its power extensions [3], [11]
yield other examples. As an illustration of the vector theory, we discuss a result from
[11] which includes and extends the example of the Dirichlet space.

A separable complex Hilbert space $C$ is chosen and fixed throughout. By a
*vector* we mean an element of $C$, and by an *operator* we mean a bounded linear trans-
formation of $C$ into itself. If c is a vector and A is an operator, we write $|c|$ and $|A|$
for their norms. The adjoint of an operator A is written $\bar{A}$, the identity operator 1. For
any vector b, $\bar{b}$ is the linear functional on $C$ such that $\bar{b}a = \langle a,b\rangle_C$ for all vectors a.

A *Herglotz function* is an operator valued analytic function $\phi(z)$ on the unit disk
such that **Re** $\phi(z) \geq 0$ and $\phi(0) = 1$. A *Herglotz family* is an indexed set of Herglotz
functions $\phi(t,z)$, $0 < t < \infty$, whose Taylor coefficients are weakly measurable functions of
positive t. Consider Loewner's equation (1) for a given Herglotz family $\phi(t,z) = 1 +$
$+ \phi_1(t)z + \phi_2(t)z^2 + \dots$ . Let $\nu$ be a real number. A generalized power series

$$(2) \qquad f(t,z) = \sum_{n=1}^{\infty} f_n(t)z^{\nu+n}$$

is a *solution* of (1) if $f_1(t)$, $f_2(t)$, ... are absolutely continuous vector valued functions

______________

[*] Research supported by NSF Grant DMS-8701395.

of positive t such that (1) holds formally a.e. Explicitly, this requires

$$(3) \quad \begin{cases} tf_1'(t) = (\nu + 1)f_1(t) \\ tf_2'(t) = (\nu + 2)f_2(t) + (\nu + 1)\phi_1(t)f_1(t) \\ \cdots \\ tf_n'(t) = (\nu + n)f_n(t) + (\nu + n - 1)\phi_1(t)f_{n-1}(t) + \ldots + (\nu + 1)\phi_{n-1}(t)f_1(t) \\ \cdots \end{cases}$$

a.e. for positive t. A vector valued function f(t) of positive t is *absolutely continuous* if there is a weakly measurable vector valued function g(t) such that $|g(t)|$ is locally integrable and

$$f(b) - f(a) = \int_a^b g(t)dt$$

for any positive numbers a and b, $a \leq b$. The integral exists in the Bochner sense [5, pp. 73, 80], and f(t) has the strong derivative $f'(t) = g(t)$ a.e. [5, p. 88].

**THEOREM 1.** *Let $\phi(t,z)$, $0 < t < \infty$, be a Herglotz family $\nu$, any real number. For a given $b > 0$ and generalized power series $h(z) = \sum_{n=1}^{\infty} c_n z^{\nu+n}$ with vector coefficients, Loewner's equation (1) has a unique solution (2) such that $f(b,z) = h(z)$.*

PROOF. There is a unique absolutely continuous function $f_1(t)$ satisfying the first equation in (3) such that $f_1(b) = c_1$ , namely, $f_1(t) = (t/b)^{\nu+1}c_1$. With $f_1(t)$ determined, the second equation in (3) has a unique solution $f_2(t)$ such that $f_2(b) = c_2$:

$$f_2(t) = (t/b)^{\nu+2}c_2 - t^{\nu+2}\int_t^b s^{-\nu-3}(\nu + 1)\phi_1(s)f_1(s)ds \ .$$

Repetition of this process yields the result.                                            □

The term *Kreĭn space* is used as in Bognár [1] for a scalar product space which is isomorphic to the direct sum of a Hilbert space and the anti-space of a Hilbert space. Let $\nu$ and $\sigma_1, \sigma_2, \ldots$ be real numbers. The *Grunsky space* $G_\sigma^\nu$ is the Kreĭn space of generalized power series

$$f(z) = \sum_{n=1}^{\infty} a_n z^{\nu+n}$$

with vector coefficients such that $\sum_{n=1}^{\infty} |(\nu + n)\sigma_n| \, |a_n|^2 < \infty$. Two series are identified if their coefficients coincide for all n such that $(\nu + n)\sigma_n \neq 0$. The scalar product in $G_\sigma^\nu$ is defined by

$$\left\langle \sum_{n=1}^{\infty} a_n z^{\nu+n}, \sum_{n=1}^{\infty} b_n z^{\nu+n} \right\rangle_{G_\sigma^\nu} = \sum_{n=1}^{\infty} (\nu + n)\sigma_n \bar{b}_n a_n$$

for any elements of the space. A *null element* of $G_\sigma^\nu$ is a series $\sum_{n=1}^{\infty} a_n z^{\nu+n}$ with vector coefficients such that $a_n = 0$ for all indices $n$ such that $(\nu + n)\sigma_n \neq 0$. Technically, such elements are representatives for the zero coset in $G_\sigma^\nu$. With a mild hypothesis, Loewner's equation is well behaved with respect to null elements of a Grunsky space.

**THEOREM 2.** *Let* $\phi(t,z)$ *be a Herglotz family, and let* $f(t,z)$ *be a solution of Loewner's equation* (1) *of the form* (2). *Let* $G_\sigma^\nu$ *be a Grunsky space such that* $\sigma_1 \cdots \sigma_n \neq 0$ *whenever* $\sigma_n \neq 0$. *If* $f(b,z)$ *is a null element of* $G_\sigma^\nu$ *for some* $b > 0$, *then* $f(t,z)$ *is a null element of* $G_\sigma^\nu$ *for all* $t > 0$.

PROOF. Assume that there is an integer $r$ such that $\sigma_n \neq 0$ if $1 \leq n \leq r$ and $\sigma_n = 0$ otherwise. The case $\sigma_n \neq 0$ for all $n$ is handled similarly, and the case $\sigma_n = 0$ for all $n$ is trivial.

If $\nu$ is not a negative integer, then an element $\sum_{n=1}^{\infty} c_n z^{\nu+n}$ of $G_\sigma^\nu$ is null if and only if $c_1 = \cdots = c_r = 0$. Let

$$f(b,z) = \sum_{n=1}^{\infty} f_n(b)z^{\nu+n}$$

satisfy this condition. By the first equation in (3), $f_1(t)$ is $t^{\nu+1}$ times a constant vector. Since $f_1(b) = 0$, $f_1(t)$ vanishes identically. In a similar way, the second equation in (3) then implies that $f_2(t)$ vanishes identically. An inductive argument shows that $f_n(t)$ vanishes identically for $1 \leq n \leq r$, and so $f(t,z)$ is a null element of $G_\sigma^\nu$ for all $t > 0$.

Suppose that $\nu$ is a negative integer, say $\nu = -p$. In this case, an element $\sum_{n=1}^{\infty} c_n z^{\nu+n}$ of $G_\sigma^\nu$ is null if and only if $c_n$ vanishes for $n = 1, \ldots, r$, $n \neq p$. Assume that

$$f(b,z) = \sum_{n=1}^{\infty} f_n(b)z^{\nu+n}$$

is such a series. If $p > r$, we can argue as above that $f(t,z)$ is a null element of $G_\sigma^\nu$ for $t > 0$. Let $1 \leq p \leq r$. As before, $f_n(t)$ is identically zero if $1 \leq n < p$. The $p$-th equation in (3) implies that $f_p(t)$ is constant. Continuing, we obtain

$$tf'_{p+1}(t) = (-p + p + 1)f_{p+1}(t) + (-p + p)\phi_1(t)f_p(t) = f_{p+1}(t) .$$

Since $f_{p+1}(t)$ vanishes for $t = b$, it vanishes identically. Continuing in this way, we see that $f_n(t)$ vanishes identically for $p < n \leq r$, and $f(t,z)$ is a null element of $G_\sigma^\nu$ for all $t > 0$. $\quad\square$

The main estimates apply to solutions of the equation

$$(4) \qquad t\frac{\partial}{\partial t}f(t,z) = \phi(t,z)z\frac{\partial}{\partial z}f(t,z) + z^\nu[\phi(t,z) - 1]c(t),$$

where $c(t)$ is a locally square summable vector valued function of positive $t$, $\nu$ is any real number, and $\phi(t,z)$ is a Herglotz family. Intuitively, one can think of (4) as describing a system driven by an input $c(t)$. A generalized power series (2) is a *solution* of (4) if its coefficients are absolutely continuous vector valued functions of positive $t$ and (4) holds a.e. in the formal sense. The scalar version of the next result is given in [11].

**THEOREM 3.** *Let $f(t,z)$ be a solution of (4), and let $G_\sigma^\nu$ be a Grunsky space such that $\sigma_1 \geq \sigma_2 \geq \ldots \geq 0$. Let $a$ and $b$ be positive numbers with $a \leq b$. If $f(b,z)$ belongs to $G_\sigma^\nu$, then $f(a,z)$ belongs to $G_\sigma^\nu$ and*

$$\langle f(a,z), f(a,z)\rangle_{G_\sigma^\nu} - \langle f(b,z),f(b,z)\rangle_{G_\sigma^\nu} \leq 2\sigma_1 \int_a^b |c(s)|^2 s^{-1}ds.$$

We first make a preliminary observation.

**LEMMA.** *Let $\mu$ be a nonnegative operator valued measure on the unit circle $\Gamma = \{\zeta : |\zeta| = 1\}$. Then*

$$(5) \qquad \int_\Gamma \bar{f}(\zeta)d\mu(\zeta)f(\zeta) \geq 0$$

*for every polynomial $f(\zeta)$ with vector coefficients.*

PROOF OF LEMMA. If $f(\zeta) = a_o + a_1\zeta + \ldots + a_n\zeta^n$, we interpret the integral in (5) as

$$\int_\Gamma \bar{f}(\zeta)d\mu(\zeta)f(\zeta) = \sum_{j,k=0}^{n} \int_\Gamma \zeta^{j-k} \bar{a}_k d\mu(\zeta)a_j.$$

The lemma thus asserts the nonnegativity of Toeplitz matrices formed from the moments of $\mu$. By considering the Poisson means

$$F_r(e^{i\theta}) = \frac{1}{2\pi} \int_\Gamma \frac{1 - r^2}{|\zeta - re^{i\theta}|^2} d\mu(\zeta),$$

we reduce the result to the case $d\mu(e^{i\theta}) = F(e^{i\theta})d\theta$, where $F(e^{i\theta})$ is a bounded weakly measurable nonnegative operator valued function on $\Gamma$. In this case, the result is a standard property of Toeplitz operators [10, § 6.2]. $\quad\square$

PROOF OF THEOREM 3. It is sufficient to give the proof in the case that $\sigma_n = 0$ if $n > r$ for some r. Since then

$$\langle f(t,z) , f(t,z)\rangle_{G_\sigma^\nu} = \sum_{n=1}^{r} (\nu + n)\sigma_n |f_n(t)|^2 =$$

$$= \sum_{n=1}^{r} (\sigma_n - \sigma_{n+1}) \sum_{j=1}^{n} (\nu + j)|f_j(t)|^2 ,$$

it can further be assumed that $\sigma_n = 1$ if $n \leq r$.

The condition that (2) is a solution of (4) says that for all $n = 1, 2, 3, \ldots$,

$$tf_n'(t) = (\nu + n)f_n(t) + (\nu + n - 1)\phi_1(t)f_{n-1}(t) +$$

$$+ \ldots + (\nu + 1)\phi_{n-1}(t)f_1(t) + \phi_n(t)c(t)$$

a.e., where $\phi(t,z) = 1 + \phi_1(t)z + \phi_2(t)z^2 + \ldots$ . By the Herglotz representation, for every positive t there is a nonnegative operator valued measure $\mu_t$ on the unit circle $\Gamma$ such that

$$\phi(t,z) = \int_\Gamma \frac{1 + z\zeta}{1 - z\zeta} d\mu_t(\zeta)$$

in the unit disk. Define a sequence of polynomials in $\zeta$ whose coefficients are vector valued functions of t by $s_o(t,\zeta) = c(t)$,

$$s_n(t,\zeta) = (\nu + n)f_n(t) + (\nu + n - 1)f_{n-1}(t)\zeta + \ldots + (\nu + 1)f_1(t)\zeta^{n-1} + c(t)\zeta^n,$$

$n = 1, 2, 3, \ldots$ . Then

$$tf_n'(t) = \int_\Gamma d\mu_t(\zeta)[s_n(t,\zeta) + \zeta s_{n-1}(t,\zeta)]$$

$$(\nu + n)f_n(t) = s_n(t,\zeta) - \zeta s_{n-1}(t,\zeta)$$

a.e. for all $n = 1, 2, \ldots$, and

$$t\frac{d}{dt}\langle f(t,z) , f(t,z)\rangle_{G_\sigma^\nu} =$$

$$= \sum_{n=1}^{r} (\nu + n)\overline{f}_n(t) t f'_n(t) + \sum_{n=1}^{r} t\overline{f'}_n(t)(\nu + n)f_n(t) =$$

$$= \sum_{n=1}^{r} \int_{\Gamma} [s_n(t,\zeta) - \zeta s_{n-1}(t,\zeta)]^{-} d\mu_t(\zeta) [s_n(t,\zeta) + \zeta s_{n-1}(t,\zeta)] +$$

$$+ \sum_{n=1}^{r} \int_{\Gamma} [s_n(t,\zeta) + \zeta s_{n-1}(t,\zeta)]^{-} d\mu_t(\zeta) [s_n(t,\zeta) - \zeta s_{n-1}(t,\zeta)] =$$

$$= \sum_{n=1}^{r} 2 \int_{\Gamma} \overline{s}_n(t,\zeta) d\mu_t(\zeta) s_n(t,\zeta) - \sum_{n=1}^{r} 2 \int_{\Gamma} \overline{s}_{n-1}(t,\zeta) d\mu_t(\zeta) s_{n-1}(t,\zeta) =$$

$$= 2 \int_{\Gamma} \overline{s}_r(t,\zeta) d\mu_t(\zeta) s_r(t,\zeta) - 2 |c(t)|^2$$

a.e. for positive t. The result follows, since by the lemma

$$\langle f(t,z), f(t,z) \rangle_{G_\sigma^\nu} - 2 \int_{t}^{b} |c(s)|^2 s^{-1} ds$$

is a nondecreasing function of positive t.                                              □

More can be said when $c(t) = 0$ identically in Theorem 3, based on the theory of contractive transformations in Krein spaces. A continuous transformation T of a Krein space $H$ into itself is *contractive* if $\langle Tf, Tf \rangle_H \leq \langle f, f \rangle_H$ for all f in $H$, and *bicontractive* if both T and its adjoint $T^*$ are contractive. Bicontractive transformations were introduced and studied in Potapov [8] and Ginsburg [4]. Kreĭn and Smul'jan [6] give a number of characterizations, including the following. Let $H = H_+ \oplus H_-$, with $H_+$ a Hilbert space and $H_-$ the anti-space of a Hilbert space. A continuous and contractive transformation T of $H$ into itself is bicontractive if and only if the restriction of $P_- T$ to $H_-$ maps $H_-$ onto itself. In this case, the restriction of $P_- T$ to $H_-$ is an invertible transformation of $H_-$ onto itself.

**THEOREM 4.** *Let* $\phi(t,z)$, $0 < t < \infty$, *be a Herglotz family, and let* $G_\sigma^\nu$ *be any Grunsky space with* $\sigma_1 \geq \sigma_2 \geq \ldots \geq 0$. *Then there exist continuous and bicontractive transformations* $T(a,b)$, $0 < a \leq b < \infty$, *of* $G_\sigma^\nu$ *into itself such that*

$$T(a,b) : f(b,z) \rightarrow f(a,z)$$

*for any solution* (2) *of Loewner's equation* (1) *such that* $f(b,z)$ *belongs to* $G_\sigma^\nu$.

PROOF. Fix positive numbers a and b, $a \leq b$, and let h(z) be any element of $G_\sigma^\nu$. By Theorem 1, (1) has a unique solution (2) such that $f(b,z) = h(z)$. By Theorem 3,

$k(z) = f(a,z)$ belongs to $G_\sigma^\nu$ and

$$\langle k(z), k(z)\rangle_{G_\sigma^\nu} \leq \langle h(z), h(z)\rangle_{G_\sigma^\nu}.$$

By Theorem 2, the coset in $G_\sigma^\nu$ determined by $k(z)$ depends only on the coset determined by $h(z)$. Hence we have a well-defined linear transformation $T(a,b) : h(z) \to k(z)$ of $G_\sigma^\nu$ into itself. The closed graph theorem can be used to show that $T(a,b)$ is continuous, and by construction, $T(a,b)$ is contractive. It remains to show that the transformation is bicontractive.

Let $H = G_\sigma^\nu$, and consider the decomposition $H = H_+ \oplus H_-$ obtained with $H_-$ chosen as the elements of $G_\sigma^\nu$ of the form $f_-(z) = \sum_{\nu+n<0} a_n z^{\nu+n}$ and $H_+$ the elements $f_+(z) = \sum_{\nu+n\geq 0} a_n z^{\nu+n}$. Let

$$P_- : \sum_{n=1}^{\infty} a_n z^{\nu+n} \longrightarrow \sum_{\nu+n<0} a_n z^{\nu+n}$$

be the projection of $H$ on $H_-$. It is sufficient to show that the restriction of $P_- T(a,b)$ to $H_-$ is an invertible transformation of $H_-$ onto itself. By the proof of Theorem 1, there are operators $M_{nj}(b,a)$, $1 \leq j < n < \infty$, such that

$$T(a,b) : \sum_{n=1}^{\infty} a_n z^{\nu+n} \longrightarrow \sum_{n=1}^{\infty} b_n z^{\nu+n}$$

if and only if

$$b_1 = (a/b)^{\nu+1} a_1$$

$$b_2 = M_{21}(b,a)a_1 + (a/b)^{\nu+2} a_2$$

$$b_3 = M_{31}(b,a)a_1 + M_{32}(b,a)a_2 + (a/b)^{\nu+3} a_3$$

$$\cdots$$

The restriction of $P_- T(a,b)$ to $H_-$ thus has block triangular form with invertible diagonal elements. It is therefore invertible on $H_-$, and the conclusion follows. $\qquad\square$

In the scalar case, $T(a,b)$ is substitution by a normalized Riemann mapping function which maps the unit disk into itself. There is a rich theory in this situation. It is treated in [11] and will be taken up in future work.

**REFERENCES**

1.    **Bognár, J.** : *Indefinite inner product spaces*, Springer-Verlag, New York, 1974.

2.    **de Branges, L.** : A proof of the Bieberbach conjecture, *Acta Math.* **154**(1985), 137-152.

3.    **de Brange, L.** : Powers of Riemann mapping functions, in *The Bieberbach Conjecture* (Proceedings of the Symposium on the Occasion of its Proof, Purdue University, 1985) eds. A. Baernstein II, D. Drasin, P. Duren, and A. Marden, pp. 51-67, Amer. Math. Soc., Providence, R.I., 1986.

4.    **Ginsburg, Ju.P.** : *J-nonexpansive operators on Hilbert space*, Dissertation, Odessa, 1958.

5.    **Hille, E. ; Phillips, R.S.** : *Functional analysis and semi-groups*, Amer. Math. Soc. Coll. Publ., vol. **31**, Providence, R.I., 1957.

6.    **Kreĭn, M.G. ; Smul'jan, Ju.L.** : Plus-operators in a space with indefinite metric, *Mat. Issled.* 1(1966), no. 1, 131-161; *Amer. Math. Soc. Translations (2)* **85**(1969), 93-113.

7.    **Löwner, K.** : Untersuchungen über schlichte konforme Abbildungen des Einheitskreises. I, *Math. Ann.* **89**(1923), 103-121.

8.    **Potapov, V.P.** : The multiplicative structure of J-contractive matrix functions, *Trudy Moscov. Mat. Obšč.* 4(1955), 125-236; *Amer. Math. Soc. Translations (2)* **15**(1960), 131-243.

9.    **Rogosinski, W.** : On the coefficients of subordinate functions, *Proc. London Math. Soc. (2)* **48**(1943), 48-82.

10.   **Rosenblum, M. ; Rovnyak, J.** : *Hardy classes and operator theory*, Oxford University Press, New York, 1985.

11.   **Rovnyak, J.** : Coefficient estimates for Riemann mapping functions, *Journal d'Analyse Mathématique* **52**(1989), 53-93.

**James Rovnyak**
Department of Mathematics
Mathematics-Astronomy Building
University of Virginia
Charlottesville, Virginia 22903
U.S.A.

Operator Theory:
Advances and Applications, Vol. 43
© 1990 Birkhäuser Verlag Basel



# A RKHS OF ENTIRE FUNCTIONS AND ITS MULTIPLICATION OPERATOR. AN EXPLICIT EXAMPLE

Franciszek Hugon Szafraniec

The theory of unbounded subnormal operators differs in many aspects from its bounded counterpart (see [2], [3] and [4] for systematic studies of the subject). In this report we wish to make transparent one of these aspects giving an explicit example of a subnormal operator in a Reproducing Kernel Hilbert Space of entire functions, which comes from some indeterminate Stieltjes moment problem [1]. This leads to an analytic subnormal operator having normal extensions in two, of quite different nature, $L^2$ spaces; one of the spaces involves a measure absolutely continuous with respect to the planar Lebesgue one, while the other is concentrated on a countable number of circles around the origin, whose radii go both to zero and to infinity. In other words, relating them to the bounded case, the first of these spaces looks like the Bergman one, the other reminds the Hardy space.

**1. Basic facts.** Let $\{a_n\}_{n=0}^{\infty}$ be a sequence of positive numbers. Define

$$(1) \qquad K(z,w) = \sum_{n=0}^{\infty} a_{2n}^{-1} \bar{z}^n w^n, \qquad z, w \in \mathbf{C}.$$

Then K is a positive definite kernel and the corresponding RKHSpace $H_K$ is composed of functions analytic in the disc $\Delta$ of convergence of the power series (1) considered in the single variable $\bar{z}w$. The Riesz representation theorem and the reproducing kernel property imply (for the detailed proof, though in more general circumstances, see [5]) the following

**RKHS TEST.** *A function* $f : \Delta \rightarrow \mathbf{C}$ *belongs to* $H_K$ *if and only if for any* $\zeta_1, \ldots, \zeta_n \in \mathbf{C}$ *and* $z_1, \ldots, z_n \in \Delta$,

$$\left| \sum_i \zeta_i f(z_i) \right|^2 \leq C \sum_{i,j} \zeta_i \bar{\zeta}_j K(z_i, z_j)$$

*with some* $C = C(f) \geq 0.$

This test gives us at once, by (1), that

1° *all monomials and, consequently, all polynomials in z are numbers of* $H_K$.

Given a measure m on $[0,+\infty)$, define a measure $\mu$ on **C** as

(2)
$$\mu(\sigma) = (2\pi)^{-1} \int_0^{2\pi} d\phi \int_0^{+\infty} \chi_\sigma(re^{i\phi}) m(dr), \quad \sigma \text{ a Borel set on } \mathbf{C},$$

where $\chi_\sigma$ stands for the characteristic function of $\sigma$.

Suppose $\{a_n\}_{n=0}^{\infty}$ is a Stieltjes moment sequence, that is

$$a_n = \int_0^{+\infty} r^n m(dr), \quad n = 0,1,\ldots .$$

Then

2° $H_K$ *is (isometricaly imbedded as) a closed subspace of* $L^2(\mu)$.

Assuming, moreover, $\inf\{a ; m([a,+\infty)) = 0\} = +\infty$ we get that

3° *radius of* $\Delta$ *is equal to* $+\infty$.

Set $g(z) = zf(z)$ and define the operator M of multiplication by the independent variable z in $H_K$ by $D(M) = \mathbf{C}[z]$ (cf.1°) and $Mf = g$, and the maximal multiplication operator $M_{\mathbf{max}}$ by $D(M_{\mathbf{max}}) = \{f \in H_K ; g \in H_K\}$ and $M_{\mathbf{max}}f = g$. Then

4° *M is closable and* $M^- = M_{\mathbf{max}}$.

Define also the multiplication operator N in $L^2(\mu)$ by

$$D(N) = \{f \in L^2(\mu) ; g \in L^2(\mu) \text{ and } Nf = g\}.$$

Then, by 3°,

5° *N is a normal extension of M (as well as of* $M_{\mathbf{max}}$*).*

Moreover

6° *N is minimal in the sense that*

(3)
$$L^2(\mu) = \mathbf{clolin}\{E(\sigma)f ; f \in H_K, \sigma \text{ a Borel subset of } \mathbf{C}\}$$

*where E is the spectral measure of* N.

(Minimality of 6° is equivalent to the fact that the only closed subspaces of $L^2(\mu)$ which reduces N and contains $H_K$ is $L^2(\mu)$ itself).

**2. The example.** Introduce the notation: for $0 < q < 1$ define

$$(a;q)_\infty = \prod_{k=0}^{\infty} (1 - aq^k),$$

$$(a;q)_0 = 1, \quad (a;q)_n = (1 - a)(1 - aq)\cdots(1 - aq^{n-1}), \quad n = 1, 2, \ldots .$$

Set

$$a'_n = (q^\alpha;q)_n q^{-n\alpha-n(n-1)/2}(1 - q)^{-n}, \quad n = 0, 1, \ldots .$$

According to (1) and (2) we wish to replace the sequence $\{a'_n\}$, which in fact is a Stieltjes moment sequence, cf. [1], by another Stieltjes moment sequence $\{a_n\}$ such that $a'_n = a_{2n}$ for $n = 0, 1, \ldots$ . Due to [1], we can represent $\{a_n\}$ as

$$a_n = \int_0^{+\infty} r^n m_i(dr), \quad i = 1, 2.$$

where $m_1$ and $m_2$ are two different measures. Define

$$a = \left\{ \int_0^{+\infty} t^{\alpha-1}[(-(1 - q)t;q)_\infty]^{-1}dt \right\}^{-1}$$

and

$$b = \left\{ \sum_{k=-\infty}^{\infty} q^{\alpha k}(-(1 - q)q^k;q)_\infty^{-1} \right\}^{-1},$$

$\alpha$ being real. After simple change of the variable in (4.4) of [1] we get

$$m_1(dr) = a(-(1 - q)r^2;q)_\infty^{-1} r^{2\alpha-1} dr,$$

while, on the other hand, the formula (4.6) of [1] leads us to a measure $m_2$ which is constant on the intervals $(q^{\frac{1}{2}(k+1)}, q^{\frac{1}{2}k})$ for $k = 1, 2, \ldots$ and $(q^{\frac{1}{2}k}, q^{\frac{1}{2}(k+1)})$ for $k = \ldots$ $\ldots, -2, -1, 0$, and has jumps of size

$$b(-(1 - q)r^2;q)_\infty^{-1} r^{2\alpha}$$

at $r = q^{\frac{1}{2}k}$, $k = 0, \pm 1, \pm 2, \ldots$ .

Thus we get two multiplication operators $N_1$ and $N_2$ in $L^2(\mu_1)$ and $L^2(\mu_2)$ respectively, where $\mu_i$ corresponds to $m_i$ via (2) (the explicit expression of $\mu_i$ can be easily get from those of $m_i$). The operators $N_1$ and $N_2$ are not $H_K$-equivalent which means there is no unitary operator $U : L^2(\mu_1) \rightarrow L^2(\mu_2)$ such that $U$ is the identity on $H_K$ and $UN_1 = N_2U$. As we have said before (cf. 6°), they are minimal in the sense of (3). However, none of them satisfies the condition

$$(4) \qquad N_i = (N_i \mid D_i)^-, \quad \text{where } D_i = \text{lin}\{N_i^{*n}f : f \in D(M), \ n \geq 0\}$$

(even more, there is no normal extension of M which would satisfy (4), cf. [4]).

Conditions (3) and (4) coincide in bounded case and both express minimality of a normal extension.

**3. The limit case** $q \to 1^-$. Since $\{a_{2n}^{-\frac{1}{2}} z^n\}_{n=0}^{\infty}$ is a basis for $H_K$, the linear mapping $U : \ell_+^2 \to H_K$ which maps the n-th basic vector $e_n$ into $a_{2n}^{-\frac{1}{2}} z^n$ becomes a unitary operator and $S = U^* M U$ is a weighted shift with weights $\{a_{2(n+1)}/(a_{2n})^{\frac{1}{2}}\}_{n=0}^{\infty}$ and domain $D(S) = \mathrm{lin}\{e_n\}_{n=0}^{\infty}$. It is a matter of direct calculation that, for $\alpha = 1$,

$$S e_n \to (n + 1)^{\frac{1}{2}} e_{n+1} \quad \text{as } q \to 1^-.$$

This means that $S \to S_0$ on $D(S)$, where, under suitable unitary isomorphism, $S_0$ is nothing else but the famous creation operator

$$2^{-\frac{1}{2}}(x - (d/dx))$$

which is a weighted shift with weights $(n + 1)^{\frac{1}{2}}$, when considered with respect to the Hermite functions normalized in the appropriate way (cf. [2 and 3]). The creation operator enjoyes the uniqueness extension property, cf. [4], while our operator apparently does not.

**4. The main reference.** The operator theoretical environment of this example can be found in [4].

## REFERENCES


1.  **Askey, R.:** Ramanujan's extension of the gamma and beta functions, *Amer. Math. Monthly*, **87**(1980), 346-359.

2.  **Stochel, J.; Szafraniec, F.H.:** On normal extensions of unbounded operators. I, *J. Operator Theory*, **14**(1985), 31-55.

3.  **Stochel, J.; Szafraniec, F.H.:** On normal extensions of unbounded operators. II, Institute of Mathematics, Polish Academy of Sciences, preprint No. **349**, November 1985; to appear in *Acta Sci. Math. (Szeged)* **53**(1989).

4.  **Stochel, J.; Szafraniec, F.H.:** On normal extensions of unbounded operators. III. Spectral properties, *Publ. RIMS, Kyoto Univ.* **25**(1989).

5.  **Szafraniec, F.H.:** Intepolation and domination by positive definite kernels, in *Complex Analysis : Fifth Romanian - Finnish Seminar. Part 2*, Proc. Bucharest 1981, *Lecture Notes Math.* **1014**, pp. 291-295, Springer, Berlin-Heidelberg-New York, 1983.



**F.H. Szafraniec**

Instytut Matematyki,
Uniwersytet Jagielloński,
ul. Reymonta 4, PL-30059 Kraków
Poland.

Operator Theory:
Advances and Applications, Vol. 43
© 1990 Birkhäuser Verlag Basel



# JOINT SPECTRAL PROPERTIES FOR PAIRS OF PERMUTABLE SELFADJOINT TRANSFORMATIONS

F.-H. Vasilescu

## 1. INTRODUCTION

Let $X$ be a complex Banach space and let $T_j : D(T_j) \subset X \to X$ $(j = 1, 2)$ be linear transformations in $X$. Then the composite operator $T_1 T_2$ is defined on the linear space

$$D(T_1 T_2) = \{x \in D(T_2) \; ; \; T_2 x \in D(T_1)\}$$

in an obvious manner and we have, in general, $T_1 T_2 \neq T_2 T_1$ on their joint domain of definition.

In this note we shall say that $T_1$, $T_2$ are *permutable* (or *permute*) if

(1.1) $$T_1 T_2 x = T_2 T_1 x, \quad x \in D(T_1 T_2) \cap D(T_2 T_1).$$

Let us observe that if $T_1, T_2$ permute, then for every pair $(z_1, z_2) \in \mathbf{C}^2$ the transformations $z_1 - T_1$, $z_2 - T_2$ also permute and

$$D((z_1 - T_1)(z_2 - T_2)) \cap D((z_2 - T_2)(z_1 - T_1)) = D(T_1 T_2) \cap D(T_2 T_1),$$

which shows that this concept has a certain invariance under translations.

Relation (1.1) is discredited by the fact that it does not generally insure equalities of the form $f_1(T_1) f_2(T_2) = f_2(T_2) f_1(T_1)$, where $f_1, f_2$ are scalar functions for which the expressions $f_1(T_1), f_2(T_2)$ make sense. This happens, in particular, to some pairs of selfadjoint operators which permute but whose spectral measures do not commute (see for instance [7]).

We think, and we shall try to prove it in the following, that it is not relation (1.1) to be blamed for such unpleasant phenomena but, rather, the joint spectral properties of the pair $(T_1, T_2)$. The discussion from this work provides a motivation to introduce a joint spectrum for pairs of permutable linear transformations, by extending (a particular case of) the definition of J.L. Taylor [6]. We focus our attention only on the case of pairs of linear transformations because we are particularly interested in pairs of permutable selfadjoint operators. The case of several permutable transformations will be treated in future work.

For a linear transformation $T : D(T) \subset X \to X$ we denote by $\rho(T)$ the *resolvent set* of $T$, that is, the set of those points $z \in \mathbf{C}$ such that $z - T : D(T) \to X$ is bijective. The *spectrum* of $T$ is the set $\sigma(T) := \mathbf{C} \setminus \rho(T)$. Finally, $N(T)$ and $R(T)$ are the *null-space* and the *range* of $T$, respectively.

## 2. PERMUTABLE SELFADJOINT TRANSFORMATIONS

It has been known for a long time that two permutable selfadjoint transformations do not necessarily have commuting spectral measures. A first example in this sense is due to E. Nelson [4]. Such phenomena have lately been systematically studied by K. Schmüdgen in a series of papers starting with [7].

In this section we shall present an approach to this problem from the point of view of the two-dimensional spectral theory. Nevertheless, most of the proofs are based on standard techniques of the theory of selfadjoint operators ([1], [5] etc.).

Let H be a Hilbert space with the scalar product $\langle \cdot , \cdot \rangle$ and the norm $\| \cdot \|$. If T is a closed operator acting in H, then its domain of definition $D(T)$ can be given a Hilbert space structure by means of the scalar product

$$(2.1) \qquad \langle x,y \rangle_T = \langle x,y \rangle + \langle Tx,Ty \rangle, \qquad x, y \in D(T).$$

The norm induced by (2.1), namely

$$(2.2) \qquad \| x \|_T = ( \| x \|^2 + \| Tx \|^2)^{\frac{1}{2}}, \qquad x \in D(T),$$

is usually called the *graph-norm* of $D(T)$.

**2.1. LEMMA.** *Let* $A_1$, $A_2$ *be permutable selfadjoint operators in H. If* $D(A_2)$ *is endowed with the graph-norm, then the linear map*

$$(2.3) \qquad A_1 : D(A_1 A_2) \cap D(A_2 A_1) \to D(A_2)$$

*is closed and symmetric in* $D(A_2)$.

PROOF. Let $\langle \cdot , \cdot \rangle_2$ be the scalar product (2.1) with T replaced by $A_2$. If $x, y \in D(A_1 A_2) \cap D(A_2 A_1)$, then

$$\langle A_1 x,y \rangle_2 = \langle A_1 x,y \rangle + \langle A_2 A_1 x, A_2 y \rangle =$$

$$= \langle x, A_1 y \rangle + \langle A_2 x, A_2 A_1 y \rangle = \langle x, A_1 y \rangle_2,$$

since $A_1$, $A_2$ permute. This shows that $A_1$ is symmetric (not necessarily densely defined) in $D(A_2)$.

Let us prove that $A_1$ is also closed. Indeed, if $\{x_k\}_{k\geq1} \subset D(A_1A_2) \cap D(A_2A_1)$ is a sequence such that $x_k \to x$ and $A_1x_k \to y$ ($k \to \infty$) in $D(\bar{A}_2)$, then $x_k \to x$, $A_2x_k \to A_2x$, $A_1x_k \to y$ and $A_2A_1x_k \to A_2y$ ($k \to \infty$) in H. This shows that $x \in D(A_1)$, $y = A_1x$, $A_2x \in D(A_1)$ and $A_1A_2x = A_2A_1x$, that is, $A_1$ is closed in $D(A_2)$.

**2.2. LEMMA.** *Let $A_1$, $A_2$ be a selfadjoint operators in H that permute and let $A_{12}$ denote the operator (2.3). If $\rho(A_{12}) \cap (\mathbf{C} \setminus \mathbf{R}) \neq \emptyset$, then the spectral measures of $A_1$ and $A_2$ commute.*

PROOF. Let $w \in \rho(A_{12}) \cap (\mathbf{C} \setminus \mathbf{R})$. Hence $w - A_{12}$ is bijective. It is clear that

$$(w - A_1)^{-1} \,|\, D(A_2) = (w - A_{12})^{-1}.$$

Therefore, if $x \in D(A_2)$, then

$$(w - A_1)^{-1}A_2x = (w - A_1)^{-1}A_2(w - A_{12})(w - A_{12})^{-1}x =$$

$$= A_2(w - A_{12})^{-1}x = A_2(w - A_1)^{-1}x.$$

From this fact we easily infer the equality

$$(w - A_1)^{-1}(w - A_2)^{-1}x = (w - A_2)^{-1}(w - A_1)^{-1}x, \quad x \in H,$$

which is equivalent to the commutation of the spectral measures of $A_1$ and $A_2$ (see for instance [5, Theorem VIII.1.3]).

**LEMMA 2.3.** *Let $A_1$, $A_2$ and $A_{12}$ be as in the previous lemma. The operator $A_{12}$ is selfadjoint if and only if the spectral measures of $A_1$ and $A_2$ commute.*

PROOF. If $A_{12}$ is selfadjoint, then $\sigma(A_{12}) \subset \mathbf{R}$ and the assertion follows from the preceding lemma.

Conversely, suppose that the spectral measures of $A_1$ and $A_2$ commute. First of all we show that $D(A_1A_2) \cap D(A_2A_1)$ is a dense subspace in $D(A_2)$ (endowed with the graph-norm). Notice that if $z \in \mathbf{C} \setminus \mathbf{R}$, then

$$(2.4) \qquad\qquad (z - A_1)^{-1}D(A_2) \subset D(A_1A_2) \cap D(A_2A_1),$$

which follows straightforward from the equality

$$(2.5) \qquad (z - A_1)^{-1}A_2 x = A_2(z - A_1)^{-1}x, \qquad x \in D(A_2).$$

Relation (2.5) is a consequence of the commutation of the spectral measures of $A_1$ and $A_2$. Let us sketch its proof.

Let $E_j$ be the spectral measure of $A_j$ $(j = 1, 2)$. If $f_1, f_2$ are bounded Borel functions on $\mathbf{R}$, then $f_1(A_1)f_2(A_2) = f_2(A_2)f_1(A_1)$, where $f_j(A_j)$ is the integral of $f_j$ with respect to $E_j$. In particular,

$$(z - A_1)^{-1}A_2 x = (z - A_1)^{-1}\left(\lim_{n\to\infty} \int_{-n}^{n} t\, dE_2(t)x\right) =$$

$$= \lim_{n\to\infty} \int_{-n}^{n} t\, dE_2(t)(z - A_1)^{-1}x = A_2(z - A_1)^{-1}x$$

for all $x \in D(A_2)$.

By (2.4), it will suffice to prove that the space $(z - A_1)^{-1}D(A_2)$ is dense in $D(A_2)$. Let $y \in D(A_2)$ be such that $\langle y,(z - A_1)^{-1}v\rangle_2 = 0$ for all $v \in D(A_2)$ (the scalar product on $D(A_2)$ is denoted as in Lemma 2.1). This means that

$$\langle y,(z - A_1)^{-1}v\rangle + \langle A_2 y, A_2(z - A_1)^{-1}v\rangle = 0.$$

In other words,

$$\langle(\bar{z} - A_1)^{-1}y,v\rangle + \langle A_2(\bar{z} - A_1)^{-1}y, A_2 v\rangle = 0,$$

that is $A_2(\bar{z} - A_1)^{-1}y \in D(A_2)$ and

$$A_2^2(\bar{z} - A_1)^{-1}y + (\bar{z} - A_1)^{-1}y = 0.$$

Since $1 + A_2^2$ is injective, we deduce that $(\bar{z} - A_1)^{-1}y = 0$, and so $y = 0$.

We have proved so far that $A_{12}$ is closed, symmetric (Lemma 2.1) and densely defined. Let us show that $A_{12}$ is actually selfadjoint. Let $x \in D(A_{12}^*)$, let $y = (\bar{z} - A_{12}^*)x$ and let $x_1 = (\bar{z} - A_{12})^{-1}y$ (the existence of $(\bar{z} - A_{12})^{-1}$ follows from (2.4)). Then $(\bar{z} - A_{12}^*)(x_1 - x) = 0$. But $N(\bar{z} - A_{12}^*) = R(z - A_{12})^{\perp} = \{0\}$. Therefore $x = x_1 \in D(A_{12})$. Since obviously $D(A_{12}) \subset D(A_{12}^*)$, we must have equality, that is, $A_{12}^* = A_{12}$.

Before proceeding further, we define a concept of joint spectrum for pairs of permutable linear transformations in arbitrary Banach spaces (see [8], [3] for other definitions).

**DEFINITION 2.4.** Let $X$ be a Banach space and let $T = (T_1, T_2)$ be a pair of permutable linear transformations. Let also $z = (z_1, z_2) \in \mathbf{C}^2$. Consider the sequence

$$(2.6) \qquad 0 \to D(T_1 T_2) \cap D(T_2 T_1) \xrightarrow{\;\delta^0(z-T)\;} D(T_2) \oplus D(T_1) \xrightarrow{\;\delta^1(z-T)\;} X \to 0,$$

where

$$\delta^0(z - T)x = (z_1 - T_1)x \oplus (z_2 - T_2)x, \qquad x \in D(T_1 T_2) \oplus D(T_2 T_1)$$

and

$$\delta^1(z - T)x_2 \oplus x_1 = (z_1 - T_1)x_1 - (z_2 - T_2)x_2, \qquad x_2 \oplus x_1 \in D(T_2) \oplus D(T_1).$$

Since $T_1$, $T_2$ permute, (2.6) is a complex of vector spaces.

We say that the pair $z - T = (z_1 - T_1, z_2 - T_2)$ is *nonsingular (singular)* if the complex (2.6) is exact (not exact). Let $\sigma(T)$ be the set of those points $z \in \mathbf{C}^2$ such that $z - T$ is singular. The set $\sigma(T)$ will be called the *joint spectrum* of the pair $T$.

Definition 2.4 is inspired from the corresponding definition for bounded linear operators in [6] (for $n = 2$).

The revelance of the joint spectrum is illustrated by the following result.

**THEOREM 2.5.** *Let* $A = (A_1, A_2)$ *be a pair of selfadjoint operators in the Hilbert space H. The following conditions are equivalent:*

*(1) the spectral measures of* $A_1$ *and* $A_2$ *commute;*

*(2)* $A_1$, $A_2$ *permute and* $\sigma(A) \subset \mathbf{R}^2$.

PROOF. We first prove that $(2) \Longrightarrow (1)$. Let $z = (z_1, z_2) \in (\mathbf{C} \setminus \mathbf{R})^2$, let $x \in H$ and let $x_j = (z_j - A_j)^{-1}x$ $(j = 1, 2)$. Since

$$\delta^1(z - A)(x_2 \oplus x_1) = (z_1 - A_1)x_1 - (z_2 - A_2)x_2 = 0,$$

it follows from the exactness of (2.6) the existence of a vector $y \in D(A_1 A_2) \cap D(A_2 A_1)$ such that $x_1 = (z_2 - A_2)y$ and $x_2 = (z_1 - A_1)y$. Then

$$y = (z_2 - A_2)^{-1}x_1 = (z_2 - A_2)^{-1}(z_1 - A_1)^{-1}x = (z_1 - A_1)^{-1}x_2 =$$

$$= (z_1 - A_1)^{-1}(z_2 - A_2)^{-1}x.$$

Therefore the operators $(z_1 - A_1)^{-1}$, $(z_2 - A_2)^{-1}$ commute, that is, the spectral measures of $A_1$ and $A_2$ commute ([5, Theorem VIII.1.3]).

We now show that $(1) \Longrightarrow (2)$. It is easily seen that if the spectral measures of $A_1$ and $A_2$ commute, then $A_1$, $A_2$ permute. Let $z = (z_1, z_2) \notin \mathbf{R}^2$. With no loss of generality we may assume that $z_1 \notin \mathbf{R}$. Then it is easily seen that the operator $\delta^0(z - A)$ is injective and the operator $\delta^1(z - A)$ is surjective.

The complex (2.6) is also exact in the middle. Indeed, let $x_2 \oplus x_1 \in D(A_2) \oplus \oplus D(A_1)$ be such that $(z_2 - A_2)x_2 = (z_1 - A_1)x_1$. By Lemma 2.3, the operator $z_1 - A_{12}$ has a bounded inverse. Let

$$x = (z_1 - A_{12})^{-1}x_2 \in D(A_1 A_2) \cap D(A_2 A_1).$$

Obviously, $(z_1 - A_1)x = x_2$. We also have $(z_2 - A_2)x = x_1$. Indeed, by (2.5),

$$(z_2 - A_2)x = (z_2 - A_2)(z_1 - A_{12})^{-1}x_2 = (z_2 - A_2)(z_1 - A_1)^{-1}x_2 =$$

$$= (z_1 - A_1)^{-1}(z_2 - A_2)x_2 = (z_1 - A_1)^{-1}(z_1 - A_1)x_1 = x_1.$$

Hence $x_2 \oplus x_1 = \delta^0(z - T)x$, and this completes the proof of the theorem.

## 3. SOME PROPERTIES OF THE JOINT SPECTRUM

In this section X will be a fixed complex Banach space.

**LEMMA 3.1.** *Let* $T = (T_1, T_2)$ *be a pair of permutable closed operators in* X. *Then the joint spectrum* $\sigma(T)$ *is a closed subset of* $\mathbf{C}^2$.

PROOF. If $D(T_j)$ is the domain of definition of $T_j$, as in the previous section we have a graph-norm on $D(T_j)$ given by

$$\|x\|_j = (\|x\|^2 + \|T_j x\|^2)^{\frac{1}{2}}, \qquad x \in D(T_j),$$

where $\|\cdot\|$ is the norm of X, which defines a Banach space structure on $D(T_j)$ $(j = 1, 2)$.

Then we may define on $D(T_2) \oplus D(T_1)$ the norm

$$\|x_2 \oplus x_1\|_{21} = (\|x_2\|_2^2 + \|x_1\|_1^2)^{\frac{1}{2}}, \qquad x_2 \in D(T_2),\ x_1 \in D(T_1).$$

We also have a norm on $D(T_1 T_2) \cap D(T_2 T_1)$ given by

$$\|x\|_{12} = (\|x\|_1^2 + \|T_2 x\|_1^2)^{\frac{1}{2}}, \qquad x \in D(T_1 T_2) \cap D(T_2 T_1).$$

Endowed with the above norms, both $D(T_2) \oplus D(T_1)$ and $D(T_1 T_2) \cap D(T_2 T_1)$ become Banach spaces.

Let $D_{12} := D(T_1 T_2) \cap D(T_2 T_1)$ and let $z = (z_1, z_2) \in \mathbf{C}^2$. Then the operator

$\delta^0(z - T) : D_{12} \to D(T_2) \oplus D(T_1)$ from (2.6) is continuous. Indeed, a simple calculation shows that

$$\| \delta^0(z - T)x \|_{21}^2 \le 2(2 + \| z \|^2) \| x \|_{12}^2, \qquad x \in D_{12},$$

where $\| z \|^2 = | z_1 |^2 + | z_2 |^2$. Analogously, the operator $\delta^1(z - T) : D(T_2) \oplus D(T_1) \to$ $\to X$ is continuous, since

$$\| \delta^1(z - T)(x_2 \oplus x_1) \| \le (2 + \| z \|^2)^{\frac{1}{2}} \| x_2 \oplus x_1 \|_{21}$$

for all $x_2 \oplus x_1 \in D(T_2) \oplus D(T_1)$.

Consequently (2.6) is a complex of Banach spaces and continuous maps. If $w = (w_1, w_2) \in \mathbf{C}^2$, we also have

$$\| (\delta^0(z - T) - \delta^0(w - T))x \|_{21}^2 = \| \delta^0(z - w)x \|_{21}^2 =$$

$$= \| (z_1 - w_1)x \|_2^2 + \| (z_2 - w_2)x \|_1^2 \le \| z - w \|^2 \| x \|_{12}^2,$$

that is,

$$\| \delta^0(z - T) - \delta^0(w - T) \| \le \| z - w \|.$$

Similarly,

$$\| (\delta^1(z - T) - \delta^1(w - T)(x_2 \oplus x_1) \| = \| \delta^1(z - w)(x_2 \oplus x_1) \| =$$

$$= \| (z_1 - w_1)x_1 - (z_2 - w_2)x_2 \| \le \| z - w \| \, \| x_2 \oplus x_1 \|_{21},$$

and so

$$\| \delta^1(z - T) - \delta^1(w - T) \| \le \| z - w \|.$$

Therefore, if the complex (2.6) is exact for a certain $z \in \mathbf{C}$, by the stability of the exactness under small perturbations (see [6, Theorem 2.1]), the complex (2.6) should also be exact when $z$ is replaced by $w$, provided $\| z - w \|$ is sufficiently small.

**REMARK 3.2.** Let $A = (A_1, A_2)$ be a pair of permutable selfadjoint operators. Then we have the following dichotomy: either $\sigma(A) = \mathbf{C}^2$ or $\sigma(A) \subset \mathbf{R}^2$. Indeed, if $\sigma(A) \ne \mathbf{C}^2$, since $\mathbf{C}^2 \setminus \sigma(A)$ is an open set (by Lemma 3.1), we can find points $z = (z_1, z_2) \in (\mathbf{C} \setminus \mathbf{R})^2$ such that $z \notin \sigma(A)$. Then we proceed as in the first part of the proof of Theorem 2.5 and deduce that $(z_1 - A_1)^{-1}$ and $(z_2 - A_2)^{-1}$ commute, which implies the commutation of the spectral measures of $A_1$ and $A_2$. Then, by Theorem 2.5, we must have $\sigma(A) \subset \mathbf{R}^2$.

For pairs $A = (A_1, A_2)$ consisting of permutable selfadjoint operators such that

$\sigma(A) = \mathbf{C}^2$ we do not expect to have, in general, $f_1(A_1)f_2(A_2) = f_2(A_2)f_1(A_1)$ even for holomorphic functions $f_j$ defined in neighbourhoods of $\sigma(A_j)$ and analytic at infinity ($j = 1, 2$). When $\sigma(A) \subset \mathbf{R}^2$, such equalities follow from the existence of the functional calculus with bounded Borel functions.

**LEMMA 3.3.** *Let* $T = (T_1, T_2)$ *be a pair of permutable closed operators in* X, *let* $D_{12} = D(T_1 T_2) \cap D(T_2 T_1)$ *and let* $T_{12} = T_1 | D_{12}$. *Then the operator* $T_{12}$ *is closed in* $D(T_2)$ *(endowed with the graph-norm) and we have the inclusion* $\mathbf{pr}_1(\sigma(T)) \subset \sigma(T_1) \cup \sigma(T_{12})$, *where* $\mathbf{pr}_1 : \mathbf{C}^2 \longrightarrow \mathbf{C}$ *is the projection on the first coordinate.*

PROOF. That $T_{12}$ is closed follows in the proof of Lemma 2.1. Let us obtain the desired inclusion. Note that the diagram

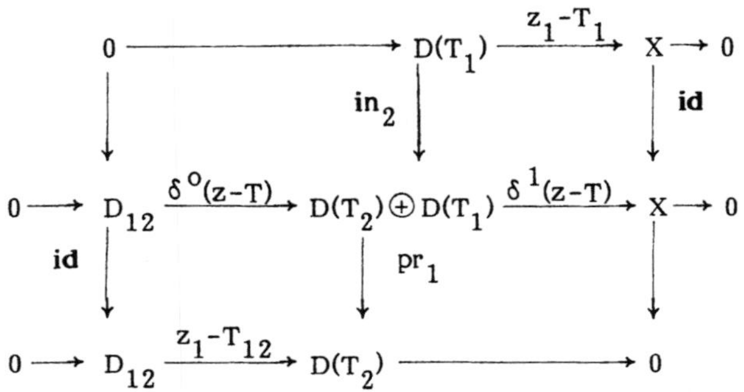

is commutative, where **id** is the identity, $\mathbf{in}_2 : D(T_1) \longrightarrow D(T_2) \oplus D(T_1)$ is the natural inclusion and $z = (z_1, z_2) \in \mathbf{C}^2$ is arbitrary.

If $z_1 \notin \sigma(T_1) \cup \sigma(T_{12})$, then the complexes from the first and the last row of the above diagram are exact. This forces the complex in the middle to be exact too for every $z_2 \in \mathbf{C}$. Therefore, if $z = (z_1, z_2) \in \sigma(T)$, then $z_1 \in \sigma(T_1) \cup \sigma(T_{12})$.

**REMARKS 3.4.** $1^\circ$ The converse inclusion in Lemma 3.3 is also true. In other words, the projection property for the joint spectrum still holds in this context. One possible way to prove it is to use some elements of spectral theory for morphisms in quotient Banach spaces (see [9], [10] for some details). Such considerations are beyond our scope so that we omit the details.

$2^\circ$ The above results, in particular Lemma 3.3, show that not only the transformations $T_1$ and $T_2$ but also the operator $T_{12}$ (or the operator $A_{12}$ from Lemmas 2.1–2.3) should be taken into consideration for the joint spectral theory of $T = (T_1, T_2)$.

$3°$ One may wonder whether similar considerations are of some significance for other important pairs of linear transformations. An example in this respect is provided by the pair of areolar derivatives $(\overline{\partial}_1, \overline{\partial}_2)$ in an open set $\Omega \subset \mathbf{C}^2$ $((\overline{\partial}_j f)(z) := (\partial f / \partial \overline{z}_j)(z)$, $z \in \Omega$, $j = 1, 2)$. Since $\overline{\partial}_j f_w(z) = w_j f_w(z)$, where $f_w(z) = \exp(w_1 \overline{z}_1 + w_2 \overline{z}_2)$, every point $w = (w_1, w_2) \in \mathbf{C}^2$ is in the "point spectrum" of the pair $(\overline{\partial}_1, \overline{\partial}_2)$. Nevertheless, if $\Omega$ satisfies some conditions, a sequence similar to (2.6) may be constructed for the pair $(\overline{\partial}_1, \overline{\partial}_2)$, and its exactness at the second and the third step can be shown, by using the methods developed in [2, Chapter IV].

## REFERENCES


1.    **Dunford, N. ; Schwartz, J.T.** : *Linear operators. Part II*, Interscience Publishers, New York, London, 1963.

2.    **Hörmander, L.** : *An introduction to complex analysis in several variables*, D. Van Nostrand Company, Princeton, 1966.

3.    **Huang, Danrun ; Zhang, Dianzhan** : Joint spectrum and unbounded operators, *Acta Math. Sinica* 2(1986), 260–269.

4.    **Nelson, E.** : Analytic vectors, *Ann. of Math.* **70**(1959), 572–615.

5.    **Reed, M. ; Simon, B.** : *Methods of modern mathematical physics. Vol. 1*, Academic Press, New York, London, 1972.

6.    **Taylor, J.L.** : A joint spectrum for several commuting operators, *J. Funct. Anal.* 6(1970), 172–191.

7.    **Schmüdgen, K.** : On commuting unbounded self-adjoint operators, *Acta Sci. Math. (Szeged)* **47**(1984), 131–146.

8.    **Vasilescu, F.-H.** : *Analytic functionnal calculus*, Editura Academiei and D.Reidel Publishing Co., Bucharest and Dordrecht, 1982.

9.    **Vasilescu, F.-H.** : Spectral capacities in quotient Fréchet spaces in *Constantin Apostol Memorial Issue*, Birkhäuser Verlag, Basel, 1988, pp. 243–263.

10.    **Waelbroeck, L.** : Quotient Banach spaces, in *Banach Center Publications*, vol.8, Warszaw, 1982, pp.553–562.



**F.-H. Vasilescu**

Department of Mathematics, INCREST
Bdul Păcii 220, 79622 Bucharest
Romania.

Operator Theory:
Advances and Applications, Vol. 43
© 1990 Birkhäuser Verlag Basel



# HOLOMORPHIC FUNCTIONS TAKING THEIR VALUES
# IN A QUOTIENT BORNOLOGICAL SPACE

**Lucien Waelbroeck**

I will discuss some aspects of the use in Operator Theory of homomorphic functions taking their values in quotient spaces.

I must add that in this paper, a topological vector space will be called $E = (E, I_E)$ where $E$ is the vector space and $I_E$ is the topology. In a similar way, a bornological space will be $E = (\underline{E}, B_E)$.

A holomorphic function near to a compact set $X \subset \mathbf{C}^n$ or to a compact subset $X$ of a complex manifold will be called $f \in O([X])$.

## 1. HOW I ARRIVED AT THE BORNOLOGICAL DOMAIN

In 1953 or 1954 [20], I constructed a holomorphic function calculus. If $(a_1, \ldots, a_n)$ are regular elements of a complete commutative unital locally convex algebra, I defined their joint spectrum, $\mathbf{sp}(a_1, \ldots, a_n)$, and constructed a homomorphism $O([\mathbf{sp}(a_1, \ldots, a_n)]) \to A$ which maps the variables $z_i$ onto $a_i$ and the constant 1 to the unit of $A$.

Simultaneously, Shilov, [17], Arens and Calderon [2] constructed a holomorphic function calculus. They used the Cauchy-Weil formula (Weil's formula, not the Taylor Cauchy-Weil formula). They proved that if the algebra $A$ is Banach, semi-simple, then the Cauchy-Weil formula gives a homomorphism $O([sp(a_1, \ldots, a_n)]) \to A$.

My construction was different. Let $X_1, \ldots, X_n, Y_1, \ldots, Y_N$ be compact subsets of $\mathbf{C}$, and $\Delta = \{z \in \mathbf{C}^n | \quad i = 1, \ldots, n: z_i \in X_i, \ \forall r = 1, \ldots, N: P_r(z) \in Y_r\}$. Then the mapping, $O([X_1 \times \ldots \times X_n \times Y_1 \times \ldots \times Y_N]) \to O([\Delta])$ which maps the function $F(x, y) \in$ $\in O(X_1 \times \ldots \times X_n \times Y_1 \times \ldots \times Y_N)$ onto $F(z, P(z))$ is surjective and its kernel is generated by the polynomials $y_r - P_r(z)$. I proved the result using K. Oka's [16] and H. Cartan's [3] articles.

Now I use Gunning's and Rossi's book, [9], Chapter I, Section F, Lemma 7, and what I call "Lemma 7". This is not stated by Gunning and Rossi, but it can be proved using Lemma 7 and one page of proof.

Let $X_i = \mathbf{sp}\, a_i$, and $Y_r = \mathbf{sp}\, P(a_r)$. Using the $n + N$ Cauchy integral formula, we map $O([\mathbf{sp}\, a_1 \times \ldots \times \mathbf{sp}\, a_n \times \mathbf{sp}\, P_1(a) \times \ldots \times \mathbf{sp}\, P_N(a)])$ into $A$. This is a continuous homomorphism. It vanishes on the ideal generated by the polynomials $y_1 - P_1(z), \ldots \ldots, y_N - P_N(z)$. The algebra $O(\Delta)$ is the quotient of $O(\mathbf{sp}\, a_1 \times \ldots \times \mathbf{sp}\, a_n \times \mathbf{sp}\, P_1(a) \times \ldots \times \mathbf{sp}\, P_N(a))$ by the ideal generated by the functions $y_r - P_r(z)$. I did not give the "Arens and Calderon" technique. I knew its existence but I did it not dare to use it. I placed it in an appendix.

Let $A$ be a non semi-simple Banach algebra, whose spectrum (its Gelfand space) is not assumed connected, $\hat{A} = X_0 \cup X_1$. Using the Cauchy-Weil formula, Shilov finds an element $e \in A$ such that $x \in X_i$, $\hat{e}(m) = 1$.

The element is such that $m : \hat{e}(m)^2 = \hat{e}(m)$, $e^2 - e \in R(A)$, the Gelfand radical of $A$. He ends his proof constructing $r' \in R(A)$ such that $e - r'$ is idempotent in $A$.

In a similar way, Arens and Calderon solve equations in the Banach algebra. Consider an equation $F(a_1, \ldots, a_n, x) = 0$ where $a_i \in A$, and $F$ a holomorphic function near to $\|a_1\| \mathbf{D} \times \ldots \times \|a_n\| \mathbf{D} \times r \mathbf{D}$. They construct a function $\phi \in C(\hat{A})$, $m : |\phi(m)| \leq r$ such that for all $m \in \hat{A}$, then $F(\hat{a}_1(m), \ldots, \hat{a}_n(m), \phi(m)) = 0$, and

$$m \in \hat{A} : \frac{\partial F}{\partial x_i}(\hat{a}_1(m), \ldots, \hat{a}_n(m), \phi(m)) \neq 0.$$

They find an element $b \in A$ such that $m \in \hat{A}$, $\hat{b}(m) = \phi(m)$. The element solves the equation modulo Gelfand's radical. They next find $r \in R(A)$ such that $F(a, b - r) = 0$. The equation is solved.

Using my Doctorate, the element $b$ constructed by Arens and Calderon, modulo the radical, or Shilov's idempotent modulo the radical is in fact a solution of the equation, or an idempotent of the algebra.

The description above applies to Banach algebras; I did not only consider Banach algebras. I considered "regular elements" of a locally convex algebra: $a \in A$ is regular if $\exists\, M \in \mathbf{R}_+$, such that if $s \in \mathbf{C}$, $|s| > M$ then $\exists\, (a - s)^{-1}$ and $\{(a - s)^{-1} \mid |s| > M\}$ is bounded in the locally convex unital algebra.

I assumed that the algebra is complete, and that the multiplication is joint continuous, Bourbaki criticized me here and was correct.

H. Cartan was Bourbaki's word-speaker. Instead of considering a complete algebra, I could have considered a quasi-complete algebra. More important, if $A$ is quasi-complete and the multiplication is separately continuous, Banach-Steinhaus proves that the product of two bounded subsets is bounded. To end the construction, Bourbaki also assumed that the multiplication is hypocontinuous (if $\{a_i\}$ and $\{b_i\}$ are two directed systems of elements of $A$, $a_i \to a$, $b_i \to b$, and either $(a_i)$ or $(b_i)$ is bounded, then $a_i \cdot b_i \to a \cdot b$).

If E is barreled and quasi-complete, $L_s(E)$, the algebra of continuous linear mappings $E \to E$ with its simple (also called strong) topology has a hypocontinuous composition and is quasi-complete. My Doctorate applies to regular elements of a commutative closed subalgebra of $L_s(E)$.

I knew my proofs better that Bourbaki. The Cauchy sequences I used were "Cauchy in the bornological sense". It would be sufficient to assume that the algebra is quasi-complete and the multiplication is separately continuous.

The above explains how I arrived at the b-spaces and the b-algebras.

A boundedness $B_X$ on a set $\underline{X}$ is a set of "bounded subsets" of X such that if $B_1$ and $B_2 \in B_X$, and $B \subseteq B_1 \cup B_2$, then $B \in B_X$, and every $\{x\}$, $x \in X$ belongs to $B_X$. The set $B_X$ is the set of bounded subsets of X. Let $\underline{E}$ be a vector space over **C**, and **D** be the unit disc of **C**. A vector boundedness $B_E$ on $\underline{E}$ is a boundedness on E such that $DB_1 +$ $+ DB_2 \in B_E$ if $B_1 \in B_E$ and $B_2 \in B_E$. A subset B of E is completant if it is absolutely convex, does not contain any non zero subspace of E, and $E_B$, the vector space absorbed by B with its Minkowski functional is a Banach space. A b-space is a vector space with a vector boundedness and every bounded subset is contained in a bounded completant subset. We see that a directed set I, $\leq$ exists, for all $i \in I$, we have a Banach space $E_i$, if $i \leq j$, then $\underline{E}_i \subseteq \underline{E}_j$ and the inclusion $E_i \to E_j$ is bounded. The b-space $E = (\underline{E}, B_E)$ is a union of Banach spaces $E = \cup_i E_i$, a bounded set B of E is contained and bounded in one of the Banach spaces $E_i$.

A b-algebra is a b-space $A$ on which a multiplication is given, and the product $B_1 \cdot B_2$ of two bounded sets is bounded.

One has two obvious classes of b-algebras: Let $(\underline{A}, I_A)$ be an algebra, with a quasi-complete locally convex topology and a separately continuous multiplication. Then $A_b$, the same set of elements, with its von Neumann boundedness is a b-algebra.

Let $E = (\underline{E}, I_E)$ be a quasi-complete locally convex space, and let $L(E)$ be the set of linear continuous mappings $E \to E$ with its equicontinuous boundedness. Then $L(E)$ is a b-algebra.


## 2. ABOUT MY THÈSE D'AGRÉGATION

Germans present a Habilatätsschrift. I presented a Thèse d'Agrégation. Bourbaki asked more generality. I put my results in a bornological language. I defined b-spaces, b-algebras, and b-ideals. I did not investigate them. Their properties would come when they would be part of Mathematics.

Simultaneously with me, Overeraert and Ch. Housel has met b-spaces. Houzel organized a Seminar in 1962-63 [11]. I was not invited. Their language is not the mine. I cannot translate my Thèse in their language.

I defined what I call the SPectrum of $(a_1, \ldots, a_n)$ where the elements $a_i$ belong to the centrum of an associative unital b-algebra $A$. To prove that the spectrum is not trivial, I introduced a b-ideal $\alpha$ and proved that the ideal is trivial $(\alpha = A)$ if the spectrum is trivial.

I had introduced a b-ideal. Once it was there, I left it. I studied the SPectrum of $(a_1, \ldots, a_n)$ modulo $\alpha$. And if $w \in SP(a_1, \ldots, a_n; A \mid \alpha)$ I defined a holomorphic function calculus $O(w) \to A$ modulo $\alpha$. The b-algebra $O(w)$ is the algebra of holomorphic functions of $S_w = (z \in \mathbf{C}^n \mid w(z) > 0)$, such that $\exists N : w(s)^N f(s)$ is bounded. The function $w$ is Lipschitz on $\mathbf{C}^n$.

$N$ given, I define $_N O(w)$, the set of $f \in O(w)$ such that $w(s)^N f(s)$ is bounded. This is a Banach space in an obvious way, and $O(w) = \cup_N \, _N O(w)$.

The following is not in my Thèse, but could be in it. For every $N \in \mathbf{N}$, a linear bounded mapping exists $\phi_N : _N O(w) \to A$ that $\phi_{N+1} \mid _N O(w) - \phi_N$ is linear bounded $_N O(w) \to \alpha$ and for each $N \in \mathbf{N}$ the mapping $(f,g) \to \phi_{2N}(f \cdot g) - \phi_N(f) \cdot \phi_N(g)$ is bilinear bounded $_N O(w), \times _N O(w) \to \alpha$.

I consider the following application important. In my Thèse, I asked whether $w \in SP(z, O(z, w))$ where $w$ is Lipschitz and $-\log w$ is p.s.h. In his Doctorate, I. Cnop [4] proved that this was the case.

In a discussion with J.-P. Ferrier (probably in 1969), I told him that since Cnop had proved the result, the difference $f(z) - f(y)$ in $O(w \times w)$ belongs to the ideal generated by the elements $z_i - y_i$ in $O(w \times w)$, where $f \in O(w)$ and $w$ is Lipschitz, $-\log w$ is p.s.h. This is true because the ideal generated by $z_i - y_i$ is a b-ideal. We can construct $f(z)$ and $f(y)$ in $O(w \times w)$, and $z \equiv y$ modulo the ideal. J.-P. Ferrier has used the remark [6].

Apply "classical" theory. The difference belongs to the closure of the ideal. Applying the "quotient" viewpoint, we see that $f(z) - f(w)$ belongs to the ideal.

In a similar way, I may remember the reader of a Banach problem. In 1977, in Warsaw, Y. Domar observed that all "primitive" ideals of $A(\mathbf{D})$ whose hulls are contained in $\mathbf{D}^\circ$ are closed (a primitive ideals is contained in a single maximal ideal). He asked what one can say about ideals whose hulls are contained in $\mathbf{D}^\circ$. It is clear that their hulls are finite.

I answered the next day that these ideals are closed, are intersections of primitive ideals. The essential part of the proof is the fact that if $A$ is a Banach space and $\alpha$ is an ideal whose hull is not connected, **Hull** $\alpha = X_0 \cup X_1$ (the sets $X_i$ are compact and disjoint), then $\exists\, e \in A$, $e^2 - e \in \alpha$ for which $\hat{e}(m) = i$ for $m \in X_i$. This result would follow from Shilov's theorem when $\alpha$ is closed [21], [23].

## 3. QUOTIENT BORNOLOGICAL SPACES

My Thèse d'Agrégation (1960, [21]) was done in a non defined category. The category itself was defined in 1962, but whatever paper was soiled and then was not publishable. The expression "The category of quotient bornological spaces" was coined then.

If $E$ is a b-space, then $F$ is a b-subspace if $\underline{E} \supseteq \underline{F}$ and $B_E \supseteq B_F$. We study $E$ by $F$ whatever that means and say that $E$ mod $F$ is a q-space, write $E\,|\,F$ instead of $E$ mod $F$. The couple $E\,|\,F$ is a "q-space".

Strict morphisms are easy to define. If $u_1 : E \to E'$ is a linear bounded mapping such that $u_1\,|\,F$ is a linear bounded mapping $F \to F'$, then $u_1$ induces a strict morphism. If $u_1$ and $v_1$ are two mappings, each inducing a strict morphism, then $u_1$ and $v_1$ induce the same strict morphism $E\,|\,F \to E'\,|\,F'$ iff $u_1 - v_1$ is a linear bounded mapping $E \to F'$.

The class of strict morphisms is a category that I call $\tilde{q}$. We need more that $\tilde{q}$.

"Pseudo-isomorphisms" exist in the category $\tilde{q}$. These are strict morphisms $u : E\,|\,F \to E'\,|\,F'$ induced by $u_1 : E \to E'$, "bornological surjective" mapping, $(\forall\, B' \in B_E$ $\exists\, B \in B_E : u_1(B) = B')$, and $u^{-1}F' = F$ in the bornological sense ($B \in B_F$ if $B \in B_E$ and $u_1 B \in B_{F'}$).

Pseudo-isomorphisms are not strict isomorphisms. I construct the category $q$ in such a way that pseudo-isomorphisms become isomorphisms of $q$, and prove the following:

A category $q$ exists, which contains $\tilde{q}$, pseudo isomorphisms of $\tilde{q}$ are isomorphisms of $q$, a functor $\tilde{\Phi} : \tilde{q} \to$ **Cat** extends to a function $\Phi : q \to$ **Cat** iff $\Phi(s)$ is an isomorphism of **Cat** whenever $s$ is a pseudo-isomorphism. The extension is unique if it exists.

F.-H. Vasilescu has defined morphisms $E\,|\,F \to E'\,|\,F'$ in another way. His quotient are quotient Fréchet spaces. The graph of a morphism is a Fréchet vector subspace $G(u)$ of $E \times E'$ such that $F \times F'$ is a Fréchet subspace of $G(u)$, the projection $G(u) \to E$ is surjective and has the kernel $F'$. The projection $G(u) \to E$ induces a pseudo-isomorphism $G(u)\,|\,F \times F' \to E\,|\,F$. The second projection $G(u) \to F'$ induces a strict morphism $G(u)\,|\,F \times F' \to E' \times F'$, [18].

At this meeting, Vasilescu has told me that in his Doctorate, Dixmier considered "Julia operators", as morphisms of one quotient Banach space into another. I assume that Julia had considered such operators before Dixmier.

I could use Vasilescu's definition. The morphism $u : E|F \to E'|F'$ would be a b-subspace $G(u)$ of $E \times E'$ such that $F \times F' \in \mathbf{Lat}(G(u))$, the projection $G(U) \to E$ would have as kernel $F'$, the projection would be bornologically surjective, $G(u)|F \times F' \to E|F$ would be a pseudo-isomorphism, the other projection $G(u) \to E'$ induces a morphism $G(u)|F \times F' \to E'|F'$.

I prefer strict morphisms and pseudo-isomorphisms. The theory is category. We need functors. We begin with strict functors $\tilde{\Phi} : \tilde{q} \to \mathbf{Cat}$. We check that $\tilde{\Phi}(s)$ is an isomorphism of $\mathbf{Cat}$ whenever $s$ is a pseudo-isomorphism. The strict functor can be continued in a unique way to a functor $\Phi : q \to \mathbf{Cat}$.

In 1962 I was very happy, I had found the category $q$ of q-spaces and its morphisms. And then I found that much mathematics was necessary before placing it in our science. Around 1969, G. Noël came to me for a Doctorate and I gave him my 1962 paper. He gave another definition, which I call the "Miracle Functor" ([24], Chapter IV). His definition may be easier than the mine. In 1962, I was convinced that one can construct the tensor product of two q-spaces. He did it. I wondered whether the tensor product of two b-spaces is a b-spaces. He found two Banach spaces E and E' such that $E \otimes_q E' \neq E \hat{\otimes} E'$, [13], [14], [29].

It is in 1971 that I began again to study quotient bornological spaces. The present definition comes from then. But proving theorems is without aim.

## 4. SHEAVES

I want to define holomorphic functions taking its values in a q-space. Holomorphic functions are defined locally, a function on U is holomorphic if it is locally holomorphic. In other words, $O(\,\cdot\,,\,\cdot\,)$ should be simultaneously a sheaf in function of the open subset U of a complex manifold and a functor in the q-space $E|F$.

But what is a sheaf taking its values in the category $q$ ?

The category $q$ is abelian, is stable under direct sums and under direct products. It is possible to define direct limits and inverse limits (inductive limits and projective limits) in this category.

Presheaves are known. Let $X = (\underline{X}, I)$ be a topological space. $\mathbf{Open}_X$ is a category whose objects are open subsets of X and if U and V are open in X, a morphism $F_{VU} : V \to U$ exists iff $U \supseteq V$ and the morphism is unique if it exists.

A presheaf on X taking its values in the category $\mathbf{Cat}$ is a contravariant functor

**Open$_X$ $\to$ Cat.**

My definition of a sheaf comes from Godement's book, [8]. If **Cat** is a subcategory of **Ens** (the category of sets and mappings), a presheaf on X taking its values in **Cat** on X is a sheaf as soon as sets $U_i$ are open, $U = \underset{i}{\cup} U_i$ and

**a.** If x and y belong to $F(U)$ and $\forall i : F_{U_i U}(x) = F_{U_i U}(y)$ then x = y.

**b.** If for each i, $x_i \in F(U_i)$ and $\forall i,j : F_{U_i \cap U_j, U_i}(x_i) = F_{U_i \cap U_j, U_j}(x_j)$, then an element $x \in F(U)$ exists such that $\forall i : F_{U_i U}(x) = x_i$.

I must change the definition. The category **q** is not a subcategory of **Ens** (sets and mappings). It is abelian and stable under general direct sums and inverse sums.

**a'.** For every i, we have a morphism $F_{U_i U} : F(U) \to F(U_i)$ and therefore a direct product morphism $\Pi F_{U_i U} : F(U) \to \Pi F(U_1)$. This morphism is $\delta_0$.

**b'.** For each i, j, we have a restriction morphism $F(U_i) \to F(U_i \cap U_j)$ hence a morphism $F(U_i) \to \underset{j}{\Pi} F(U_i \cap U_j)$, and a morphism $\Pi F(U_i) \to \Pi F(U_i \cap U_j)$. In a similar way, we define a morphism $\underset{j}{\Pi} F(U_j) \to \Pi F(U_i \cap U_j)$. Of course, $\Pi F(U_i) = \Pi F(U_j)$. The morphism $\delta_1 : \Pi F(U_i) \to \Pi F(U_i \cap U_j)$ is the difference between the two given morphisms.

The morphisms $\delta_0$ and $\delta_1$ can be composed and $\delta_1 \circ \delta_0 = 0$. The presheaf F is a sheaf if all the above complexes $(0, \delta_0, \delta_1)$ are left exact.

In the category **Mod** of modules, one can associate a sheaf $F^=$ to a presheaf F and a presheaf morphism $F \to F^=$. If $\Phi : F \to G$ is a morphism of presheaves and G is a sheaf, then $\Phi$ factors in a unique way through the associated morphism $^= : F \to F^=$.

In the category **q**, to every presheaf F taking its values in **q**, we can associate a sheaf $F^= : \mathbf{Open}_X \to \mathbf{q}$ and a presheaf morphism $^= : F \to F^=$.

I need some sheaves taking values in the category **q** before defining the sheaf $O(\cdot, E \mid F)$.


## 6. EXAMPLES

### a. Continuous functions

Let $X = (\underline{X}, d_X)$ be a metrisable topological space, let $U \subset X$ be open, and $E \mid F$ be a q-space. We let $C(U, E \mid F) = C(U, E) \mid C(U, F)$. If $f : E \mid F \to E' \mid F'$ is a strict morphism, induced by $f_1 : E \to E'$, we define the morphism $\tilde{C}(U, f) : C(U, E \mid F) \to C(U, E' \mid F')$ induced by $C(U, f_1)$. In this way, we have defined a strict functor $\tilde{C}(U, \cdot) : \tilde{\mathbf{q}} \to \mathbf{q}$.

The Bartle and Graves theorem shows that this functor can be extended to a functor $q \to q$, because $\tilde{C}(U, s)$ is an isomorphism of $q$ as soon as $s$ is a pseudo-isomorphism of $\tilde{q}$. (A pseudo-isomorphism $E \mid F \to E' \mid F'$ is isomorphic to a morphism $E \mid F \to (E/G) \mid (F/G))$.

We shall call $C(U, \cdot)$ the extension of $\tilde{C}(U, \cdot)$. This is a functor $q \to q$. It happens to be exact $q \to q$.

Using partitions of the unity we see also that $C(\cdot, E \mid F)$ is a sheaf.

### b. Sheaf $E(\cdot, \cdot)$

I have two sheaves of functions of class $C^\infty$ taking its values in a q-space. I shall call them $E(\cdot, E \mid F)$ and $C^\infty(\cdot, E \mid F)$. In my situation, $E(U, E \mid F) \neq C^\infty(U, E \mid F)$. In the quotient Fréchet situation, $E(\cdot, E \mid F) = C^\infty(\cdot, E \mid F)$.

We let $E(U, E \mid F) \simeq \varprojlim E(V) \otimes_q E \mid F$ where V ranges over the relatively compact open subsets of a smooth manifold U. The b-space $E(V)$ is nuclear. $E(V) \otimes_q E$ is (isomorphic to) a b-space. And $f \in E(U, E)$ iff $\forall x \in U$ an open neighbourhood V of x exists and a completant bounded subset B of E exists such that $f \mid V \in E(V, E_B)$ where $B_1 \subset E(U, E)$ is bounded if $\forall x \in U$, an open neighbourhood V of x and a bounded completant subset B of E exist such that $B_1 \mid V$ is bounded in $E(V, E_B)$.

It is known that $E(V) \otimes_q E \mid F \simeq E(V) \otimes_q E \mid E(V) \otimes_q F$. We can show that $E(U, E \mid F)$ depends in an exact way on function of $E \mid F$, also that $E(\cdot, E \mid F)$ is a sheaf, using a partition of the unity.

$\otimes_q$ is the tensor product in the category $q$.

### c. The space $C^r(U, E \mid F)$

For a long time I have tried to define $C^r(U, E \mid F)$ where $r \in \mathbf{N}$. This is difficult. It is easier to define $C^r(U, E \mid F)$ when $r \in \mathbf{R}_+ \setminus \mathbf{N}$. In the preparation of his Doctorate, [1] B. Aqzzouz has found the theory of W. Kaballo, [12], the results of Frampton and Tromba [7] who proved that $C^r(U) \simeq c_o$ when $r \in \mathbf{R}_+ \setminus \mathbf{N}$, and U is a compact manifold. Of course, Whitney had defined $C^r(X)$ when X is a compact subset of U, and proved that $C^r(X)$ is a direct summand of $C^r(U)$. The space $C^r(U) \simeq c_o$ and is therefore a $L_\infty$-space. The space $C^r(X)$ is a direct summand of $c_o$, so is a $L_\infty$-space. Using the work of Kaballo, he takes the $\varepsilon$-product of $C^r(X)$ by $E \mid F$. This gives $C^r(X, E \mid F)$. The space $C^r(V, E \mid F)$ is the inverse limit (projective limit) of spaces $C^r(X, E \mid F)$ with X compact in V. Of course, $C^r(V, E \mid F) \simeq \varprojlim C^r(X, E \mid F)$ with X compact in V. We can prove that $C^r(V, \cdot)$ is an exact functor, and that $C^r(\cdot, E \mid F)$ is a sheaf, both statements are proved using a partition of the unit.

Whitney's results follow from papers [28], [29] when $r \in \mathbf{N}$. I have attended a talk of L. Schwarz at the Séminaire Bourbaki, part of my knowledge comes from that talk.

### d. The sheaf $C^\infty(\,\cdot\,, E \,|\, F)$

In the category $\mathbf{q}$, inverse (projective) limits exist. For all $r \in \mathbf{R}_+ \setminus \mathbf{N}$, we have defined $C^r(U, E \,|\, F)$. When $r' \geq r$, we have a natural transformation $C^{r'}(U, E \,|\, F) \rightarrow$ $\rightarrow C^r(U, E \,|\, F)$. At the inverse limit, we have a sheaf $C^\infty(U, E \,|\, F)$. In my category, the functor is left exact, and is not exact. In the category $\mathbf{qFre}$, the sheaves $E(\cdot, E \,|\, F)$ and $C^\infty(\,\cdot\,, E \,|\, F)$ are equal. The functor $C^\infty(U, \cdot)$ is then exact.

### 7. The functor-sheaf $O(\,\cdot\,,\,\cdot\,)$

The sheaf $O(\,\cdot\,,\,\cdot\,)$ has several definitions. They are not isomorphic to the presheaf defined by F.-H. Vasilescu in a seminar at OT X, i.e. to the presheaf defined in his talk at OT XII. I have also defined holomorphic functions, taking its values in a quotient Banach space using the tensor product [22].

If $F_1$ and $F_2$ are sheaves, if $u : F_1 \rightarrow F_2$ is a morphism of sheaves, then $\mathbf{Ker}\, u(\cdot)$ is a sheaf. Of course, $\overline{\partial}$ is a morphism of sheaves. We can define $O(\,\cdot\,, E \,|\, F)$ as the kernel of any of the following morphisms of sheaves:

a. The kernel of $\overline{\partial} : C^\infty(\,\cdot\,, E \,|\, F) \rightarrow C^\infty(\,\cdot\,, E \,|\, F) \times \mathbf{C}^n$.

b. The kernel of $\overline{\partial} : E(\,\cdot\,, E \,|\, F) \rightarrow E(\,\cdot\,, E \,|\, F) \times \mathbf{C}^n$.

c. The kernel of $\overline{\partial} : C^r(\,\cdot\,, E \,|\, F) \rightarrow C^{r-1}(\,\cdot\,, E \,|\, F) \times \mathbf{C}^n$.

d. The kernel of $\overline{\partial} : D'(\,\cdot\,, E \,|\, F) \rightarrow D'(\,\cdot\,, E \,|\, F) \times \mathbf{C}^n$.

I have not defined $D'(\,\cdot\,, E \,|\, F)$. In this section, $D'(U, E \,|\, F) \simeq \varprojlim D'(V) \otimes_q E \,|\, F$, where V is relatively compact in U. I believe that the $O(\,\cdot\,, E \,|\, F)$ is also isomorphic to the kernel of $\overline{\partial} : q(D(\,\cdot\,), E \,|\, F) \rightarrow q(D(\,\cdot\,), E \,|\, F) \otimes_q \mathbf{C}^n$.

In the quotient Fréchet situation, when $n \geq 2$ and $E \,|\, F$ does not have the u.c.p. then $O(\,\cdot\,) \hat{\otimes} E \,|\, F$ is not a sheaf. In the general q-situation, $O_1(\,\cdot\,, E \,|\, F)$ is not a sheaf when $O_1(U, E \,|\, F) \simeq \lim O(V) \otimes_q E \,|\, F$ where V is relatively compact in U.

But in any case, the sheaf associated to the presheaf defined is $O(\,\cdot\,, E \,|\, F)$.

### 8. THE UNIQUE CONTINUATION PROPERTY

I must speak of the unique continuation property (the u.c.p.). In classical situations, the u.c.p. is valid:

Let $U \subset \mathbf{C}^n$ be open connected, and $V \subset U$ open, non empty. Let E be a locally convex space or a convex bornological space. Then the restriction mapping $O(U, E) \to O(V, E)$ is injective. The space has the u.c.p.

It is unfortunate that many q-spaces (and q-Fréchet) spaces do not have the u.c.p.

If $E|F$ is a genuine q-Banach space, i.e. if F is a Banach space contained in E in a continuous, not a bicontinuous way, then $E|F$ does not have the u.c.p.

Let $(e_n)$ be a sequence of elements in E that is contained in F and is not bounded in F. From it we can extract a subsequence, that I shall again call $(e_n)$ such that $e_n = O_F(2^n)$ but for all $a < 2$, $e_n \neq O(a^n)$. The series $\sum e_n z^n$ converges on the disc D in E but has a singularity on the circle $(z \mid |z| = \tfrac{1}{2})$. The restriction $O(D^\circ, E|F) \to$ $\to O(\tfrac{1}{2}D^\circ, E|F)$ is not monic. $E|F$ does not have the u.c.p.

Some genuine quotient bornological spaces have the u.c.p. The most known example is the one-dimension hyperfunctions $Hf(V) \simeq O(U)|O(U \setminus \mathbf{R})$. We assume that $V = U \cap \mathbf{R}$. (Classical mathematicians call $B(V)$ what I call $Hf(V)$, but for me, $B$ is the name of the boundedness.) Also the continuous germs is a q-space with the u.c.p.

We can say that $E|F$ has the u.c.p. if $\exists \alpha < 1$, $\alpha \in \mathbf{R}_+$ such that B is bounded in $O(D^\circ, F)$ if it is bounded in $O(D^\circ, E)$ and its restriction to $\alpha D$ is bounded in $O(\alpha D^\circ, F)$. We can also say that $E|F$ has the u.c.p. if some connected open subset of $\mathbf{C}^n$ exists, such that B is bounded in $O(U, F)$ if it is bounded in $O(U, E)$ and its restriction to V is bounded in $O(V, F)$. The two statements are equivalent.

In the quotient Fréchet case, $n \geq 2 : O(\,\cdot\,)\hat{\otimes}E|F$ cannot be a sheaf unless $E|F$ has the u.c.p. In the q-case, $n \geq 2$, $O_1(\,\cdot\,, E|F)$ cannot be a sheaf unless $E|F$ has the u.c.p. in both cases, when $n \geq 2$. And in both cases, we use the set $U = \{(z_1, z_2) \in \mathbf{C}^n \mid \tfrac{1}{2} < |z_1|^2 + |z_2|^2 < 1, \mathbf{Re}\, z_1 > 0\}$ and its holomorphic hull $\tilde{U} = \{(z_1, z_2) \in \mathbf{C}^N \mid |z_1|^2 + |z_2|^2 < 1, \mathbf{Re}\, z_1 > 0\}$. And it is not difficult to show that in quotient Fréchet case, $O(U, E|F) \simeq O(U)\hat{\otimes}E|F$ then $E|F$ has the u.c.p. In the q-case, if $O(U, E|F) \simeq O_1(U, E|F)$, then $E|F$ has the u.c.p.

## 9. THREE OTHER PROBLEMS

### a. Conjecture that $O(U, E|F) \simeq O_1(U, E|F)$ if U is pseudo-convex

Begin with what I call Dolbeault's lemma:

Let $\omega$ be a form belonging to $\Lambda^{p,q}E(\mathbf{C}^n, E)$ such that on a neighbourhood of $X = X_1 \times \ldots \times X_n$ we have $\bar{\partial}\omega|U \in \Lambda^{p,q+1}(U, F)$. Then a neighbourhood $U'$ of X exists,

$\tilde{\omega} \in \Lambda^{p,q-1}E(U',E)$, $\phi \in \Lambda^{p,q}E(U', F)$ such that $\omega \mid U' = \bar{\partial}\,\tilde{\omega} + \phi$.

Using this lemma, we can prove that $O([X], E \mid F) \simeq O([X]) \otimes_q E \mid F$ if $X$ is a direct product of compact subsets of $\mathbf{C}$, hence $O_1(U, E \mid F) \simeq O(U, E \mid F)$.

Using the Cauchy-Fantappiè formula, I can prove that $O_1(U, E \mid F) \simeq O(U, E \mid F)$ when $U$ is a domain of holomorphy contained in $\mathbf{C}^n$.

Before going to Timişoara I could not prove that $O(U, E \mid F) = O_1(U, E \mid F)$ if $U$ is a Stein manifold. During the OT meeting, using Aqzzouz $\varepsilon$-product of a b-space and a q-space (his Doctorate, [1]), I have probably solved the problem.

### b. A category of quotient webbed spaces should exist

M. Dewilde has defined the category of webbed spaces (in French, espaces à réseaux). It is the smallest category that contains **Ban** and is stable under countable direct limits and countable inverse limits [5]. The category of webbed spaces has the "closed graph" property: if E and F are two webbed spaces, if $u : E \to F$ is a linear mapping with a closed graph, then u is continuous.

I conject that a category of quotient webbed spaces exists, is abelian, is stable under countable direct limits and countable inverse limits and the tensor product of two quotient webbed spaces can be defined.

In Analysis, the problems we meet use countable direct limits and countable inverse limits. The statements of theorem about quotient webbed spaces would be easier to be understood than those of theorems about quotient bornological spaces.

It is also possible that one could prove a lemma about quotient webbed spaces. I think of a very important lemma applying to quotient Fréchet spaces. Let $(E_n)$, $(F_n)$ be sequences of Fréchet spaces, where each $F_n$ is a Fréchet subspace of $E_n$, for each n we have a morphism $E_{n+1} \to E_n$ mapping $F_{n+1}$ into $F_n$, with a dense range. Then $\varprojlim (E_n \mid F_n) \simeq \varprojlim E_n \mid \varprojlim F_n$. (See an application of the lemma in [10], Chapter IX.)

### c. If E|F has the u.c.p. it is possible that $O_1( \cdot , E \mid F) \simeq O( \cdot , E \mid F)$

We assume that $E \mid F$ has the u.c.p. Let $f \in \sigma\beta(X, O(U, E \mid F))$. Let U be open in $\mathbf{C}^n$. For some arcs j in $\mathbf{C}^n$, beginning at $z_0 \in U$, ending at $z_1 \in \mathbf{C}^n$, we continue f to $z_1$, and have a continuation of f at $z_1$. Let $j_0$ and $j_1$ be two arcs from $z_0$ to $z_1$ and let G be a homotopy from $j_0$ to $j_1$. Assume that f can be extended along all the arcs of G. Then the continuation of f along the arcs of the homotopy does not depend on the parameter of the homotopy. Using this technique, we obtain a domain $U_1$ which contains U. All holomorphic functions on U can be extended to $U_j$. <u>If we can prove that $U_1$ is pseudo-convex, we prove that</u>

$O( \cdot , E \mid F)$ would be a sheaf if $E \mid F$ has the u.c.p.

## REFERENCES


1.  **Aqzzouz, B.** : *Le ε-produit d'un b-espace et d'un q-espace*, Doctorate at the Université Libre de Bruxelles, May, 1988.

2.  **Arens, R. ; Calderon, A.P.** : Analytic functions of several Banach elements, *Ann. of Math.*, **62**(1955), 204-216.

3.  **Cartan, H.** : Idéaux de fonctions analytiques de plusierus variables complexes, *Bull. Soc. Math. France* **78**(1950).

4.  **Cnop, L.** : Spectral study of holomorphic functions with a bounded growth, *Ann. Inst. Fourier (Grenoble)*, **22**(1979), 293-309.

5.  **Dewilde, M.** : *Reseaux dans les espaces linéaires à semi-normes*, Mém. Soc. Roy. de Liège, v. **18**(1969).

6.  **Ferrier, J.P.** : *Spectral theory and complex analysis*, North Holland, Noord- -Holland Notas de Matematica, **4**(1973).

7.  **Frampton, J. ; Tromba, A.** : On the classical spaces of Hölder continuous functions, *J. Funct. AnaL*, **10**(1972), 336-345.

8.  **Godement, R.** : *Théorie des faisceaux*, Hermann, 1958.

9.  **Gunning, W. ; Rossi, H.** : *Analytic functions of several complex variables*, Prentice Hall in Modern Analysis, 1965.

10. **Hörmander, L.** : *The analysis of linear partial differential operator. I*, 1983, Springer-Verlag.

11. **Houzel, Ch.** (Ed.) : *Séminaire Banach*, Springer-Verlag, Lecture Notes in Mathematics, **277**(1973).

12. **Kaballo, W.**, Lifting theorems for vector-valued functions and their ε-tensor products, in *Functional analysis, Surveys and recent results*, K. Bierstedt and B. Fuchssteiner, Eds., Noord Holland.

13. **Noël, G.** : Une immersion de la catégorie des espaces bornologiques séparés dans une catégorie abélienne, *C.R. Acad. Sci. Paris, Ser. I Math.* **269**(1969), 195-197.

14. **Noël, G.** : Produits tensoriels et platitude dans les QESP et les QESPC, *Bull. Soc. Math. Belg. Ser. A* **13**(1970), 114-142.

15. **Noël, G.** : Espaces ultraplats et espaces de Lindenstrauss-Pełczinski (Bordeaux), *Coll. AnaL Fonctionelle, Memoires, SMF*, **31-32**(1973), 257-263.

16. **Oka, K.** : Fonctions de plusieurs variables complexes. VII. Sur quelques notions arithmétiques, *Bull. Soc. Math. France* **58**(1952).

17. **Shilov, G.** : *Math. Sb.* **32**(1953), 354-364.

18. **Vasilescu, F.-H.** : Spectral theory in quotient Fréchet spaces. I. *Rev. Roumaine Math. Pures Appl.* **32**(1987), 561-579.

19. **Vasilescu, F.-H.** : Spectral theory in quotient Fréchet spaces. II, *J. Operator Theory* **21**(1989),

20.   **Waelbroeck, L.** : Le calcul symbolique dans les algèbres commutatives, *J. Math. Pure Appl.* **33**(1954), 147-186.

21.   **Waelbroeck, L.** : *Etudes spectrales des algèbres commutatives complètes*, Acad. Royale de Belgique, 1960, Memoires, tome **31**, fascicule 7 et dernier.

22.   **Waelbroeck, L.** : Fonctions à valeurs dans les quotients banachiques, *Bulletin Acad. Royale de Belgique, Classe des Sciences* (1981), 319-327.

23.   **Waelbroeck, L.** : Holomorphic functional calculus and quotient Banach spaces, *Studia Math.* **75**(1982), 273-286.

24.   **Waelbroeck, L.** : Quasi-Banach algebras, ideals and holomorphic function calculus, *Studia Math.* **75**(982), 287-292.

25.   **Waelbroeck, L.** : A first set of notes over quotient bornological spaces ended in 1986.

26.   **Waelbroeck, L.** : A second set of notes ended in 1988.

27.   **Waelbroeck, L.** : Quotient Fréchet spaces, *Rev. Roumaine Math. Pures Appl.*, to appear.

28.   **Whitney, H.** : On analytic extensions of differentiable functions defined on closed sets, *Trans. Amer. Math. Soc.* **36**(1934), 63-89.

29.   **Whitney, H.** : On ideals of differentiable functions, *Amer. J. Math.* **70**(1948), 635-658.

**Lucien Waelbroeck**
Departement de Mathématique
Université Libre de Bruxelles
Campus Plaine, C.P. 214
Bld. du Triomphe, B-1050 Bruxelles
Belgique.

Operator Theory:
Advances and Applications, Vol. 43
© 1990 Birkhäuser Verlag Basel



# ON CAUSAL NETS OF ALGEBRAS

Manfred Wollenberg

## 1. INTRODUCTION

In the algebraic approach to quantum field theory over curved space-time the main object is a family of $C^*$-algebras (or even von Neumann algebras) $A(O)$ indexed by open bounded sets $O$ from the space-time manifold. The isometries g of this manifold are represented by $*$-automorphisms $\alpha_g$ of the whole algebra $A = C^*(\underset{O}{\cup} A(O))$. These objects $A(O)$, $\alpha_g$ satisfy some properties such as isotony, covariance, and locality (for details see [3], [7], [9], [10]). In the case that the space-time is the Minkowski manifold the structure of these algebras and automorphisms has been intensively studied mostly under additional assumptions (see e.g. [4], [5], [8]). The aim of the present paper is to extend some of these results to more general manifolds. For this goal a slight abstraction of the algebraic approach to quantum field theory is given. It has the sense to simplify and unify the essential ideas and to abstract from details which are unimportant for some questions. Thus it can be applied to more general situations.

In Section 2 the notion of a causal manifold is introduced. An essential ingredient is the so-called causal map $\perp$. It is similar to the orthogonality relation used in [2], [4] for the definition of quasilocal algebras. The notion of a causal net of algebras (over causal manifolds) is given in Section 3. Roughly speaking, causal nets of algebras are quasilocal algebras with additional structure. Causal nets of algebras over lattices have been first introduced in [1] (there are slight differences in notation, notions, and assumptions compared with those used here).

In the last section properties of the algebras and automorphisms of these causal nets of algebras are derived (under natural additional assumptions). It is proved that the algebras $A(O)$ are not abelian, are not finite-dimensional, and satisfy the Wightman inequality and extended locality. Further it is shown that the whole algebra $A$ is simple and that the automorphism are in general not inner. As already mentioned these results are well known for Minkowski manifolds. The proofs for the more general manifolds given here are in some parts similar and in other parts very different to the ones for the Minkowski manifold. Naturally, the simple properties described here present only a choice giving a flavour of this theory.

## 2. CAUSAL MANIFOLDS

Let $M$ be a connected $C^\infty$-manifold of dimension $n \geq 1$. Let $G$ be a Lie group of transformations acting on $M$. Further, let the symbol $\perp$ denote a map from the open subsets of $M$ into the open subsets of $M$, $\perp : M \supset O \rightarrow O^\perp \subset M$. We say the triple $\{M, G, \perp\}$ is a causal manifold if it satisfies the following five assumptions:

(i) $O \subset (O^\perp)^\perp$  ($O^\perp$ is called the causal complement of $O$).

(ii) $O \cap O^\perp = \emptyset$ .

(iii) $(O_1 \cup O_2)^\perp = O_1^\perp \cap O_2^\perp$ .

(iv) $(gO)^\perp = gO^\perp$ for all $g \in G$ ($g$ commutes with $\perp$).

Open sets $O$ with $O = (O^\perp)^\perp$ are called causally complete. In general there does not exist such sets. We require the existence of a large family of such sets.

(v) There is a family $T$ of causally complete subsets of $M$ which is a basis for the topology of $M$. For the elements $O$ of this family it holds $O^\perp \neq \emptyset$ and $\overline{O} \cap \overline{O^\perp} \neq \emptyset$.

Special examples of causal manifolds are:

(a) $M$, $G$ arbitrary. $\perp : M \supset O \rightarrow \text{int}\,(M \setminus O)$.

(b) $M$ is a globally hyperbolic manifold, $G$ a subgroup of the group of isometries, $O^\perp$ is the interior of the set of all points which cannot be reached by causal or lightlike curves from $O$.

Simple properties of causal manifolds are: $\emptyset^\perp = M$, $O_1 \subseteq O_2$ implies $O_2^\perp \subseteq O_1^\perp$. For each open set $O \neq \emptyset$ there is an open set $\tilde{O} \neq \emptyset$ such that $\tilde{O} \subset O$ and $\tilde{O}^\perp \cap O \neq \emptyset$.

The last property implies that we can find for each nonempty open set $O_1$ an infinite sequence of nonempty open sets $O_m$ from $M$ such that $O_j \subset O_{j-1}$ and $O_{j-1} \cap O_j^\perp \neq \emptyset$ for all $j = 2, 3, \ldots$ . We will use this property in the last section.

## 3. CAUSAL NETS OF ALGEBRAS

A quadruple $\{A, H, \gamma, \alpha(\cdot)\}$ is called a causal net of algebras (for the causal manifold $\{M, G, \perp\}$) if it satisfies the following five assumptions:

(i) To each open bounded set $O \subset M$ there is assigned a von Neumann algebra $A(O)$ on a Hilbert space $H$ (independent of $O$) such that $C^*(V_O A(O)) = A$. For unbounded regions $\tilde{O}$ (if they exist) we define $A(\tilde{O}) := (V_{O \subset \tilde{O}} A(O))'' \cap A$ (net property).

(ii) $O_1 \subseteq O_2$ implies $A(O_1) \subseteq A(O_2)$ (isotony).

(iii) There is a representation of $G$ by $*$-automorphisms of $A$, $G \ni g \rightarrow \alpha_g(\cdot)$, such that $\alpha_g(A(O)) = A(gO)$ for all open $O \subset M$ and $g \in G$. For $A \in A$ the map $G \ni g \to$ $\to \alpha_g(A)$ is weakly continuous.

(iv) There is a $*$-automorphism $\gamma$ on $A$ such that $\gamma^2 = 1$, $\gamma \cdot \alpha_g = \alpha_g \gamma$, $\gamma A(O) = A(O)$ for all open sets $O \subset M$ and $g \in G$. Thus $\gamma$ graded the algebras $A(O)$, $A$. This means $A(O) = A_+(O) \oplus A_-(O)$, $(A = A_+ \oplus A_-)$, $\gamma(A_\pm) = \pm A_\pm$ for $A_\pm \in A_\pm(O)$ $(A_\pm \in A_\pm)$. Further, for $O_1 \subset O_2^\perp$ we have $A_1 A_2 = -A_2 A_1$ for $A_i \in A_-(O_i)$ and $A_1 A_2 = A_2 A_1$ in all other cases, i.e. $A_i \in A_+(O_i)$ or $A_1 \in A_+(O_1)$, $A_2 \in A(O_2)$ or $A_2 \in A_+(O_2)$, $A_1 \in A(O_1)$ (causality).

(v) $A$ is a factor, i.e. $A'' \cap A' = \mathbf{C}1$, and $A \neq \mathbf{C}1$.

The last assumption $A \neq \mathbf{C}1$ was made to avoid this trivial case which satisfies all other assumptions.

This framework is very wide. If the group $G$ is small we have not enough relations between the algebras $A(O_1)$, $A(O_2)$ belonging to different regions $O_1$, $O_2$. Thus we often need some additional assumptions.

(vi) Weak additivity. A causal net $\{A, H, \gamma, \alpha(\cdot)\}$ over $\{M, g, \perp\}$ is called weakly additive if the group $G$ acts transitively on $M$ and for each open set $O \neq \emptyset$ the relation $(\cup_g A(gO))'' = A''$ holds.

(vii) Weak primitive causality. A causal net $\{A, H, \gamma, \alpha(\cdot)\}$ over $\{M, G, \perp\}$ is called weakly primitive causal if for each causally closed set $O$ from $T$ and $\tilde{O} \supset \bar{O}$ it holds $(A(\tilde{O}) \cup A(O^\perp))'' = A''$.

(viii) Split property for $A_+$. We say the algebra $A_+$ of the causal net $\{A, H, \gamma, \alpha(\cdot)\}$ over $\{M, G, \perp\}$ satisfies the split property if for each open bounded set $O_1$ with $O_1$, $O_1^\perp \neq \emptyset$ there are open bounded sets $O_2$, $O_3$ such that $O_3 \subset O_1 \subset O_2$, $O_1^\perp \cap O_2 \neq \emptyset$, $O_2^\perp \neq \emptyset$, and $A_+(O_3) \subset N_2 \subset A_+(O_1) \subset N_1 \subset A_+(O_2)$ holds where $N_1$, $N_2$ are type $I_\infty$ - factors.

## 4. SIMPLE PROPERTIES OF CAUSAL NETS OF ALGEBRAS

In this section simple algebraic properties of causal nets are investigated. As already mentioned these properties are known in case that $M$ is the Minkowski manifold, $G$ is the Poincare group, $\gamma$ is the identity automorphism, and the causal net satisfies spectrality and/or some other conditions.

We begin with some preliminary results which are simple but useful for the proof of the following results.

**LEMMA.** (i) *Let $B$ be a unital $C^*$-algebra and $\gamma$ a $*$-automorphism of $B$ with*

$\gamma^2 = 1$. *Let $B = B_+ \oplus B_-$ be the corresponding decomposition (see (iv) in Section 3). Then $B_+ = \mathbf{C}1$ implies $B_- = \{0\}$ or $B_- = \mathbf{C}V$ where $V = V^*$ and $V^2 = 1$.*

(ii) *Let $B = A(O)$, $O \neq \emptyset$, where $A(O)$ belongs to a causal net of algebras. Then $B_+ = \mathbf{C}1$ implies that there is a nonempty set $\tilde{O} \subset O$ such that $A(\tilde{O}) = \mathbf{C}1$.*

PROOF. (i) Assume $W \in B_-$. Then $W^* \in B_-$ and $W^*W$, $WW^* \in B_+$. This implies that $W^*W = c_1 1$ and $WW^* = c_2 1$ where $c_i$ are nonnegative numbers. The case that the numbers $c_i = 0$ for all $W$ leads to $B_- = \{0\}$ which is clearly a solution. Now we assume that $W \neq 0$ and therefore at least one number is not zero. Assume $c_1 > 0$. We put $U = (c_1)^{-\frac{1}{2}} W$. Thus $U^*U = 1$, i.e. $U$ is a partial isometry with $UU^* = P$. On the other hand we have $UU^* = c_1^{-1}c_2 1$. This gives $P = 1$ and $c_2 = c_1$. Therefore $U$ is unitary. Now $U \cdot U \in B_+$ giving $U \cdot U = c1$. Because of the unitarity of $U$ we get $c = e^{ia}$, $a \in \mathbf{R}$. Setting $V = e^{-ia/2}U$ we obtain $V^2 = 1$ and $V^* = V$. Let $X \in B_-$. Then $XV \in B_+$ and $XV = b1$. Thus $X = bV$.

(ii) Because of (i) we have that $B_- = A_-(O)$ is equal to $\{0\}$ or to $\mathbf{C}V$. We have to exclude that the second case is true for all subsets $A_-(\tilde{O})$, $\tilde{O} \subset O$, from the properties of $A(O)$. From Section 2 we know that there is a nonempty open set $O_1 \subset O$ such that $O_1^\perp \cap O \neq \emptyset$. Thus $A_-(O_1)$ anticommutes with $A_-(O_1^\perp \cap O)$.

First case: One of these sets $(A_-(O_1), A_-(O_1^\perp \cap O))$ is $\{0\}$. Then we are ready.

Second case: Assume both sets are equal to $\mathbf{C}V$. This gives $VV = -VV$ (because of causality) implying $V = 0$. Thus the second case is not possible.                    $\square$

Now we are in a position to prove the nonabelianess of the algebras $A(O)$.

**PROPOSITION 1.** *Suppose the causal net of algebras satisfies (vi) and (vii). Then for each open nonempty set $O$ the von Neumann algebra is not abelian.*

PROOF. Assume $A(O)$ is abelian. For $O$ there is an nonempty open set $O_1$, $\bar{O}_1 \subset O$, with $O_1^\perp \cap O \neq \emptyset$. We can assume that $O_1$ is from $T$. Because of isotony $A(O_1)$ is abelian and commutes with $A(O)$. Further, $A_+(O_1)$ commutes with $A(O_1^\perp)$. Thus $A_+(O_1)$ commutes with $(A(O) \cup A(O_1^\perp))'' = A''$ (weakly primitive causality). Since $A$ is a factor, $A_+(O_1) = \mathbf{C}1$. From the lemma we obtain that there is a nonempty open set $O_2 \subset O_1$ such that $A(O_2) = \mathbf{C}1$. Now using weak additivity we get $A'' = (\cup_g A(g\,O_2))'' = (\cup_g \mathbf{C}1)'' = \mathbf{C}1$. This contradicts assumption (v) in Section 3. Thus $A(O)$ is not abelian.    $\square$

**REMARK 1.** The corresponding result for the Minkowski manifold and $\gamma = 1$ can be found in [8, p. 39] where other assumptions have been used for the proof.

The next result says that by enlarging the regions $O$ the algebras also enlarge. Because of the importance of this statement we present two versions.

**PROPOSITION 2.** (Wightman inequality). (i) *Suppose the causal net of algebras is weakly additive with respect to the connected component $G_o$ of the identity of G. Let $O_1$, $O_2$ be open bounded sets of $M$ such that $O_1 \subset O_2$ , $O_1^{\perp} \cap O_2 \neq \emptyset$, and* **dist** *$(O_1, \partial O_2) > 0$ (the Euclidean distance in the charts). Then $A(O_1) \subsetneq A(O_2)$.*

(ii) *Suppose the causal net of algebras is weakly additive (vi) and weakly primitive causal (vii). Let $O_1$, $O_2$ be two open bounded sets such that $O_1 \subset O_2$, there is a $g \subset G$ with $gO_1 \subset O_2$, and $g^m O_1 \subset O_1^{\perp}$ for some natural number $m$. Then $A(O_1) \subsetneq A(O_2)$.*

PROOF. (i) We can find a generating neighbourhood $N$ of the identity such that $gO_1 \subset O_2$ for all $g \in N$. Assume $A(O_1) = A(O_2)$. Then we get $\alpha_g(A(O_1)) = A(gO_1) \subseteq A(O_2)$ and $\alpha_g(A(O_2)) = A(gO_2) = \alpha_g(A(O_1)) = A(gO_1)$. This gives $A(gO_2) \subseteq A(O_2) = A(O_1)$ for all $g \in N$. Since $N$ is generating for $G_o$ we obtain $A(gO_2) \subseteq A(O_2) = A(O_1)$ for all $g \in G_o$. From the weak additivity (with respect to $G_o$) we find $A'' = (\cup_g A(gO_2))'' \subseteq (A(O_2))'' = A(O_2) = A(O_1)$. Now we have $O_1^{\perp} \cap O_2 \neq \emptyset$. Further, $A(O_1^{\perp} \cap O_2) \subseteq A(O_2) = A(O_1)$. Thus $A_+(O_1^{\perp} \cap O_2)$ commutes with $A(O_1) = A''$. Since $A''$ is a factor we get $A_+(O_1^{\perp} \cap O_2) = \mathbf{C}1$. From the lemma we obtain $A(O_3) = \mathbf{C}1$ for an open bounded set $O_3 \subset O_1$ with $O_3 \neq \emptyset$. Using again weak additivity this gives $A'' = \mathbf{C}1$ contradicting assumption (iv) in Section 3. Thus $A(O_1) \subsetneq A(O_2)$.

(ii) Assume $A(O_1) = A(O_2)$. We have $\alpha_g(A(O_1)) = A(gO_1) \subseteq A(O_2) = A(O_1)$. This implies $\alpha_{g^m}(A(O_1)) = A(g^m O_1) \subseteq A(O_1)$. On the other hand we have $A(g^m O_1) \subseteq A(O_1^{\perp})$ because of $g^m O_1 \subset O_1^{\perp}$. Thus $A_+(g^m O_1)$ commutes with $A(O_1) \supseteq A(g^m O_1)$. This gives that $A_+(g^m O_1)$ is abelian. Moreover $A_-(g^m O_1)$ anticommutes with $A_-(O_1) \supseteq A_-(g^m O_1)$. Thus for every element $V \in A_-(g^m O_1)$ we have $V^2 = -V^2$, $V^*V = -VV^*$, and $(V^*V)V = V(V^*V)$ $(V^*V \in A_+(g^m O_1))$. Using this we get $V = 0$, i.e. $A_-(g^m O_1) = \{0\}$. Therefore $A(g^m O_1) = A_+(g^m O_1)$ is abelian. Applying now Proposition 1 we obtain a contradiction. Thus $A(O_1) \subsetneq A(O_2)$.                                                                $\square$

**REMARK 2.** One can prove the same assertion with the help of other conditions. But a kind of weak additivity is always necessary. The corresponding result ((i) with $\gamma = 1$) for the Minkowski manifold can be found in [8, p. 38].

An easy consequence of this proposition is

**THEOREM 1.** *Suppose the causal net of algebras is weakly additive with respect to the connected component $G_o$ of the identity of $G$. Let $O_1$ be an open nonempty set of $M$. Then $A(O_1)$ is infinitely dimensional.*

PROOF. According to the remarks in Section 2 we can find an infinite sequence of open bounded sets $O_m$ such that $O_m \subset O_{m-1}$, $O_m^\perp \cap O_{m-1} \neq \emptyset$, $O_m \neq \emptyset$, and **dist**$(O_m, \partial O_{m-1}) > 0$ for all $m = 2, 3, \ldots$ . From Proposition 2 we get $A(O_m) \subsetneq A(O_{m-1})$ for all $m = 2, 3, \ldots$ . Thus $A(O_1)$ is infinitely dimensional.           $\square$

The next result says that the algebras which belong to causally disjoint regions (in a slightly stronger sense) have only scalars in their intersection.

**PROPOSITION 3.** (Extended locality). *Suppose the causal net of algebras is weakly primitive causal (vii). Let $O_1$, $O_2$ be causally closed sets from $T$ such that there is some $\tilde{O}$ with $\tilde{O} \supset O_1$ and $\tilde{O} \subset O_2^\perp$. Then $A(O_1) \cap A(O_2) = \mathbf{C}1$.*

PROOF. Assume $A \in A_+(O_1) \cap A(O_2) = A_+(O_1) \cap A_+(O_2)$. Thus $A$ commutes with $A(O_1^\perp)$ and with $A(\tilde{O}) \subset A(O_2^\perp)$. This implies that $A$ commutes with $(A(\tilde{O}) \cup A(O_1^\perp))'' = A''$ because of weak primitive causality. Thus $A = c1$ because $A''$ is a factor. This gives $(A(O_1) \cap A(O_2))_+ = A_+(O_1) \cap A_+(O_2) = \mathbf{C}1$. From the lemma we get $(A(O_1) \cap A(O_2))_-$ is equal to $\{0\}$ or to $\mathbf{C}V$. Let us assume the second case. We have $V \in A(O_1) \cap A(O_2) \subset A(O_1) \cap A(O_1^\perp)$. Thus $V \in A_-(O_1) \cap A_-(O_1^\perp)$. Therefore $V$ anticommutes with itself. This gives $V^2 = -V^2$. On the other hand we have $V^2 = 1$ (see the lemma). So we obtain $(A(O_1) \cap A(O_2))_- = \{0\}$ and therefore $A(O_1) \cap A(O_2) = \mathbf{C}1$.           $\square$

**REMARK 3.** The corresponding result for the Minkowski manifold and $\gamma = 1$ can be found in [8, p. 80] under other assumptions.

Next we prove that the global algebra $A$ is simple.

**THEOREM 2.** *Suppose the causal net of algebras satisfies the split property for $A_+$ and $A = C^*(\cup_m A(O_m))$ with $O_m$ a sequence of open bounded sets such that $O_m^\perp \neq \emptyset$. Then $A$ is simple.*

PROOF. 1. First we note that we can find a sequence $\tilde{O}_m$ such that $\tilde{O}_m \subset \tilde{O}_{m+1}$ and $\tilde{O}_m^\perp \cap \tilde{O}_{m+1} \neq \emptyset$, and $A = C^*(\cup_m A(\tilde{O}_m))$. Second we note that the last property implies that $A_+ = C^*(\cup_m A_+(A(\tilde{O}_m))$. This follows from the simple observation that if $A_s \in A$ converges in norm to $A$ then $A_{s,+} = \frac{1}{2}(\gamma(A_s) + A_s)$ and $A_{s,-} = \frac{1}{2}(A_s - \gamma(A_s))$

tend in norm to $A_+$, $A_-$, respectively.

**2.** From the split property and **1** we get that $A_+ = C^*(\underset{m}{\cup} N_m)$ where $N_m$ are type $I_\infty$-factors such that $N_m \subset N_{m+1}$ and for each $N_m$ there is a type $I_\infty$-factor $M_m$ such that $M_m \subset N_n$ and $M_m$ commutes with $N_{m-1}$.

Now we use the following statement (Corollary 2.6.20 in [2]): Let $B$, $\{B_m\}$ be a quasilocal algebra with $\gamma = 1$ and $B_m$ isomorphic to $L(H_m)$ where $H_m$ is a separable infinte-dimensional Hilbert space. It follows that $B$ is simple.

We apply it to our situation with $B = A_+$, $B_m = B_m'' = N_m$ using that the proof of this corollary does not really requires that $B$, $\{B_j\}$ is a quasilocal algebra. It needs only the properties of $N_j$ noted before this statement. Thus we get that $A_+$ is simple.

**3.** Assume $I$ is an ideal of $A$. Then we have the decomposition (with respect to $\gamma$) $I = I_+ \oplus I_-$. Further $I_+$ is an ideal for $A_+$: Then $A_+ I_+$, $I_+ A_+ \subseteq A_+ \cap I = I_+$. Since $A_+$ contains the identity we get $A_+ I_+ = I_+ A_+ = I_+$.

Now we have that $A_+$ is simple. Thus $I_+ = \{0\}$ or $I_+ = A_+$. The second case implies $A_- I_+ = A_- A_+ = A_-$ and $I_+ A_- = A_+ A_- = A_-$. On the other hand we have $A_- I_+ \subset I_- \subset$ $\subset A_-$. This gives $A_- \subset I_- \subset A_-$, i.e. $I_- = A_-$. Thus $I = A$. Next we consider the case $A_+ = \{0\}$. Assume $W \subset I_-$. Then we get $W^* W = 0$ and $WW^* = 0$ because $W$, $W^*$ are elements of $A_-$ and $A_- I_- \subseteq I_+$. But this implies $W = 0$ and therefore $I_- = \{0\}$. Thus we have only the ideals $I = \mathbf{C} 1$ and $I = A$, i.e. $A$ is simple.                            $\square$

**REMARK 4.** Under the split assumptions for $A$ (not only for $A_+$) this theorem is an easy consequence of the mentioned corollary from [2] or of Proposition 10 from [6]. If $M$ is the Minkowski manifold and $\gamma = 1$ then one proves that $A$ is simple under the spectrality assumption for the causal net (see e.g. [8, p. 151]).

The last result which is proved here says that the automorphisms are in general not inner. The assumptions imply that the manifold $M$ is unbounded. Because the proof is very similar to the corresponding one for Minkowski space-time and $\gamma = 1$ we give only an outline of the proof.

**THEOREM 3.** *Suppose that for the causal net of algebras there is an element $g \in G$ such that $g^m O \subset O^\perp$ for all $m > m_0(O)$ and all open bounded sets $O \subset M$. Then $\alpha_g(\,\cdot\,)$ is not an inner automorphism.*

PROOF. **1.** Let $A \in A_\pm$, $B \in A$. Set $\alpha_m := \alpha_g m$. Then $\| \alpha_m(A)B \mp B\alpha_m(A) \| \rightarrow 0$ as $m \rightarrow \infty$. The proof uses the approximation of $A, B$ in norm by $A_s, B_s \in A(O_s)$ with $O_s$ open bounded sets, $\alpha_g \circ \gamma = \gamma \circ \alpha_g$ and $\alpha_m(A(O_s)) = A(g^m O_s)$ with $g^m O_s \subset O_s^\perp$ for large m.

**2.** We have (assuming $\alpha_g$ is inner, i.e. $\alpha_g(A) = UAU^{-1}$, $U \in A$)

$$\| U\alpha_m(A) \pm \alpha_m(A)U \| = \| U\alpha_m(A)U^* \pm \alpha_m(A) \| =$$

$$= \| \alpha_{m+1}(A) \pm \alpha_m(A) \| = \| \alpha_m(\alpha_1(A) \pm A) \| = \| \alpha_1(A) \pm A \|$$

for $A \in A_\pm$. The left side goes to zero as $m \to \infty$ for the $\pm$ sign (see 1.). Thus $\alpha_1(A) = \pm A$ for all $A \in A_\pm$. On the other hand we have that $\alpha_m(A(O)) \subseteq A(O^\perp)$ for large m. This implies $A(O) \subset A(O^\perp)$ for all open bounded sets $O$. Thus $A_+(O)$ is abelian and the elements of $A_-(O)$ anticommute with themselves implying $A_-(O) = \{0\}$. Therefore all $A(O)$ are abelian giving $A$ is abelian. This contradicts (v) of Section 3.                     $\Box$

**REMARK 5.** The proof for the Minkowski manifold and $\gamma = 1$ can be found in [4].

**Acknowledgement.** It is a pleasure to thank professor H. Baumgärtel for discussions about the topics of this paper.

## REFERENCES

1.   **Baumgartel, H.** : *Einstein causal quantum fields on lattices with Lorentz invariance*, in Topics in quantum field theory and spectral theory, Report R-Math-01/86, Karl-Weierstrass Institut fur Mathematik, Berlin, 1986.

2.   **Bratteli, O. ; Robinson, D.W.** : *Operator algebras and quantum statistical mechanics I*, Springer, New York, 1979.

3.   **Dimock, J.** : Algebras of local observables on a manifold, *Comm. Math. Phys.* **77** (1980), 219-228.

4.   **Emch, G.** : *Algebraic methods in statistical mechanics and quantum field theory*, Wiley Interscience, New York, 1972.

5.   **Haag, R. ; Kastler, D.** : An algebraic approach to quantum field theory, *J. Math. Phys.* 5(1964), 848-861.

6.   **Haag, R. ; Kastler, D. ; Kadison, R.V.** : Nets of $C^*$-algebras and classification of states, *Comm. Math. Phys.* 16(1970), 81-104.

7.   **Haag, R. ; Narnhofer, H. ; Stein, H.** : On quantum field theory in gravitational background, *Comm. Math. Phys.* **94**(1984), 219-238.

8.   **Horuzii, S.S.** : *Introduction to algebraic quantum field theory* (Russian), Nauka, Moscow, 1986.

9.   **Isham, C.** : Quantum field theory in curved space-time: a general mathematical framework, in *Differential geometric methods in mathematical physics. II*, eds. K. Bleuler et al., Springer, New York, 1978.

10.  **Kay, B.** : The double wedge algebra for quantum fields on Schwarzschild and Minkowski space times, *Comm. Math. Phys.* **100**(1985), 57-81.

**Manfred Wollenberg**
Karl-Weierstrass-Institut für Mathematik
Akademie der Wissenschaften der DDR
Mohrenstrase 39, 1086 Berlin
G.D.R.

# ON UNIFORM DUAL ALGEBRAS

F. Zarouf

## ABSTRACT

Let $A$ be a uniform dual algebra with separable predual, denoted be $Q_A$, and such that the product is separately weak$^*$-continuous. We show that there exists a positive measure $\mu$ on the Shilov boundary $\Gamma$ of $A$ such that $A$ is isometrically embedded in $L^\infty(\Gamma,\mu) = L^\infty(\mu)$ and a contractive weak$^*$-continuous homomorphism $\Phi$ from $H^\infty(\mu)$, the weak$^*$-closure of $A$ in $L^\infty(\mu)$, into $A(= Q_A^*)$ such that $\Phi(f) = f$, $f \in A$. Consequently, for any isometric representation of a uniform algebra $A$ on a separable complex Hilbert space of infinite dimension H has a weak$^*$-continuous homomorphic contractive extension to $H^\infty(\mu)$ for some positive measure $\mu$ on the Shilov boundary of $A$.

## 1. UNIFORM DUAL ALGEBRAS WITH SEPARABLE PREDUAL

Let $A$ be a unitary commutative Banach algebra and **Max**$(A)$ the maximal ideal space of $A$. Then $A$ is called a uniform algebra if the Gelfand's transform $f \to \hat{f}$ from $A$ into $C(\textbf{Max}(A))$, the Banach algebra of all complex valued continuous functions on **Max**$(A)$ is an isometry. Hence, a unitary commutative Banach algebra is uniform if and only if $A$ is a function algebra (in the sense of [9]) on a compact Hausdorff space X. That is $A$ is norm closed in $C(X)$, the constant function 1 is an $A$ and for every x, y $\in$ $\in$ X, x $\neq$ y, there exists an f $\in$ $A$ such that f(x) $\neq$ f(y).

Let $A$ be a function algebra on a compact Hausdorff space X. A closed subset F of X is called a *boundary* of X relative to $A$ if for f $\in$ $A$, the supremum norm $\| f \|_\infty =$ $= \sup\limits_{x \in X} |f(x)|$ is equal to $\sup\limits_{x \in X} |f(x)|$. We know, (see [4, Chapter I, Theorem 4.2]), that the intersection $\Gamma$ of all boundaries of X relative to $A$ is a boundary. The set $\Gamma$ is called the *Shilov boundary* of X relative to $A$ (or of $A$) and it is equal to Shilov boundary of **Max**$(A)$ relative to $A$ (cf. [9, Theorem 2.17]). The map $\theta : A \to C(\Gamma)$ defined by $\theta(f) = f|\Gamma$, the restriction of f to $\Gamma$ is an isometric isomorphism from $A$ onto $\theta(A)$ and $\Gamma$ is the minimal closed set of X satisfying the above property. In other words the Shilov boundary $\Gamma$ of $A$ is the minimal compact set on which $A$ can be realized as function algebra. If $\mu$ is a positive measure on $\Gamma$, we will denote by $L^\infty(\mu)$ the Banach algebra of all measurable essentially bounded functions on $\Gamma$, by $L^1(\mu)$ the Banach space of $\mu$-integrable functions on $\Gamma$ and by $H^\infty(\mu)$ the weak$^*$-closure of $A$ in $L^\infty(\mu)$. A *uniform dual algebra* is a uniform algebra $A$ such there exists a Banach space $Q_A$ such that $A = Q_A^*$, the dual of $Q_A$. We say the product is separately weak$^*$-continuous if for any $f \in A$, the map $g \rightarrow f{\cdot}g$ is weak$^*$-continuous.

Recall that if H is a Hilbert space, the dual of the Banach space $C_1(H)$ of trace-class operators on H coincides with the Banach space $L(H)$ of all bounded linear operators on H. The bilinear functional is given by:

$$\langle L,T \rangle = \mathbf{tr}(TL), \qquad T \in L(H), \ L \in C_1(H).$$

This duality defines on $L(H)$ a weak$^*$-topology. A subalgebra $A$ of $L(H)$ which contains 1 and is closed in the weak$^*$-topology on $L(H)$ is called a dual algebra. (For more details see [1].)

**PROPOSITION 1.1.** *Let $\mu$ be a $\sigma$-finite positive measure on a set $\Omega$. Then the Banach algebra $L^\infty(\Omega,\mu) = L^\infty(\mu)$ of all measurable essentially bounded functions on $\Omega$ is isometrically isomorphic and weak$^*$-homeomorphic to a dual algebra on a Hilbert space.*

PROOF. Let $H = L^2(\mu)$ be the Hilbert space of all functions f such that $|f|^2$ is integrable with respect to the measure $\mu$, let $L(H)$ be the Banach algebra of all bounded linear operator on H, and let $M : f \rightarrow M_f$ be the map from $A$ into $L(H)$ such that $M_f g = f{\cdot}g$, $g \in A$. It is well-known (and straightforward to prove) that M is an isometric representation of $A$ on H. It is also easy to see that M is weak$^*$-continuous. Indeed since every trace-class operator K can be written in the form:

$$K = \sum_{i=1}^{\infty} (u_i \otimes v_i),$$

where $\sum_{i=1}^{\infty} \| u_i \|^2 < \infty$, $\sum_{i=1}^{\infty} \| v_i \|^2 < \infty$ (and, as usual, $u \otimes v$ denotes the operator: $x \rightarrow (x,v)u : H \rightarrow H$). The function $k = \sum_{i=1}^{\infty} u_i \bar{v}_i$ belongs to $L^1(\mu)$ and satisfies:

$$\langle K, M_f \rangle = \int fk \, d\mu, \qquad f \in L^\infty(\mu).$$

**THEOREM 1.2.** *Let $A$ be a uniform dual algebra with a separable predual $Q_A$ and suppose that the product is separately weak$^*$-continuous. Then there exists a finite positive measure $\mu$ on the Shilov boundary $\Gamma$ of A such that:*

(a) **supp** $\mu = \Gamma$ *and hence A is isometrically embedded in* $L^\infty(\mu)$,

(b) *There exists a contractive weak*$^*$-continuous homomorphism $\Phi$ from $H^\infty(\mu)$ onto A ( = $Q_A^*$) such that:* $\Phi(f) = f$, $f \in A$.

The proof uses the following lemma:

**LEMMA 1.3.** *(The notations are as in the above theorem.) For every $\ell \in Q_A$, there exists a measure $\mu_\ell$ on $\Gamma$ such that:*

(1) $$\int f d\mu_\ell = \langle \ell, f \rangle, \quad f \in A, \text{ and } \|\mu_\ell\| = \|\ell\|$$

*where $\langle \cdot, \cdot \rangle$ denotes the canonical bilinear form* $(Q_A^* = A)$.

PROOF. Let $\ell \in A$. The mapping $f \longrightarrow \langle \ell, f \rangle$ from $A$ into $\mathbf{C}$ is a continuous map since $|\langle \ell, f \rangle| \leq \|f\| \|\ell\|$. Since $A$ is a subalgebra of $C(\Gamma)$, the Hahn–Banach theorem together with the Riesz theorem imply the existence of a measure $\mu_\ell$ on $\Gamma$ satisfying the condition of the lemma.

PROOF OF THEOREM 1.2. Let $L$ be a dense countable $(\mathbf{Q} + i\mathbf{Q})$-linear subspace of $Q_A$, where $\mathbf{Q} + i\mathbf{Q}$ is the field of complex rational numbers. Via Lemma 1.3, we select for each $\ell \in L$ a measure $\mu_\ell$ satisfying (1). There exists a positive measure on $\Gamma$ such that for every $\ell \in L$, the measure $\mu_\ell$ is absolutely continuous with respect to $\mu$ ($\mu_\ell \ll \mu$). (Indeed one can take $\mu = \sum_{\ell \in L \setminus \{0\}} \alpha_\ell |\mu_\ell| / \|\mu_\ell\|$ where $(\alpha_\ell)_{\ell \in L}$ is a summable family of (strictly) positive numbers). For any f in $A$, we have:

$$\|f\|_{L^\infty(\mu)} = \sup_{x \in \text{supp}\mu} |f(x)| \leq \|f\|_\infty = \sup_{\substack{\|\ell\| \leq 1 \\ \ell \in L}} |\langle f, \ell \rangle| = \sup_{\substack{\|\ell\| \leq 1 \\ \ell \in L}} \left| \int f d\mu_\ell \right| \leq$$

$$\leq \sup_{\substack{\|\ell\| \leq 1 \\ \ell \in L}} \|\mu_\ell\| \|f\|_{L^\infty(|\mu_\ell|)} \leq \sup_{\substack{\|\ell\| \leq 1 \\ \ell \in L}} \|f\|_{L^\infty(|\mu_\ell|)} \leq \|f\|_{L^\infty(\mu)}.$$

This proves the assertion (a).

(b) To prove the second assertion of the theorem, we construct a continuous linear map $\Phi$ from $Q_A$ into the predual, $(Q_{H(\mu)}^* = L^1(\mu)/^1 H^\infty(\mu)$, of $H^\infty(\mu)$ (where $^1 H^\infty(\mu)$ is the preannihilator of $H^\infty(\mu)$ in $L^1(\mu)$ and then will show that $\Phi^*$ is the desired homomorphism. We first define $\Phi$ on $L$ by setting

$$\Phi(\ell) = [d\mu_\ell / d\mu], \quad \ell \in L$$

where $[d\mu_\ell/d\mu]$ is the coset in $Q_{H^\infty(\mu)}$ of the density function $d\mu_\ell/d\mu$ of the measure $\mu_\ell$ with respect to the measure $\mu$. Let $\alpha$, $\beta$ be in $(\mathbf{Q} + i\mathbf{Q})$ and let $k$, $\ell$ in $L$. For every $f \in A$, we have:

$$\int f d\mu_{\alpha\ell + \beta k} = \langle \alpha\ell + \beta k, f \rangle = \alpha\langle \ell, f \rangle + \beta\langle k, f \rangle = \int f d(\alpha\mu_\ell + \beta\mu_k).$$

Therefore $\mu_{\alpha\ell + \beta k} - (\alpha\mu_\ell + \beta\mu_k)$ vanishes on $A$ and, consequently $(d\mu_{\alpha\ell+\beta k}/d\mu) - (\alpha(d\mu_\ell/d\mu) + \beta(d\mu_k/d\mu))$ belongs to the preannihilator $A = {}^{\perp}H^\infty(\mu)$ in $L^1(\mu)$. This proves:

$$\phi(\alpha\ell + \beta k) = \alpha\phi(\ell) + \beta\phi(k)$$

and so the $(\mathbf{Q} + i\mathbf{Q})$ linearity of $\phi$. Moreover $\|\phi(\ell)\| \leq \|\ell\|$.

The usual techniques of extension (which easily adjust in the case of a $(\mathbf{Q} + i\mathbf{Q})$-linear space), imply that the map $\phi$ has a unique continuous linear extension, denoted again by $\phi$, to $Q_A$. The adjoint map $\Phi$ $(= \phi^*)$ of $\phi$ is therefore a weak$^*$-continuous contractive linear map. For $f \in A$, we have:

$$\langle \ell, \Phi(f) \rangle = \langle \phi(\ell), f \rangle = \langle \ell, f \rangle, \qquad \ell \in L.$$

Then $\Phi(f) = f$ and $\Phi$ is an extension of the canonical injection of $A$ into $L^\infty(\mu)$. It remains to show that $\Phi$ is an homomorphism. Let $f \in A$ and $g \in H^\infty(\mu)$. There exists a net $(g_\alpha)$ in $A$ which is convergent to $g$. Since $\Phi$ is weak$^*$-continuous and $\Phi|A$ is a homomorphism and the product is separately weak$^*$-continuous in $A$ $(= Q_A^*)$, we have:

$$\Phi(fg) = \lim_\alpha \Phi(fg_\alpha) = \lim_\alpha \Phi(f)\Phi(g_\alpha) = \Phi(f)\lim_\alpha \Phi(g_\alpha) = \Phi(f)\Phi(g).$$

(The limit is taken in the weak$^*$-topology.) If $f, g \in H^\infty(\mu)$ we have $f = \lim_\alpha f_\alpha$ for some net $(f_\alpha)$ in $A$ so $fg = \lim_\alpha f_\alpha g$. By the same argument as above we obtain $\Phi(fg) = \Phi(f)\Phi(g)$ and so $\Phi$ is an homomorphism. The proof of theorem is therefore complete.

**COROLLARY 1.4.** *A uniform dual algebra $A$ with separately weak$^*$-continuous product is isometrically isomorphic weak$^*$-homeomorphic to a quotient of a weak$^*$-closed subalgebra of $L^\infty(\mu)$ (by one of its weak$^*$-closed ideals).*

**REMARK 1.5.** We do not know of any uniform dual algebra whose product is not separately weak$^*$-continuous. Furthermore it may be the case, as far as we know, that every uniform dual algebra is a weak$^*$-closed subalgebra of $L^\infty(\mu)$ for some $\mu$.

## II. DUAL ALGEBRA GENERATED BY ISOMETRIC REPRESENTATION

Let $A$ be a uniform algebra, $\Gamma$ its Shilov boundary. We know that $A$ can be isometrically embedded into $C(\Gamma)$. Let H be a separable complex Hilbert space of infinite dimension, let $L(H)$ be the Banach space of all bounded linear operators on H and let $\Phi : A \longrightarrow L(H)$ be an isometric representation, so that $\Phi(1) = I$, where I is the identity of H. We denote by $B$ the dual subalgebra generated by $\Phi$ in $L(H)$ (this means $B$ is the weak$^*$-closure of $\Phi(A)$ in $L(H)$). We denote by $Q_B = C_1(H)/^\perp B$ the predual of $B$ where $^\perp B$ is the preannihilator of $B$ in $C_1(H)$. We will write [L] for the coset in $Q_B$ of element $L \in C_1(H)$.

**THEOREM 2.1.** *Let A be a uniform algebra and $\Phi$ an isometric representation of A on H. Then there exists a finite positive measure $\mu$ with support $\Gamma$ such that:*

(a) *A is isometric to a subalgebra of $L^\infty(\Gamma, \mu) = L^\infty(\mu)$.*

(b) *The representation $\Phi$ has a unique weak$^*$-continuous contractive homomorphic extension $\Phi$ to $H^\infty(\mu)$.*

PROOF. Let $L$ be a $(\mathbf{Q} + i\mathbf{Q})$-countable subspace of $C_1(H)$ such that $[L] = ([L])_{\ell \in L}$ is dense in $Q_B$. For each $[L] \in [L]$ the mapping:

$$f \longrightarrow <[L], \Phi(f) > = \mathbf{tr}(\Phi(f)L)$$

from $A$ into $\mathbf{C}$ is a continuous linear form. According to Lemma 1.4 there exists a measure $\mu_{[L]}$ on $\Gamma$ such that:

$$<[L], \Phi(f) > = \int f d\mu_{[L]}, \qquad f \in A \text{ and } \| \mu_{[L]} \| = \| [L] \|.$$

Set $\mu = \sum_{[L] \in [L] \backslash \{0\}} \alpha_{[L]}(| \mu_{[L]} | / \| \mu_{[L]} \|)$ where $(\alpha_{[L]})_{[L] \in [L]}$ is a summable family of (strictly) positive numbers. By the same arguments used in the proof of Theorem 1.1, we have the first and the second part of theorem.

**REMARK 2.2.** We have only used the fact that the representation $\Phi$ is an isometry to prove the assertion (a). Thus the above procedure can be applied to obtain an extension of an arbitrary representation $\Phi$ of a uniform algebra to a certain algebra $H^\infty(\mu)$. In fact, even if $\Phi$ is not isometric, it may turn out (cf. III, Example B) that **supp** $\mu = \Gamma$ and hence that $A$ is isometrically embedded in $H^\infty(\mu)$.

The following examples illustrate Theorem 2.1.

### III. EXAMPLES

**Example A**

This example illustrates how the above techniques lead quickly to the construction of the scalar-spectral measure and of the borelian functional calculus (hence of the spectral measure) associated to an arbitrary normal operator.

Let $N : H \to H$ be a normal operator, let $X = \sigma(N)$ be its spectrum, let $C(X)$ be the complex valued continuous functions on X and let M(X) be the space of all measures on X (it is the dual of the Banach space $C(X)$). It is well-known that the functional calculus $\Phi : f \to f(N)$ is an isometric representation from $C(X)$ into $L(H)$. Denote by $A$ the dual algebra generated by $\Phi(C(X))$ and by $Q_A$ its predual.

For any $[L] \in Q_A$, according to Lemma 1.3, there exists a measure $\mu_{[L]}$ on X such that:

$$\int f d\mu_{[L]} = \langle [L], f(N) \rangle, \qquad f \in C(X) \text{ and } \| \mu_{[L]} \| = \| [L] \| .$$

In this case the measure $\mu_{[L]}$ is unique and so the map $[L] \to \mu_{[L]}$ from $Q_A$ into M(X) is a linear isometry. This enables us to simplify the procedure used in the proofs of Theorem 1.2 and 2.1. Indeed, selecting a countable subset $L$ in the unit sphere of $Q_A$ we obtain easily a finite positive measure $\mu$ on X such that $\mu_{[L]} \ll \mu$, for all $[L]$ in $L$ and, hence, (via the density of $L$ and the fact that the map $[L] \to \mu_{[L]}$ is a linear isometry) for all $[L]$ in $Q_A$.

Denote by $d\mu_{[L]}/d\mu$ the density function of $\mu_{[L]}$ relative to $\mu$. The map $\phi$:
$: Q_A \to L^1(\mu)$ defined by:

$$\phi([L]) = (d\mu_{[L]}/d\mu)$$

is thus a linear isometry. The borelian functional calculus for N is now immediately at hands.

**THEOREM 3.1.** *The map $\Phi = \phi^*$ (adjoint of $\phi$) is an isometric $^*$-isomorphism weak$^*$-homeomorphism from $L^\infty(\mu)$ onto A such that:*

$$\Phi(f) = f(N), \qquad f \in C(X).$$

PROOF. The equality $\Phi(f) = f(N)$ is a consequence from:

$$\langle [L], \Phi(f) \rangle = \langle \phi([L]), f \rangle = \int f d\mu_{[L]} = \langle [L], f(N) \rangle, \qquad [L] \in Q_A.$$

To end the proof it suffices to prove that $\ker \Phi = \{0\}$. The ideal $\ker \phi$ is weak$^*$-closed and is therefore of the form $\chi_A \cdot L^\infty(\mu)$, where $\chi_\Delta$ is the characteristic function of a certain Borel set $\Delta$ of X.

Let $\Delta^c$ be the complementary set of $\Delta$ in X. Then, for $[L] \in Q_A$, we have:

$$\int f d\mu_{[L]} = \langle [L], \Phi(f) \rangle = \langle [L], \Phi(\chi_{\Delta^c} \cdot f) \rangle =$$

$$\langle \phi([L]), \chi_\Delta c \cdot f \rangle = \int \chi_\Delta c \cdot f \, d\mu_{[L]},$$

$f \in C(X)$. This implies that $|\mu_{[L]}|(\Delta) = 0$ and so $\mu(\Delta) = 0$. Then $\ker \Phi = \{0\}$ and $\Phi$ is one--to-one.

If we denote by $\mathfrak{M}$ the Borel subsets of $X$, we see that the map $E : \mathfrak{M} \rightarrow L(H)$ defined by $E(\Delta) = \Phi(\chi_\Delta)$, $\Delta \in \mathfrak{M}$ is the spectral measure of $N$ and the measure $\mu$ is a scalar-valued spectral measure of $N$ in the sense of [3].

### Example B

Let $\mathbf{T} = \{z \in \mathbf{C} \, ; \, |z| = 1\}$ be the unit circle in the complex plane and $\mathbf{A}$ be the standard disc algebra. It is well-know that $\mathbf{A}$ is a uniform algebra with Shilov boundary $\mathbf{T}$. For a finite positive measure $\mu$ on $\mathbf{T}$, $H^\infty(\mu)$ will denote the weak$^*$-closure of $\mathbf{A}$ in $L^\infty(\mu)$. Let $H$ be a Hilbert space, let $T$ be a contraction on $H$, let $A_T$ be the dual algebra generated by $T$ (that is $A_T$ is the weak$^*$-closure of $\{p(T), p \text{ polynomial}\}$ in $L(H)$) and let $Q_T \, ( = C_1(H)/^\perp A_T)$ be the predual of $A_T$.

The inequality of von Neumann $(\| p(T) \| \leq \sup_{\lambda \in D} |p(\lambda)|)$ provides a contractive representation $\Phi$ of $\mathbf{A}$ on $H$. It is easy to see that $\Phi$ is isometric if and only if the spectrum of $T$ contains $\mathbf{T}$. For $f \in \mathbf{A}$, we denote by $f(T)$ the image by $\Phi$ of $f$.

**LEMMA 3.1.** *Let* $[L]$ *be an element of* $Q_T$. *Then there exists a Borel measure* $\mu_{[L]}$ *on* $\mathbf{T}$ *such that:*

(a) *For every* $f \in \mathbf{A}$, $\langle [L], f(T) \rangle = \int f \, d\mu_{[L]}$,

(b) $\| \mu_{[L]} \| \leq \| [L] \|$,

(c) *If* $\mu_{[L]}$ *and* $\mu'_{[L]}$ *verify* (1), *then the measure* $\mu_{[L]} - \mu'_{[L]}$ *is absolutely continuous with respect to the Lebesgue measure* m.

PROOF. For (a) and (b), see the proof of Lemma 1.3. The assertion (c) is a consequence of Theorem of F. and M. Riesz ([4, Chapter II, Theorem 7.10]).

**THEOREM 3.2.** *Let* $T$ *be a contraction on a Hilbert space* H. *Then there is a positive measure* $\mu$ *on* $\mathbf{T}$ *such that:*

(a) *the representation* $\Phi$ *has a contractive weak$^*$-continuous homomorphic extension* $\Phi^\mu$ *to* $H^\infty(\mu)$.

(b) *Moreover, if* $\nu$ *is a positive measure on* $\mathbf{T}$ *and* $\Phi^\nu$ *is a weak$^*$-continuous extension of* $\Phi$ *to* $H^\infty(\nu)$, *then:*

*The measures* $\mu$ *and* $\nu$ *are mutually absolutely continuous (and hence* $H^\infty(\nu) = H^\infty(\mu)$) *provided* $T$ *is not unitary).*

PROOF. The first part of the theorem is a direct consequence of Theorem 2.1. The second part uses the following well-known lemma:

**LEMMA 3.3.** *Let $\nu$ be a Borel measure on* $\mathbf{T}$ *and let* $\nu = \nu_a + \nu_s$ *be its Lebesgue decomposition relative to the Lebesgue measure* $m$ *on* $\mathbf{T}$. *Then:*

$$H^\infty(\nu) = H^\infty(\nu_a) \oplus L^\infty(\nu_s)$$

PROOF. Let $f \in {}^1H^\infty(\nu)$ (in $L^1(\nu) = L^1(\nu_a) \oplus L^1(\nu_s)$). We have $\int z^n f d\nu = 0$, $n \geq 0$. By the theorem of F. and M. Riesz, the measure $f d\nu$ is absolutely continuous with respect to $m$ and so $f \in {}^1H^\infty(\nu_a)$. This implies that ${}^1H^\infty(\nu) = {}^1H^\infty(\nu_a)$ and $H^\infty(\nu) = H^\infty(\nu_a) \oplus L^\infty(\nu_s)$.

PROOF OF THE SECOND PART OF THEOREM 3.4. Let $\mathbf{H} = \mathbf{H}_a \oplus \mathbf{H}_s$ be the decomposition of $\mathbf{H}$, where $\mathbf{H}_a$ (resp. $\mathbf{H}_s$) is the reducing subspace associated to the absolutely continuous part (resp. the singular part) of T. Let $(\nu, \Phi^\nu)$ be an other pair satisfying the condition (a) of the theorem. Then there exists a continuous linear map $\phi^\nu : Q_T \to Q_{H^\infty(\nu)}$ such that $(\phi^\nu)^* = \Phi^\nu$.

For $[L] \in Q_T$, let $\nu_{[L]}$ be a measure such that $\nu_{[L]} \ll \nu$ and $[d\nu_{[L]}/d\mu] = \phi^\nu([L])$. In particular for every $[L]$ in $[L]$ (the countable $(\mathbf{Q} + i\mathbf{Q})$-linear space dense in $Q_T$ used to build $\mu$, cf. proofs of Theorems 1.2 and 2.1) the measure $\nu_{[L]} - \mu_{[L]}$ is orthogonal to $\mathbf{A}$ and, hence, absolutely continuous with respect to $m$. It follows that $\mu \ll \nu + m$ and therefore $\mu_s \ll \nu_s$. Let $E_s$ be a Borel subset of $\mathbf{T}$ such that $m(E_s) = 0$ and $\nu_s(\mathbf{T}) = \nu_s(E_s)$. Then, $T_s^1 = \Phi^\nu(\chi_{E_s} \cdot z)|\,\mathbf{Ran}\,\Phi^\nu(\chi_{E_s})$ is a singular unitary operator (with $\nu_s$ as scalar--valued spectral measure). Then $\mathbf{Ran}\,\Phi^\nu(\chi_{E_s}) \subset \mathbf{H}_s$. On the other hand, since $\nu_a \ll m$, we easily obtain that $\mathbf{Ran}\,\Phi^\nu(\chi_{E_s^c}) \subset \mathbf{H}_a$. It follows then that $\mathbf{Ran}\,\Phi^\nu(\chi_{E_s}) = \mathbf{H}_s$. Consequently $\nu_s$ and $\mu_s$ are both scalar-valued spectral measures for $T_s^1$ and hence mutually absolutely continuous. As T is not unitary the measures $\mu_s$, $\nu_a$ and $m$ are mutually absolutely continuous. This completes the proof of the theorem.

**REMARK 3.6.** The extension (in the sense of Theorem 2.1) of an isometric representation is not, in general isometric. For example, the extension of the representation associated to a contraction of the class $C_o$ on a Hilbert space H whose spectrum is $\mathbf{T}$ (cf. [7, Chapter III, Corollary 5.3]) of A on H is not isometric.

This is the beginning of a program aiming at a new look at representations of uniform algebras based on a dual algebras approach: the measures $\{\mu_{[L]}\}_{[L] \in Q_A}$ appear as a natural generalization of the notion of elementary measures introduced in [5], and used, in particular, in [5] and [6].

I would like to thank B. Chevreau and J. Esterle for many fruitful mathematical discussions.

## REFERENCES


1. **Bercovici, H. ; Foiaş, C. ; Pearcy, C.** : *Dual algebras with applications to invariant subspaces and dilation theory*, C.B.M.S. Regional Conferences Series in Math., **56**, A.M.S., Providence, 1985.

2. **Brown, S. ; Chevreau, B. ; Pearcy, C.** : Contractions with rich spectrum have invariant subspaces, *J. Operator Theory* 1(1979), 123-136.

3. **Conway, J.B.** : *Subnormal operators*, Pitman, Boston, 1981.

4. **Gamelin, T.** : *Uniform algebras*, Prentice-Hall, Englewood Cliffs, N.J., 1969.

5. **Lebow, A.** : On von Neumann theory of spectral sets, *J. Math. Anal. Appl.* **7** (1963), 64-90.

6. **Mlak, W.** : Absolutely continuous operator valued representations of function algebras, *Bull. Acad. Sci. Polon.* **17**(1969), 547-550.

7. **Sz.-Nagy, B. ; Foiaş, C.** : *Harmonic analysis of operators on Hilbert space*, North-Holland, Amsterdam, 1970.

8. **Rudin, W.** : *Real and complex analysis*, Mc.Graw-Hill Book Company, New York, 1960.

9. **Suciu, I.** : *Function algebras*, Noordhoff, Leyden, 1975.



**F. Zarouf**

Département de Mathématiques
Faculté des Sciences
Université Mohammed V, B.P. 1014, Rabat
Maroc.

Present address:

U.E.R. de Mathématiques et d'Informatique
Université de Bordeaux I
351, Cours de la Libération
33405, Talence Cedex
France.

MIX
Papier aus verantwortungsvollen Quellen
Paper from responsible sources
FSC® C105338

If you have any concerns about our products,
you can contact us on
ProductSafety@springernature.com

In case Publisher is established outside the EU,
the EU authorized representative is:
**Springer Nature Customer Service Center GmbH**
**Europaplatz 3, 69115 Heidelberg, Germany**

Printed by Libri Plureos GmbH
in Hamburg, Germany